STATISTICS FOR SPORT AND EXERCISE STUDIES

Statistics for Sport and Exercise Studies guides the student through the full research process, from selecting the most appropriate statistical procedure, to analysing data, to the presentation of results, illustrating every key step in the process with clear examples, case studies and data taken from sport and exercise settings.

Every chapter includes a range of features designed to help the student grasp the underlying concepts and relate each statistical procedure to their own research project, including definitions of key terms, practical exercises, worked examples, clear summaries and guides to further reading and resources. The book also offers an in-depth and practical guide to using SPSS in sport and exercise research, the most commonly used data analysis software in sport and exercise departments. In addition, a companion website includes downloadable data sets and work sheets for use in or out of the classroom.

This book is a complete, user-friendly and easy-to-read introduction to the use of statistical tests, techniques and procedures in sport, exercise and related subjects.

Visit the companion website for this book at www.routledge.com/cw/odonoghue/

Peter O'Donoghue is Reader and Discipline Director for Performance Analysis in the Cardiff School of Sport, Cardiff Metropolitan University, UK. Peter started his career as a computer scientist before moving into sport and exercise science in 1997. Since then he has been active in performance analysis of sport and is Chair of the International Society of Performance Analysis of Sport.

STATISTICS FOR SPORT AND EXERCISE STUDIES

An introduction

PETER O'DONOGHUE

 Routledge
Taylor & Francis Group

LONDON AND NEW YORK

First published 2012
by Routledge
2 Park Square, Milton Park, Abingdon, Oxon OX14 4RN

Simultaneously published in the USA and Canada
by Routledge
711 Third Avenue, New York, NY 10017

Routledge is an imprint of the Taylor & Francis Group, an informa business

British Library Cataloguing in Publication Data
A catalogue record for this book is available from the British Library

Library of Congress Cataloging in Publication Data
   Statistics for sport and exercise studies : an introduction /
   edited by Peter O'Donoghue.
   p. cm.
   1. Sports sciences. 2. Sports—Statistics. 3. Exercise—Statistics.
   I. O'Donoghue, Peter.
   GV558.S73 2012
   613.7'1—dc23
   2011040739

ISBN: 978–0–415–59556–8 (hbk)
ISBN: 978–0–415–59557–5 (pbk)
ISBN: 978–0–203–13350–7 (ebk)

Typeset in Zapf Humanist 601 BT
by Swales & Willis Ltd, Exeter, Devon

MIX
Paper from
responsible sources
FSC
www.fsc.org   FSC® C004839

Printed and bound in Great Britain by
TJ International Ltd, Padstow, Cornwall

To Mum

CONTENTS

FIGURES

figures

TABLES

XVI
tables

XVII

tables

xviii

tables

PREFACE

Statistical analysis has been and continues to be an area where many students have difficulties. This textbook, like the many other statistics textbooks, can only form part of the solution to this problem. The author has read other statistics textbooks, particularly those that are specifically aimed at sport and exercise sciences. While there is a rationale for the current textbook, the author wishes to recognize the outstanding work of Ntoumanis (2001), Hinton (2004), Hinton et al. (2004), Fallowfield et al. (2005) and Newell et al. (2010). These are excellent textbooks with different emphases on statistical procedures, the use of statistical packages and statistics within a research context. The key for successful development in the area of statistics is engagement with the area. This textbook is designed to assist students who are prepared to do the work necessary to develop the knowledge, intellectual and practical skills of the area.

In designing this textbook, the author considered how statistics are used within research and identified three things that need to be done: (a) select the most appropriate statistical procedures for the given research design; (b) apply the procedures to research data; and (c) report the results of data analysis. Therefore, each of the chapters covering statistical procedures follows a structure based on these phases aided by the use of a sport or exercise science example. The book uses SPSS (SPSS: An IBM Company, Armonk, NY) as an example computerized statistical analysis package. SPSS data sheets are provided on an accompanying website to allow students to try the examples for themselves and compare their results with those in the book. The exercises with accompanying data and solutions are also found on the website.

While engaging in these exercises is strongly encouraged, the author also had to consider where and when readers might be reading the textbook. It is not always convenient for readers to be logged onto a computer that has the SPSS package installed. Some may prefer to read through chapters before engaging in the exercises. Therefore, the book was written so that it could be used without the accompanying resources. However, the full potential of the book will only be realized by those who do engage in the practical exercises.

Developing the book has also been a learning experience for the author who has enjoyed the discussions he has had with colleagues about the examples that can be used and types of analyses that the book should promote. The book has also made a couple of contributions that may be original. The reason that the author says 'may be original' is because the two developments seem so obvious that somebody somewhere must have made such proposals before. One is an alternative to the Bonferroni adjustment used to restrict the experiment-wise α level. The author has always been uncomfortable with the Bonferroni adjustment leading to p values of 1.000 being reported because probabilities of greater than 1 have been calculated! The alternative approach proposed in Chapter 11 ensures an α level for individual pairwise comparisons that leads precisely to the intended experiment-wise α level. The second contribution is post hoc testing for the chi square test of independence. Where an association between two categorical variables measured at more than two levels is found, the author proposes testing each nominal value against the others put together to determine precisely where significance occurs.

This book is a reference book and is not intended to be read from cover to cover in sequential order. Indeed, the placement of some chapters was a difficult decision to make. In particular, there is a case for covering reliability before the main descriptive and inferential procedures. The first chapter is a general introduction to the area of statistics within the context of quantitative research. The second chapter is more of a housekeeping chapter providing guidance on how to use the book. Those who feel they cannot deal with equations and formulae should read this chapter and hopefully find that they are not so difficult. Chapter 3 covers the all-important descriptive statistics before Chapter 4 considers the use of standardized scores to assist interpretation of statistics. Chapters 5 and 6 provide essential background material on probability and probability distributions respectively which are pre-requisite to Chapter 7 where hypothesis testing is introduced. Chapters 8 and 9 cover correlation and regression. Chapters 11 to 15 cover inferential statistical procedures before chapters 16 to 18 cover some of multivariate procedures that few sport and exercise science students would cover before masters level. Chapter 19 was an interesting one for the author to write because it highlights that different disciplines use the word 'reliability' to mean different things. Chapter 20 is a very brief introduction to statistical power analysis and is heavily influenced by the work of Murphy *et al.* (2009). Those particularly interested in statistical power are referred to the more in-depth coverage provided by Murphy *et al.*'s (2009) book.

ILLUSTRATION CREDITS

The figures listed below are reprinted courtesy of International Business Machines Corporation, © SPSS, Inc., an IBM Company. SPSS was acquired by IBM in October, 2009.

XXii

illustration credits

ACKNOWLEDGEMENTS

I'd like to express my sincere thanks to all the people who helped me during the writing and production of this book. In particular, I'd like to thank Richard Neil, the Discipline Director of Undergraduate Research Methods at University of Wales Institute Cardiff (Cardiff Metropolitan University), who developed some of the examples used in this textbook and proofread the chapters that use them. I'm very grateful to Laurence Llewelyn, a former colleague, from whom I learned some new tricks in SPSS and statistical analysis in general. This textbook has been motivated by the excellent and exciting work that has been done in statistics in recent years at Cardiff Metropolitan University and I'd like to express my thanks to the members of the team I have been part of: Owen Thomas, Steve-Mark Cooper, Ian Bezodis, Michael Hughes, Jon Oliver, Paul Smith, Gemma Robinson, Genevieve Williams, Kylie Wilson, Ray Ponting, Deborah Welford, Ian Mitchell, Rhodri Lloyd, Joseph Esformes, Marianne Gittoes, Cassie Wilson, Nic James, Karianne Backx, Christian Edwards, David Wasley and Kieran Kingston. I'd also like to thank Josh Wells and Simon Whitmore from Routledge for their assistance during the planning and developing of this book.

Thank you all.

CHAPTER 1

DATA, INFORMATION AND STATISTICS

INTRODUCTION

There are different definitions of what statistics is or are, with some authors using the word 'statistics' to represent statistical results while others use it to describe the processes of statistical analysis. Anderson *et al.* (1994: 2) described statistics as 'numerical facts'. Newell *et al.* (2010: 2) and Thomas and Nelson (1996: 1) defined the word statistics as representing the general area of statistical analysis incorporating its methods and application. Thomas and Nelson (1996: 92) introduced statistics as 'simply an objective means of interpreting a set of observations'. Newell *et al.* (2010: 1) stated that 'statistics is the science of collecting, analysing, presenting and interpreting data to provide the logical framework which enables objective evaluation of research questions of interest'. Fallowfield *et al.* (2005: 18–21) stated that statistics are used 'to make sense of phenomena, occurrences or behaviour'. In this first chapter, the area of statistics will be introduced, followed by a commentary on the nature of data and information, the use of statistics within research processes and the misuse of statistics.

Research methods in general and statistics in particular are considered to be difficult areas for students. In introducing the book entitled *Taking the fear out of data analysis*, Diamantopoulos and Schlegelmilch (1997: xxi) stated their intention to produce a book in a 'digestible' form for readers. In a more recent book, Hinton (2004: xvii) described how undergraduate students find formulae to be mysterious and how he personally was frustrated by statistics books explaining what to do to perform a statistical analysis but not why.

There is a school of thought that students find statistics difficult because of how the area is presented in books and how it is delivered in university programmes. This author has decided not to jump on this bandwagon in writing the current textbook. Instead, this first chapter will commence with a much-needed pep talk to aspiring sports scientists, students, staff and management. The author stands solidly with other lecturers of research methods and statistics who work hard to facilitate student learning through the design and delivery of modules in the area. The published textbooks in the area that the author has read and referred to are excellent. There is also a wealth of excellent internet resources on statistics and even material on www.youtube.com showing how to do statistical analysis. There are two serious issues to cover right at the outset of this book:

■ Why do we do research methods in general and statistics in particular in sport and exercise science degree programmes?

- Why is statistics harder than the disciplines of sports medicine, management, sociology, psychology, biomechanics and physiology that we do on sport and exercise science degrees?

Consider the first question: 'Why do we do research methods in general and statistics in particular in sport and exercise science degree programmes?' It is primarily because the degrees are scientific programmes. The 'Sc' of 'BSc' stands for 'Science' and the 'Sc' of 'MSc' stands for 'Science'. Research methods and statistics are not merely taught to prepare a student to undertake an independent research project and write up a dissertation report. To fully understand the main disciplines covered in a sport and exercise science degree, we need to know where the theories came from and the nature of the evidence supporting these theories. BSc graduates need to be capable of independent critical thought rather than blindly accepting the theories that are presented to them during their programmes of study. An understanding of the research process and the role of various data analysis techniques including statistics is essential for students to develop into successful graduates. Salkind (2004: 11) stated that you will be a far better social or behavioural scientist if you have an understanding of statistics. This also applies to sport and exercise science. So the author has a question for anybody who asks why we do research methods and statistics: why did you apply to do a BSc? There are plenty of professional qualifications in sport, exercise and leisure that may be more suitable than a university degree and this author is not discouraging or devaluing those qualifications in any way. They have their place just as academic qualifications have their place.

Consider the second question: 'Why is statistics harder than the disciplines of sports medicine, management, sociology, psychology, biomechanics and physiology that we do on sport and exercise science degrees?' The author's reply is simply that it isn't. When one looks at the amount of knowledge, intellectual ability, analytical ability and practical skills that need to be developed to successfully achieve the learning outcomes of statistics modules (or statistics elements of research methods modules), it is far less than the learning required in sports medicine, psychology, physiology, biomechanics, sociology, management and just about every other discipline that is included in a sport and exercise degree programme. There are about 20 things one needs to know how to do in order to be a good statistician at undergraduate level within a sport and exercise science degree. Remember, the term 'good statistician' is relative as there are entire degree programmes in statistics that can be done in other faculties of universities and higher education institutions. A sport and exercise graduate is not expected to match the statistical ability of a student graduating from a specialist statistics or mathematics programme. So why do students perceive statistics to be harder and less interesting than the other main disciplines?

Considering student perception of statistics, or more correctly student misperception of statistics, much of the problem comes down to lack of engagement. To achieve the learning outcomes of any module requires private study to support the learning and practical exercises done during contact hours. Research methods and statistics are no different to any other areas in this respect. Students need to engage with the area, read the recommended textbooks and other essential reading as well as practise selecting and applying statistical procedures. A student with the correct entry requirements for a programme, investing ten effort hours for every UK credit point that a module is worth, should achieve

2

the learning outcomes of that module. For some failing students, the problem originates at the very outset of the module. Salkind (2004: xvii) described high anxiety of 'fear struck' students at the start of statistics modules and the attitudes passed down by previous cohorts of students. A lecturer delivering a statistics session to a cohort of students for the first time is often confronted with a minority of students shaking their heads and displaying other negative body language as though it is all too much for them. Psyching one's self out of being able to do statistics from the outset of a module is a fundamental mistake. My advice to students feeling that statistics is beyond them is to commit to tackling this area from the outset. Students should go into the module with a determination to keep up with the content throughout every stage of the module. They should not so much be going into the module thinking 'yes I can' but rather they should believe 'yes I will!' Students should not be surprised if they come out of some sessions confused having not fully understood all the concepts introduced. This is the same as in any other module: few students would be able to successfully answer a sports psychology exam question having only attended a lead lecture on the given topic. Similarly, it is doubtful that students would fully understand hypoxia after a lead lecture on it without additional private study. Students need to put in the additional hours between sessions to develop the necessary knowledge, intellectual ability and practical skill.

So why do students find statistics less interesting than other areas? It may come down to career aspirations and scientific degree programmes not being suitable for them. Any student with a genuine interest in sport and exercise science should find statistics interesting. When we consider the research problems we are interested in and the specific questions we want to answer, the whole process of finding the answer is one that students should be enthused by. Research can be both exhilarating and frustrating (Fallowfield et al., 2005: 1), but the experience of producing evidence in answer to some research question gives immense satisfaction. Students will not have to wait until their final year dissertation to experience this feeling. In areas such as physiology, practical laboratory exercises involve the collection and analysis of data, presenting and interpreting results to either confirm or refute some theory that students have been exposed to. The fact that they have gone through the process and that data independent of their own personal opinion is providing the evidence supporting their conclusions is the most satisfying part.

It is very important that students are able to learn by analogy. There are many possible research designs that combine within-subjects factors, between-subjects factors, covariates and lots of other things covered in this book in different ways. It is impossible for any statistics module or statistics textbook to cover the statistical approach to be taken in every possible situation. Furthermore, there is often more than one way of designing a study and more than one legitimate way of analysing data collected according to a given research design. Statistical procedures are like tools (Hinton 2004: xviii) and researchers need to be capable of using these tools to solve a statistical problem within their research. It is like a child playing with Lego® bricks (Lego, Billund, Denmark). The child knows what the individual bricks are for and how they could be used in small combinations. However, the child may not have been shown how to build a car, or a castle or a ship using Lego® bricks. The child is able to build cars, castles and ships because the child is able to solve the problem of how to build the intended structure using the different Lego® bricks available.

3

THE NATURE OF DATA AND INFORMATION

Abstraction and communication

The text that you are currently reading is a means of communicating information. If a lecture is being given based on this chapter, the words you are hearing are spoken in a way to communicate information. We use representations of information to communicate the information to others. These representations of information have become confused with the information itself. Consider the number 10. You have just read '10': two characters '1' and '0' which you assume are written to represent the numerical value 10. But was the author talking about 10 to the base 10 (10_{10}), which is the number 10 as we commonly know it, or was the author talking about 10 to the base 2 (10_2) which is a binary number which we commonly refer to as 2 because we use base 10?

Abstraction is used all the time in communication in order to convey only the necessary information that is required. If I wanted to tell you how I created this chapter, I would tell you that I created a Microsoft Word file by typing the text in using a laptop computer. This is an example of abstraction because I have avoided describing the biomechanical detail of every key depression made during the typing of the chapter. I have often wondered what would happen if I were to appear in court as a witness (not the defendant I hope) and were to be asked if the evidence I was about to give would be 'the truth, the whole truth and nothing but the truth'. It would not be feasible to tell the whole truth and I would need to provide an abstract testimony that satisfactorily answered any question asked. So sorry Your Honour, I will not be telling the whole truth! We naturally use abstraction all the time in quantitative research as well. For example, when investigating the demands of a sport or exercise activity, we would reduce the full complexity of the participant's experience to a fingerprint of relevant variables such as mean heart rate response, blood lactate concentration and self reported exertion (RPE). The use of qualitative data also involves abstraction with real behaviour, thoughts, emotions, beliefs and experiences being represented by words. Some studies may abstract lived experience to a sequence of discrete events while others would represent experience as a more continuous process. In considering where research data ultimately originate, there are deeply philosophical questions about human consciousness: about the way we see things and hear things. Does anyone else see the colour blue the way the author does? We may agree an object is blue but we may not see it the same way, but we think others see it the way we do. The way we hear sounds may differ. When someone says the word 'blue', we all believe that everyone else hears the sound of the word 'blue' being spoken the same way that we do. However, we never truly know how others experience and sense the world. The use of written and spoken language helps us communicate as we all have our own ways of interpreting writings and the sound waves produced by speech.

Data types, variables, values and constants

Fallowfield *et al.* (2005: 12) distinguished between cognitive, affective and psychomotor domains, showing that different aspects of these domains are measured using tests that provide data that can be communicated. For example, perception is an aspect of the affective domain that could be represented by a 1 to 9 Likert scale. There is a vast array of data types that can be used in research (Anderson *et al.*, 1994: 8; Diamantopoulos and Schlegelmilch, 1997: 4–7). These different types of data include facts, knowledge, intentions, attitudes, motives, primary and secondary data and published statistics. Whatever the source of data, the researcher needs to use an abstract representation of the data to make the study manageable.

The author's own original degree programme was in computer science (data processing) and provided an excellent understanding of different types of data. Data structures need to be unambiguously declared so that equally unambiguous algorithms can be written to populate data structures as well as retrieve and amend data from these data structures. The algorithms and data structures developed by computer programmers need to be unambiguous and complete because computers do not possess human intelligence and really do need to be told everything to the last detail. In order to understand data structures, it is necessary to understand data types, constants and variables. In statistics, these same things need to be understood, especially with much statistical analysis being done with computerized packages such as SPSS (SPSS: An IBM Company, Armonk, NY) and Minitab (Minitab Inc., State College, PA).

Vincent (1999: 4–6) distinguished between constants and variables while Diamantopoulos and Schlegelmilch (1997: xxiii) discussed variables, values, types, constants and data sets. Data types include character strings, real numbers, cardinal numbers, integers and sets of named values. Cardinal numbers are positive whole numbers including 0 (0, 1, 2, 3, . . .), integers are positive and negative whole numbers, while real numbers include the infinite number of values between each pair of integers. There is no need to go into the more complex types used in computer programming languages in this book. It is suffice to say that there is a lot in common between developing data structures for computerized systems and data structures for statistical analysis in research.

Scales of measurement

For our purposes, the types of data are specific versions of the scales of measurement that have been discussed by authors of other textbooks (Anderson *et al.*, 1994: 2–7; Vincent, 2005: 6–7; Rowntree, 2004: 32–3; Hinton, 2004: 21–3; Fallowfield *et al.*, 2005: 19–20; Newell *et al.*, 2010: 4–5). These scales of measurement are:

- Nominal scale
- Ordinal scale
- Interval scale
- Ratio scale.

The first two scales are categorical scales of measurement as variables use a finite set of named values. The difference between nominal scale variables and ordinal scale variables is that the values of a nominal scale variable have no natural order (for example, male and female), whereas ordinal scale variables use values that have a defined order (for

5

example, 'never', 'seldom', 'sometimes', 'often' and 'always'). This allows comparison operations such as less than (<) and greater than (>) to be applied.

Interval and ratio scale variables are numerical variables used to measure quantities such as distance, mass, force, velocity as well as counts of events. Interval scale variables are characterized by a fixed interval between points on the measurement scale. For example, four is not just greater than three, it is exactly one greater than three. This measurable interval means that not only comparisons such as greater than and less than can be applied to interval scale variables, but values can also be subtracted. An example of an interval scale variable is temperature using the degrees Celsius scale: 4°C is not just greater than 1°C, it is exactly 3°C greater than 1°C.

Ratio scale variables include all the properties of interval scale variables but in addition ratio scale variables use zero to represent an absence of the concept being measured. This means that as well as subtraction, division can be done. For example, 6m is twice as long as 3m. We cannot do this with temperature measured in degrees Celsius: 4°C is not four times as warm as 1°C because 0°C does not represent a complete absence of heat. Interval and ratio scale variables could be discrete or continuous. Discrete interval scale values are integer values whereas discrete ratio scale values are cardinal numbers. For example, if we were counting sit ups performed in a minute, 40 sit ups is 20 more sit ups than 20 sit ups and twice as many sit ups as 20 sit ups. Just because the variable is not measured on a continuous numeric scale does not mean that division and subtraction cannot be done. Continuous numeric variables use real numbers as opposed to cardinal numbers and integers.

There are different types of nominal scale; for example, gender has two named values (female and male), while position in a soccer team might have four values (goalkeeper, defender, midfielder and forward). Gender sounds like a variable, but what computer science students realize is that it is best to keep it as a type in case we need several variables of this type. For example, the participant's gender and the gender of the participant's coach could be two separate variables of the type gender. Female and male are values of the type gender. Many statistics textbooks do not go to the level of detail that computer programming texts do and typically refer to gender as a variable rather than a type, especially if there is only one gender variable used in the given study. The current textbook will not go to the level of definition of computer programming textbooks either for two main reasons. First, many examples may only have a single gender variable (for example) and a proliferation of types and variables could become confusing. Second, statistical analysis packages typically allow variables to be defined using standard types or being made up of user-defined values without any need for user-defined types.

There are also different types of ordinal scale; for example, frequency of occurrence might contain five ordered values ('never', 'seldom', 'sometimes', 'often' and 'always') while level of agreement might contain five different values ('strongly disagree', 'disagree', 'undecided', 'agree' and 'strongly agree'). Interval variables might use a subrange of the set of real numbers; for example, angular displacement might range from $-\pi$ to $+\pi$. The same is true of ratio scale data where a percentage will range from 0.0 to 100.0.

Height is a variable while the constant 1.73m is a particular value of height that might be a lower limit for joining a police service. Variables such as height, body mass and body mass index (BMI) allow general relationships between variables to be defined that hold for an infinite number of values without having to specify the relationship for every combination of values. A functional relationship defining BMI in terms of height and body mass is shown in equation 1.1.

$$BMI = Body\ mass\ /\ (Height^2) \tag{1.1}$$

Units of analysis

Height is a variable and 1.80m is a value that could be assigned to this variable for a particular case. Anderson *et al.* (1994: 2) used the terms 'elements', 'variables' and 'observations' when introducing the nature of data to be analysed using statistical techniques. The elements are the cases or participants in a study. Variables are defined measurable characteristics such as height, body mass and gender, as has already been discussed. An observation is a set of values representing an element: one value per characteristic. For example, a study may include variables gender, height and body mass. We may have a particular element (participant 1) in our study who is represented by the values 'female', 1.75m and 54.2kg for gender, height and body mass respectively.

Diamantopoulos and Schlegelmilch (1997: xxiii) and O'Donoghue (2010: 144–5) used the term 'unit of analysis' rather than elements to represent the cases or participants in a study. O'Donoghue (2010: 144–5) pointed out that there was a choice of unit of analysis in many sports performance studies. For example, a study of tennis strategy could use a point, a match (combining two player performances), a player performance within a match or a typical player performance based on multiple match data as the unit of analysis.

Independent and dependent variables

As well as being classified by their type, variables can also be classified according to their role within a study. Newell *et al.* (2010: 5) distinguished between explanatory and response variables in experimental studies. Others refer to these as independent and dependent variables respectively (Vincent, 1999: 8–9; Fallowfield *et al.*, 2005: 52–3). Where nominal or ordinal variables are hypothesized to have an influence on some dependent interval or ratio scale variables, the nominal and ordinal scale variables are sometimes referred to as factors. Factors are independent variables with a finite number of values or levels. For example, gender is measured at two levels because there are two values of gender (female and male). Socio-economic group could be measured at six levels (A, B, C1, C2, D and E). Some factors are referred to as between-subjects factors because they are used to distinguish different groups of people. Both gender and socio-economic groups would be between-subjects factors. Other factors are within-subjects factors where their levels (or values) represent different times or conditions under which all participants are measured. For example, in a diurnal variation study we might use a variable time which is measured at six levels (04:00, 08:00, 12:00, 16:00, 20:00 and 00:00). This is a within-subjects factor because all participants would be tested at all six times.

7

PARAMETERS, STATISTICS AND SAMPLES

Most research is undertaken using samples drawn from populations of interest. This is due to factors of cost, decision importance, time to market and confidentiality (Diamantopoulos and Schlegelmilch (1997: 10–11).

Samples should be random to avoid bias and systematic influencing of research data (Hinton, 2004: 50). Parameters are characteristics of a population while statistics are characteristics of a sample (Vincent, 1999: 12–13; Newell *et al.*, 2010: 3–4). Both population parameters and sample statistics are values for some variables of interest. Where a sample is used to represent a population within a study, there is usually some error between the true population parameter and the sample statistic used to estimate it. Sampling error will be covered in detail in Chapter 7.

STATISTICS IN RESEARCH

The normative paradigm

Most statistical procedures are used within quantitative research that is underpinned by the assumptions of the normative paradigm (Cohen *et al.*, 2007: 7–26). The normative paradigm assumes that there is a single real World that is independent of the view of individuals. This not only applies to objects within the World but also to processes and the rules that govern human behaviour. If behaviour is deterministic and predictable, then it follows that all humans behave in a similar way. Essentially, there is an assumption that there is an average human with some variability about that average person. The average human being can be studied through an empirical approach. Human behaviour is abstracted to hard tangible facts that can be used to represent behaviour in a concise way, facilitating quantitative analysis. Quantitative research involves the following steps listed by O'Donoghue, (2010: 46):

- Abstraction – representing human behaviour with a finger-print of variables.
- Sampling – using data for a representative sample drawn from the population of interest.
- Summarizing – determining sample statistics to describe the average participant as well as the variability about this average.
- Analysis – using statistical procedures to determine if there are relationships between variables and differences between groups.
- Generalization – a large enough random sample is assumed to be representative of the population of interest allowing results to be generalized to the wider population.

Research Design

Different statistics textbooks have different purposes with some covering solely statistical analysis (Vincent, 1999; Ntoumanis, 2001; Hinton, 2004; Hinton *et al.*, 2004) and others setting statistics in a context or research design (Fallowfield *et al.*, 2005; Newell *et al.*, 2010). The approach taken in this book is to consider research design and measurement issues that are related to statistical analysis. Planning and design of a study are essential to avoiding major errors. The 'V'-shaped diagram used by O'Donoghue (2010: 51) made the point that the earlier an error occurred in a research study, the greater the amount of work that might be done before the error was detected. Newell *et al.* (2010: 11) also warned against the consequences of poor research design; 'No amount of "fancy data analysis" can save a badly designed study.' Studies should have clear aims and avoid complex designs (Thomas and Nelson, 1996: 63). Quantitative research typically commences with a process of developing a research problem and then specifying a research question. This process starts using vague concepts which are gradually operationalized with precise variables being defined. O'Donoghue (2010: 55–6) listed the following reasons for selecting a research topic of interest:

- Personal interests
- Career aspirations
- The research interests of colleagues
- Topical issues in sport and exercise
- Scientific literature in the area
- Practical application of research in the area
- Importance of the area
- The strengths and weaknesses of the researcher.

Having selected a research topic of interest, the researcher should review the literature to identify a research problem and provide a rationale for studying it. O'Donoghue (2010: 80–101) described a four-step approach that gradually transforms a vague research problem into a specific research question through focusing the research problem, selecting variables, defining the variables and forming hypotheses.

The research question may be to evaluate relationships between variables, or to compare different groups in terms of variables of interest, or to examine some variables of interest under different conditions. These basic types of quantitative research can be modified into more specialist designs if necessary. In each case, the research identifies the variables of interest and the relationships and differences to be tested for. It is essential that the researcher chooses the appropriate statistical procedures for the study designed. The researcher who tries to fit a research design to a particular statistical procedure that they have a preference for is not doing research properly. The statistical procedures should always be governed by the research design and the research design should be made to answer the specific research question.

There are some studies to determine if variables are related and other studies to determine if there are differences between groups or conditions. A correlation study assesses the relationship between variables: the variables involved will be measured on ordinal, interval or ratio scales. None of the variables would be considered to be dependent variables by the correlation procedure to be used, but there may be temporal or logical relationships between them which would lead a researcher to conceptually classify some as independent and others as dependent (Newell *et al.*, 2010: 12–15).

When comparing different groups or conditions of interest, there will typically be a classification of variables as independent and dependent. An independent variable is hypothesized

9

to have some influence on the dependent variable(s). Very often an independent variable is a nominal variable that is used to classify participants into groups or a nominal variable that is used to classify measurements into conditions. An example of a nominal variable used for grouping participants is gender. This would be an independent variable in many sport and exercise studies. Consider a study comparing the time spent watching soccer per week between males and females. The study might find that males watch more soccer than females. Making gender the dependent variable is a bit like hypothesizing that a female's gender will change if they watch too much soccer or a male's gender will change if they don't watch enough soccer. An example of a nominal variable used to classify measurements into conditions might be venue for sports contests. This could be used in a study to compare the home and away performances of a set of teams.

Some ordinal variables can also be used to arrange data values into independent samples or related samples. For example, quarter of a basketball match is ordinal but it could be used in a study the same way as a nominal within-subjects effect is used.

This section on statistics in research will contrast survey research with experimental research. Survey research could use a questionnaire to gather self-report data from respondents. Survey research is useful for diagnosing situations but it does not provide evidence of any cause–effect relationship between variables and cannot be used as evidence to support action (Kirk-Smith, 1998). Experiments, however, deliberately manipulate conditions to test the effect of some explanatory variable on a response variable. The research design issues that are specific to surveys and experiments will be discussed. First, research issues that are common to both survey research and experimental research are discussed.

General research issues

Planning alone will not prevent problems during data collection and, therefore, pilot studies should be used before commencing the main data collection activity (Newell *et al.*, 2010: 28–9). If pilot work is not done and participants arrive for the pre-test of an experiment and there is a problem with the testing procedure that means the data cannot be collected on that occasion, the participants' time will have been wasted and many will withdraw from the study. Similarly, if a questionnaire has questions that cannot be understood due to poor wording, a lot of responses will be lost if participants attempt to complete the questionnaire before it has been properly tested. Pilot work can identify problems with methods that can be rectified before the main investigation commences.

The sample size for a study depends on a number of factors (Fallowfield *et al.*, 2005: 49–50; Newell *et al.*, 2010: 29). There are statistical means of determining the number of participants required to give a good chance of a significant finding. This may require some exploratory study to be done that provides an indication of the sample statistics that might be found in a larger study. There are statistical power calculations that can be used to determine how many participants would be required for a larger clinical study to be significant assuming the same sample statistics were produced from a larger sample. Statistical power and this application of statistical power will be covered in Chapter 20. It may not be possible to identify sufficient participants satisfying the criteria for inclusion in the study. A further issue is that the time required for testing or observation of a participant may make it impossible to collect the volume of data necessary to give a chance of a significant result. For example, the effort hours to be devoted to the research project may impose a limit on the size of the sample (O'Donoghue, 2010: 143–4). Researchers also face trade offs between what they would like to do in an ideal study and what is possible given resource constraints, ethics policies and practicalities (Fallowfield *et al*, 2005: 17).

Fallowfield *et al*. (2005: 50–2) discussed the advantages of using a homogenous sample of participants. Differences between samples being compared in a study may be rendered insignificant if there is too much within-sample variance. Homogenous samples can be created by restricting the scope of an investigation and applying strict criteria for inclusion in the study. This is related to the MAXICON principle described by Thomas and Nelson (1996: 62–3) whereby researchers attempt to maximize true variance, minimize error variance and control extraneous variance.

Error variance is a term that is often used in statistics for normal variability between and within individual participants as well as genuine measurement error. This 'error' variance can be minimized by gathering data using standard procedures consistently (Fallowfield *et al*., 2005: 55–6) and using reliable measures that maximize sensitivity to changes in the response variables being measured (Fallowfield *et al*., 2005: 52–4).

Another problem is where measures do not discriminate between different levels of performance. Consider a test of rugby union penalty kicking that involves a series of three sets of penalty kicks taken from 10 different locations on a rugby pitch. There may be four locations that are straightforward for a penalty to be scored from by all but novice rugby players. There may be four other locations from which few non-international rugby players would score from. This leaves only two of the kick locations that might discriminate between different kicking abilities expected in good club rugby players. It is recommended that such a test use different locations that would give a better spread of test scores in the population of rugby players of interest to the study. This clearly requires pilot work to determine such a set of locations.

The independence assumption of some tests is taken more seriously by some than others. Consider research in the area of performance analysis of sport. There are many published papers where the unit of analysis is individual match events or phases rather than whole matches. Two recent examples at the time this textbook was written are the volleyball papers by Alexandros and Athanasios (2011) and Costa *et al*. (2011). Alexandros and Athanasios (2011) used a sample of 1,472 phases from eight volleyball matches to show that when the setter made the correct choice, it significantly increased the 'attacking efficiency ratio'. Costa *et al*. (2011) applied multinomial logistic regression to analyse the association of reception effect, defence effect, attack tempo and attack type, with the 'attack effectiveness'. The data used were 863 actions of reception, 1,191 actions of attack and 435 actions of defence from 11 matches. One criticism of both studies is that the patterns found may be largely influenced by the small number of matches used. A further criticism is that the data are not independent as each study includes multiple events from individual matches. The problem of independence could largely be overcome if each match was characterized by percentage performance indicators showing the 'attacking efficiency ratio' and 'attack effectiveness' for different tactical choices rather than using nominal variables. This would admittedly reduce the number of observations to 8 and 11 respectively for these two investigations. However, using related samples tests to compare the effectiveness of different tactical choices might still be significant. Alternatively, studies such as these might need to use a greater number of matches. There is still a philosophical question about the independence of different matches; this author remembers Blackburn Rovers winning the league having lost to Liverpool on the last day of the 1994–95 English FA Premier League season because Manchester United drew at West Ham United. Were these two matches truly independent? Are all matches and all data related because we all live in the same universe?

Survey research

Surveys can take different forms including those using questionnaires and interviews. Interview data are not of concern to the current book as they are not usually analysed using statistical procedures. Questionnaires pose interesting problems for statisticians as some can contain multiple types of questions leading to variables measured on different scales. The same questionnaire could include nominal choices, ordinal choices, ranking responses and indeed completely open text responses.

There are many different types of questionnaire ranging from self-completion questionnaires to questionnaires to be completed by the researcher while asking questions in the presence of the respondent. Questionnaires should be complete, clear, well designed, well presented and provided with sufficient supporting instructions for the respondent. Questionnaires have measurement and analysis issues that need to be recognized by those undertaking questionnaire surveys (Kirk-Smith, 1998). The way in which questions are asked and the order in which questions are asked might influence the responses received. The scales of measurement used for ordinal rating of agreement ('strongly disagree', 'disagree', 'undecided', 'agree' and 'strongly agree') might not allow sufficient discrimination between different levels of agreement. The same could be said of other ordinal concepts recorded in this way.

Where questionnaires are used in quantitative research, there should be a clear aim and purpose to the study. If the purpose of the study is to compare different types of respondent, there should be some key question allowing respondents to be grouped. This key question should be involved in every statistical test that is performed on the data. Any analysis that does not include this key question is failing to answer the research question. Despite their limitations relating to self-report data and theory building, questionnaires are useful in descriptive research (Kirk-Smith, 1998).

Experimental research

Thomas and Nelson (1996: 351–62) described 13 different types of experimental design: three true experimental designs, three pre-experimental designs and seven quasi-experimental designs. This book will not cover all of them but will discuss implications for statistical analysis in the main experimental designs. The main true experimental design is the pre-test/post-test randomized groups design. This design involves randomly assigning participants to an experimental group that is exposed to the condition of interest and a control group that is not. Both groups are tested before and after the experimental period using a valid test for the response variable. Ideally, all other conditions experienced by the participants should be the same so that any difference between the groups for the pre- to post-change in the response variable can be attributed to the condition of interest. Using a random sample and randomly assigning participants to the experimental and control groups is necessary for an experimental design to qualify as a true experiment. This is because it is impossible to ethically control all aspects of the participants' lives over the experimental period. If there is a large enough random sample and participants are randomly assigned to the experimental and control groups, then there is a good chance that variability in training, sleep patterns, diet and other factors that might influence the response variable will be balanced between the experimental and control groups.

Often randomization is not possible and experiments without random assignment of participants to groups are either pre-experimental designs or quasi-experimental designs. A pre-experimental design may test an experimental group without a control group or the control group may be some pre-existing group that was not formed at the same time as the experimental group. Quasi-experimental design can involve control but without random assignment. For example, the pre-test/post-test groups design is identical to the pre-test/post-test randomized groups design except participants are not randomly assigned to groups. The statistical procedures used for this design and the pre-test/post-test randomized groups design are the same because in each case we are testing whether an independent grouping variable has a significant influence on some response variable of interest. The ex-post facto quasi experimental design (Thomas and Nelson, 1996: 358–60) involves different statistical procedures because the grouping variable is actually the dependent variable. For example, data may be gathered on a set of participants over a period of two years. At the end of the experimental period, the researcher determines which participants have experienced some illness or other condition of interest. These participants form one group while the remainder form another group that they are compared with. The research question is whether any of the variables gathered over the two-year experimental period are associated with the risk of developing the condition of interest. Therefore, the measured variables are independent variables within the study which are hypothesized to influence which group the participants find themselves in. Thus the grouping variable is the dependent variable in this case.

There are various issues that need to be considered when planning an experiment. First, participants may improve their performance between a pre-test and a post-test due to increased familiarity with the test (Newell et al., 2010: 29). The Solomon four group design attempts to overcome this by using two experimental groups and two control groups (Thomas and Nelson, 1996: 356). One experimental group is tested before and after the experimental period while the other only performs the post-test. Similarly, one control group is tested before and after the experimental period while the other only performs the post-test. The statistical procedures necessary for the Solomon four group design are quite complex and this author has not been able to work out how one single statistical test can determine if there is a significant experimental effect or not. The best the author could come up with was one statistical test applied to the two groups that performed the pre-test and another statistical test applied to the two groups that did not. An alternative way of addressing familiarization is for all participants to undergo a training programme to become familiar with the testing procedure prior to the main study commencing. This should minimize any additional pre- to post-test improvement due to familiarization.

Standardization and control is important in experimental design even though it is not possible to totally control every aspect of the participants' lives during the experimental period (Fallowfield et al., 2005: 25; Newell et al., 2010: 20). The same procedure for testing should be applied to all participants during the pre-tests and post-tests. Participants should also be given clear advice on preparation they need to do prior to pre- and post-testing; for example, there may be diet and exercise constraints that need to be adhered to or else test results might be invalid. Where tests are performed outdoors, every effort should be made to test all participants using similar weather conditions.

The timing of the pre-tests and post-tests may also impact on the test performance. Pre- and post-tests should be applied at the same time of day to avoid any differences due to diurnal variation (Waterhouse et al., 2005). The author has noticed that experiments in many student dissertations carry out pre-testing in October and post-testing in December or January. If the participants are students, it is possible that they are healthier and fitter in October than during December and January. This is not only due to climate but also due to pressure of university work towards the end of the first semester.

There may be psychological effects of the experimental and control conditions that may influence performance during post-tests. For example, if a participant genuinely believes that the treatment they have been exposed to is beneficial, their post-test performance may be enhanced. Alternatively, some participants in the control group may view performing the post-test without being exposed to the experimental treatment as a challenge. Their desire to prove that they can improve even without benefitting from the experimental treatment may lead to an improved post-test performance. These psychological factors can be overcome by using a placebo with the control group (Fallowfield et al., 2005: 39–40) with all participants being unaware, or 'blind', as to whether they are in the experimental or the control group. If the testing that is done before and after the experimental period involves some judgement of performance by the researcher, it is also necessary that the researcher is not aware of which participants are in which group. This is called a double blind study (Fallowfield et al., 2005: 41–2). Placebos, blind studies and double blind studies are possible where participants are taking some product that is hypothesized to influence the response variable. However, if participants are performing some mental or physical training programme, it will be obvious to them whether they are performing the programme or not.

The internal validity of an experiment is the extent to which changes in the response variable can be attributed to the experimental treatment (Thomas and Nelson, 1996: 344). The internal validity of a study is reduced by maturation of participants or other events affecting the participants during the experimental period, withdrawal of participants from the study, increased familiarization with the tests used, limited reliability of testing and participants not being randomly assigned to groups (Thomas and Nelson, 1996: 345).

The external validity of an experiment is the extent to which the results of an experiment can be generalized beyond the sample used in the investigation (Thomas and Nelson, 1996:

14

344). Participants should be representative of the population of interest. If the population of interest is senior male, club-level field hockey players, then participants should be drawn from several clubs. If the participants are a convenience sample from one particular field hockey club, there may be training patterns that are specific to that club that influence the results. Worse still, if the control group uses participants from one club and the experimental group uses participants from another club, any difference in the response variable over the experimental period may be due to differences in other activity between the two clubs. Psychological factors to do with the experimental treatment might also reduce the external validity of the study. Blind and double blind studies and the use of placebos can counter such psychological factors.

The ecological validity of a study is the extent to which it represents the real world of the concepts of interest. For example, we may be interested in the effect of some training programme for club-level soccer players on the specific fitness required to perform intermittent high-intensity activity during soccer match play. We will need some test variable to represent the ability to perform intermittent high-intensity activity. Timing of repeated sprints performed during a friendly soccer match has greater ecological validity than a laboratory test of intermittent high-intensity activity. There is a trade off between experimental control and ecological validity because it is impossible to control the activity within friendly soccer matches that might be played before and after the experimental period.

One of the issues with the pre-test/post-test randomized groups design is that the experimental and control groups are made up of different participants and, therefore, inter-individual differences may be responsible for any difference in the pre-test/post-test change between the groups. Therefore, some experimental designs expose all participants to the experimental and control conditions, testing the response variable after each. This reduces the chance of inter-individual differences impacting on the results of the study. One such design is the crossover design where half of the participants experience control conditions first and then experimental conditions with the remaining participants experiencing the conditions the other way round (Fallowfield et al., 2005: 39). This helps avoid order effects if participants perform the test better on the second occasion than the first or vice versa (Newell et al., 2010: 19). There are other experimental designs where all participants experience experimental and control conditions. One of these is the reversal design where participants experience the experimental and control conditions more than once in alternation (Thomas and Nelson, 1996: 358). Crossover and reversal designs work well if there are no carry-over effects of treatments. A carry-over effect is where the status of a participant is affected for a time after exposure to an experimental condition or if the experimental condition leads to other changes that are not quickly reversible (Fallowfield et al., 2005: 40–1; Newell et al., 2010: 19). This could interfere with control conditions and, in such cases, independent group designs might be preferable. Where carry-over effects are possible and the researcher is concerned about inter-individual variability impacting on the results of the study, a matched pairs design could be used (Fallowfield et al., 2005: 41). In this experimental design, pairs of similar participants are identified using specified criteria. A random selection is done to place one member of each pair in the experimental group and the other member of the pair into the control group.

Measurement issues

The previous sections of this chapter have already touched on measurement issues in questionnaires and experimental testing. The main measurement issues to be considered for all research variables are validity, objectivity and reliability. Validity is the extent to which a variable measures the concept it is supposed to be measuring as well as the importance of that concept. Objectivity is concerned with the extent to which a variable is independent of researcher opinion. This usually requires an objective measurement procedure that is free of researcher judgement. Reliability is concerned with the consistency of measurement for a given variable. Chapter 19 will cover these at length with detail of statistical procedures for assessing validity and reliability. At this stage, it is worth noting that validity, objectivity and reliability are related measurement issues. This author takes the view that validity is the most important of the three. If a variable is a 'so what' or 'who cares' variable that is meaningless to practitioners, then it does not matter how objective and reliable it is, the entire study will be compromised. In saying this, a variable cannot be valid unless it is reliable. Objective measurement procedures assist reliability in that they promote consistent understanding of the variable. Where a variable is vaguely defined, there will not only be problems with reliability but also with replicability of the study. Research needs to be open to public scrutiny with studies being replicable so that others can confirm or refute the findings.

A statistically significant result is not necessarily a meaningful result (Thomas and Nelson, 1996: 109). An experiment may reveal that a group doing a new training programme improves some relevant fitness variable significantly more than a group doing a traditional training programme. The statistical test used might produce a p value of less than 0.05 meaning that there is less than a 5 per cent chance that this difference was a chance occurrence due to sampling error. However, although we are more than 95 per cent confident that this is a real effect of the training programme, it may be such a small effect that it is not meaningful in real athletic terms. Alternatively, it may also be possible that a small difference may be a meaningful difference in real athletic terms even though it is not statistically significant. Effect sizes are used to estimate the proportion of the variance in the data that is attributed to the factor of interest (Vincent, 1999: 133–4). For example, if the factor of interest is a categorical variable arranging participants into groups doing alternative training programmes and an effect size of 0.25 is produced, then 25 per cent of the variance is down to the different training programmes with the remaining 75 per cent being explained by other factors. This was specifically for ω^2 from a t-test and some other effect sizes are not interpreted this way. Fallowfield *et al.* (2005: 56–8) suggested that rather than using standard criteria for effect sizes, what we really need to know is what size of change would be considered important in practice. Hinton (2004: 101–3) also stated that we need to have a feel for what sort of effect size is important.

MISUSE OF STATISTICS

Naïve use

There is naïve use of statistics as well as intentional misuse of statistics. Naïve use of statistics, as the term suggests, is where the researcher makes a mistake through lack of

16

knowledge or ability rather than intentionally using statistics in a misleading way. Naïve use of statistics is where a researcher uses an inappropriate statistical test, interprets the results of a statistical test incorrectly or fails to use a statistical test at all, meaning that relationships or differences observed could be chance occurrences. An example of naïve use of statistics is where a researcher does a test-retest reliability study for some fitness test score. The researcher uses a paired samples t-test to compare the test and retest scores finding no significant difference ($p > 0.05$). The problem here is that the paired samples t-test is typically used to test for differences with the null hypothesis being that there is no difference between the paired samples. An alpha level of 0.05 means that the fitness test score will be considered reliable if there is a p value of greater than 0.05. The p value is the chance of making a mistake if we are claiming there is a difference. So although the chances of such a mistake are greater than 1 in 20, if the p value is between 0.05 and 0.50, we are still more likely to be talking about a difference rather than a similarity between the paired samples – it is just not a significant difference! Remember in reliability studies we are tying to demonstrate a similarity not difference.

Another form of naïve use of statistics is where an association between two variables may be influenced by some third variable that has not been considered by the researcher. For example, a student may find that in English FA Premier League soccer matches between teams with differing recovery days since their previous matches, that the team with fewer recovery days has a greater chance of winning. The fact is that in many of the matches between teams with differing recovery days from their previous matches, the team with fewer recovery days may have been playing a mid-week European Champions League match while the other team had not played since the previous weekend. The teams with fewer recovery days from their previous matches did not win because they had fewer recovery days. They won because they were England's top four club sides and that is why they played in Europe. It is also possible that this type of misuse of statistics could be intentional where the researcher is fully aware of some other variable responsible for the result found but deliberately does not mention this in the methods or results of the study.

Intentional misuse of statistics

Intentional misuse of statistics can come in two forms, one more serious than the other. The most serious intentional misuse of statistics is where a knowledgeable researcher with a good statistical ability blatantly uses an inappropriate statistical test or no statistical test at all so that they can draw the conclusions that they desire from their study. This is an unethical use of statistics.

The less serious form of misusing statistics is where the research reports factual information in a transparent but misleading way. This sounds like a contradiction in terms but in fact it isn't. They have chosen to truthfully emphasize some results over others. It is up to readers to be able to recognize this. For example, it may be announced that the rate at which unemployment is increasing has decreased in the last month. This means that unemployment is still rising, but not by as much as it was in the previous month. There will be some who interpret the statement as meaning unemployment is falling and that may be why the statement has been worded this way.

17

There are many other forms of misleading uses of statistics that have given statistics a bad press (Wood, 2003: 3). Vincent (1999: 14–15) described other misuses of statistics including improper generalization and inadequate definition of terms. Improper generalization occurs where very small case evidence is used. For example, if we were to analyse Rafael Nadal's first serve to the advantage court against Roger Federer in the 2011 French Open final, we would see that he won 75 per cent of these points when the serve was played to Federer's forehand. These serves go to the left of the service box from Nadal's viewpoint and Federer is right handed. This is a higher percentage of points won than when he served to any other area of the service box. Therefore, Nadal should serve to Federer's forehand more often. The problem here is that an improper generalization has been made. Indeed, Rafael Nadal won three out of four points when the first serve was placed in the left third of the advantage service box (75%), which is greater than the 12 out of 18 (66.7%) won when it was placed in the middle of the service box and the 21 out of 35 (60%) won when it was placed in the right third. Generalizing to this extent from just four points is inappropriate especially as if one of the three won points had been lost, the success rate would be down to 50 per cent. A further issue with the way this statistic has been presented touches on the definition of terms mentioned by Vincent (1999: 14–15). The data used to produce the percentages do not include serves played to the left, middle or right of the advantage count that were faults, second serves are not included and serves to the deuce court are not included. The sentence 'Therefore, Nadal should serve to Federer's forehand more often' is very misleading.

SUMMARY

This chapter has introduced statistics and the use of statistics within scientific research. Good research design is essential to a successful research project and the statistical procedures used should be dictated by the research design. Most quantitative research uses samples to represent the populations of interest to the study. Large random samples are more likely to be representative of the wider population than smaller convenience samples.

CHAPTER 2

USING THIS BOOK

INTRODUCTION

Having introduced the area of statistics within its research context in Chapter 1, the purpose of this chapter is to give the reader specific advice on how to use this book. This chapter comes in two parts. The first part discusses how the book should be used within different modules by readers. This book is a reference book and it is anticipated that many readers will target specific parts of the book that support their needs. This is the case for academic staff in universities as well as students. The book covers a range of areas from descriptive statistics to multivariate analysis methods. This does make it unlikely that the whole of the book would support one single module. Certainly, the author would not recommend a single module to cover all of these areas at his own university. The second part discusses the terminology and notation used within the book. In particular, the notation used to describe the use of the SPSS interface, its menu structure and pop-up windows are outlined in this chapter. Occasionally, the book uses equations where they help explain statistical concepts. The second part of the chapter also includes a brief description of equation notation used to assist readers.

GUIDANCE FOR READERS

Recommended use of material within different modules

Different academic programmes teach statistical techniques at different stages. University programmes in the UK typically use three years where students cover level four to six material. Masters programmes cover level seven material. The author works in Cardiff School of Sport at Cardiff Metropolitan University where an undergraduate scheme consists of a family of individual undergraduate degree programmes with some programme-specific modules, some modules being common to several programmes and some modules being common to all programmes. The research methods module is a 20-credit point level five module done by all students in the undergraduate scheme. The module covers research processes in general, qualitative research, action research as well as quantitative research. Therefore, quantitative methods in general and statistics in particular will make up less than half of the material in the module. The statistical procedures are embedded within quantitative research exercises including problem selection, specification, study design, data gathering, analysis and interpretation of results. This approach has been successful as students perform statistical tests using a sport or exercise research question. The statistical procedures and the situations in which they are applied are covered in lead lectures while entering the data into SPSS and analysing the data in SPSS are done during practical exercises in the microcomputer laboratory. The quantitative research and statistics material is covered over a period of six weeks of the 24-week module with part of the assessment being the completion of a portfolio of the quantitative research that has been done within the module. Therefore, the material covered during this level five module is limited to the following analyses which are shown with the corresponding chapters of this book in parentheses:

- Descriptive statistics (Chapter 3)
- Hypothesis testing (Chapter 7) underpinned by probability (Chapter 5) and data distributions (Chapter 6)
- Correlations (Chapter 8)
- t-tests (Chapter 10)
- One-way ANOVA (Chapter 11)
- Mann–Whitney U-test (Chapter 14)
- Wilcoxon Signed ranks test (Chapter 14)
- Chi square tests (Chapter 15)

In the final year of the undergraduate scheme, there is a level six dissertation support module where some of the more advanced statistical procedures are covered. This module provides discipline-specific support sections including sessions on the commonly used data analysis procedures in those disciplines. For example, performance analysis students would have a session on non-parametric correlations (Chapter 8) and non-parametric tests of difference, recapping on the Mann–Whitney U-test and Wilcoxon signed ranks test and also including the Kruskal–Wallis H-test and the Friedman test (Chapter 14). This session would also revisit chi square tests (Chapter 15). Performance analysis students would also have a separate session on reliability assessment (Chapter 19). Physiology students would have a session on research design and data analysis in experimental research. This would recap on the t-tests (Chapter 10), Pearson's r (Chapter 8) and the one-way ANOVA (Chapter 11) and also cover repeated measures ANOVA tests (Chapter 11), some of the factorial ANOVA tests used in experimental research (Chapter 12) and linear regression (Chapter 9).

With around 400 students undertaking final year dissertations each year, it is not possible to cover all the statistical procedures that every student might require within their particular research project. Therefore, it is sometimes necessary for students to develop the knowledge and practical ability to use some procedures not covered in the level five research methods module or the level six dissertation support module. Supervisors may direct students to the specific material they need to cover and tasks they can use to develop the analytical capability they will require for the particular project. This book and other textbooks in the area will be able to assist supervisors and students in situations like this.

The research methods module within the postgraduate scheme is a level seven common and core module to a family of masters degree programmes. This module is a 20-credit point module done over a 10-week teaching block commencing with three weeks of general content on issues of scientific inquiry, epistemology, ontology and ethical principles and procedures. After this initial three weeks, there are two streams of sessions: one on quantitative research and one on qualitative research. These are timetabled at different times permitting students to attend one or both of the streams. Engagement with either stream on its own in addition to the first three weeks of the module is sufficient for students to achieve the learning outcomes. This book supports the quantitative stream through the coverage of multivariate statistical procedures such as multiple linear regression (Chapter 9), analysis of covariance (Chapter 11), multivariate ANOVA tests (Chapter 13), principle components analysis (Chapter 18), discriminant function analysis (Chapter 16), binary logistic regression (Chapter 16) and cluster analysis (Chapter 17). Statistical power (Chapter 20) and reliability tests such as Bland and Altman's (1986) 95 per cent limits of agreement (Chapter 19) are also covered within this level seven module. At the beginning of the quantitative stream there is a recap on the purpose of tests covered during undergraduate programmes and the situations where these tests should be applied. However, the series of practical exercises relating to these tests in the current book is not formally repeated at masters level. This book will provide a useful source of revision exercises for those masters students who do need to revisit undergraduate material before the quantitative stream of the postgraduate research methods module moves onto the more advanced procedures.

Within both undergraduate and postgraduate programmes at Cardiff Metropolitan University, students undertake exercises within other modules that require statistical analysis to be undertaken. One example of this is a physiology laboratory report where correlations and paired samples t-tests may be required. Another example is a performance analysis coursework to develop and test a hand notation system for a sport of the student's choice. This requires statistical analysis of inter-operator reliability data. In both cases and any other modules that use statistical analysis, this book can be used by the students to determine the appropriate statistical test to apply and how to apply it.

This section has described how the book is to be used within the undergraduate and postgraduate programmes at Cardiff Metropolitan University. Other universities and higher education institutions may cover research methods and statistics at different stages from the stages at which they are covered at Cardiff Metropolitan University. This will be for sound reasons relating to the particular curriculum or for practical reasons if sport and exercise students take research methods in a common module with non-sports students.

Chapter structure

Different chapters of this book have different purposes and cover different content. For example, the current chapter covers the terminology, notation and structure of the book. Chapters 3 and then Chapters 8 to 18 cover particular types of statistical analysis with examples and exercises. Therefore, these chapters follow a common structure although some differences in content lead to some variation in that structure. For example, Chapter 18 on data reduction covers one type of analysis called Principal Components Analysis while Chapter 14 on non-parametric tests covers four different tests. Each chapter that describes statistical procedures has the following general structure:

- Introduction
- Section or series of sections on the analysis procedure(s) covered

- Summary
- Exercises
- Project Exercise.

Each chapter section covering an analysis procedure is further broken down into the following structure:

- Purpose of the test
- Assumptions
- An example
- How to do the example in SPSS
- Reporting results.

There may be some variation to this structure; for example, some analysis procedures may be associated with follow-up procedures that are used in the event of significant results being found.

Slides

Rather than providing a set of slides for each chapter in the book, there is a set of slides for each statistical procedure that is covered. This allows individual lecturers using the book in their teaching to organize sessions for individual procedures if they prefer. If they wish to cover more than one procedure within a given session, they can use all of the PowerPoint presentations for that session or even combine them, changing the initial slides to combine the learning outcomes.

The slides follow the structure of the book with learning outcomes and a session overview being followed by slides on the purpose and assumptions of the statistical procedure, followed by an example which is illustrated using SPSS before guidelines for presenting results are provided. The exercises are followed by slides on the solutions to the exercises that are not shown in this text of the book itself. There is a slide on the project exercise described in the chapter which can be discussed within sessions. Most lecturers basing statistics modules on this book will probably not use the slides showing the use of SPSS. Instead, they will probably go into the SPSS package and illustrate the procedures to the students in a more interactive way. That is certainly the way the author will be using the book during teaching. The main benefit of the slides is that after the session, students can check steps on the slides if they are provided on virtual learning environments such as Blackboard™ (Blackboard Inc., Washington DC) or Moodle (Moodle Trust, Perth, Western Australia).

There are some chapters that cover more theoretical content, such as Chapters 1, 7 and 20. The slides for these chapters have been organized individually for the most appropriate presentation of that material. Each set of slides comes with a set of notes. These notes are not a verbatim script to be read out by the lecturer but key points and details that need to be presented when a slide is being shown. Each lecturer can present each slide in their own way based on their own teaching style, their learning environment and the nature of the group they are delivering to.

Exercises

At the end of all of the chapters, except Chapters 1 and 2, there are exercises to help readers check their understanding of the statistical procedures covered in the chapter. This book provides the exercise questions and identifies the files containing the data for the exercise and the Word file containing the solution to the exercise. Readers of this book can obtain these files from the internet site associated with the book. The relevant files are marked with the download icon ⬇ and can be found at www.routledge.com/cw/odonoghue. This material can be used in the way that best suits each student's development in the area. For example, some readers may prefer to attempt an exercise and not look at the solution until after they have completed the exercise. Other students may wish to look at the solution after each stage of the analysis process to make sure they are on the right course. Other students may prefer to look at the solution before attempting the exercise if they are unsure of where or how to start. In fact, recalling my own experience of learning statistics, it is fair to say that there are some basic tests that the author would have done blind of the solutions but other more complex tests where the author would have needed to look at the solutions regularly to make sure he was doing the right thing and knowing why he was doing the right thing.

Project exercises

The project exercises do not have provided data sets. Instead, they are ideas for research projects that involve data collection, analysis and interpretation that students may wish to consider as the basis of a dissertation idea. Alternatively, if statistical analysis is taught within a context of a research exercise, as it is within the level five research methods module at Cardiff Metropolitan University, lecturers might consider these project exercises as good examples to use within their teaching so that students see statistics being applied to real data that they have personally collected to answer a research question of interest.

SPSS

Comparing SPSS and Microsoft Excel

The software package used to illustrate statistical analysis is Statistics Package for the Social Sciences (SPSS) Version 18.0 (SPSS: An IBM Company, Armonk, NY). Many readers will have used spreadsheet packages such as Microsoft Excel to store and analyse data. There are two fundamental differences between SPSS and Excel. The first is that in Excel, data can be placed in any cells the operator wishes to use. For example, the rows could represent variables and columns could represent participants or vice versa. Indeed, a participant's data could be spread over more than one row and/or column. We could also use a single column to include data for more than one variable, stacked on top of each other. In SPSS, however, the columns are always the variables and the rows are always the units of analysis (participants, matches or cases). The second difference between SPSS and Excel is that SPSS uses separate windows to display the data and the results of the analysis of that data. In Excel, data, charts and functions evaluating summary statistics can all be included on a single spreadsheet.

Creating an SPSS data sheet

This textbook is not a specific textbook on the SPSS package and some of the detail of how to open the package and other very detailed file processing operations will not be covered. Readers wishing to read about file options are referred to the book by Ntoumanis (2001: 2–13). There are two ways of creating an SPSS data sheet. First, data could be entered into the datasheet in the SPSS package. Second, the data could be keyed into a spreadsheet package, such as Microsoft Excel, and then loaded into SPSS. The author prefers the latter approach because Excel is excellent for the typical pre-processing of data that is required prior to statistical analysis. For example, we may enter variables representing the frequency of events which then need to be pre-processed to determine percentage occurrences. This may identify some impossible values such as percentages that are greater than 100 per cent. This type of data cleaning activity is best done in Excel before using SPSS.

If data are being entered directly into an SPSS datasheet, the user should use the 'Variables' tab to access a form that allows them to define their variables first. This is shown in Figure 2.1. The user can give names to the variables, select the type of variable and define the number of decimal places to use with each numerical scale variable. The scale of measurement is very important because some analysis facilities in SPSS will only use independent variables or dependent variables that use the expected scales. Where we have nominal or ordinal variables, we can use numeric codes to represent the values of the variable. For example, '1' could indicate female and '2' could indicate male. This eases the process of entering data because '1' and '2' are easier to type than female and male. Furthermore, errors are reduced during data entry because 'Female' and 'female' are considered to be two different values by SPSS. Even if data are loaded in from an Excel spreadsheet, the variables should be defined to assist analysis later on.

Figure 2.2 shows an SPSS datasheet presented by the data view window. The gender variable is currently using the values '1' and '2' but if the View Label facility was selected, these would appear as 'Female' and 'Male'. The View Label menu item allows us to toggle between these two presentation modes for labelled variables.

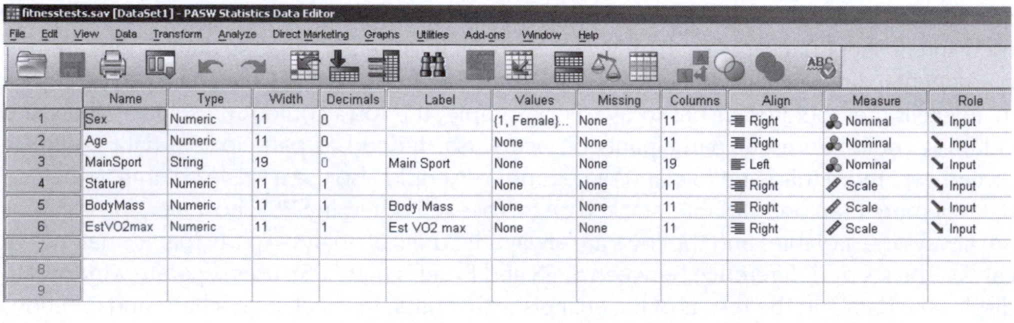

	Name	Type	Width	Decimals	Label	Values	Missing	Columns	Align	Measure	Role
1	Sex	Numeric	11	0		{1, Female}...	None	11	Right	Nominal	Input
2	Age	Numeric	11	0		None	None	11	Right	Nominal	Input
3	MainSport	String	19	0	Main Sport	None	None	19	Left	Nominal	Input
4	Stature	Numeric	11	1		None	None	11	Right	Scale	Input
5	BodyMass	Numeric	11	1	Body Mass	None	None	11	Right	Scale	Input
6	EstVO2max	Numeric	11	1	Est VO2 max	None	None	11	Right	Scale	Input
7											
8											
9											

Figure 2.1 The variable definition form in SPSS.

using this book

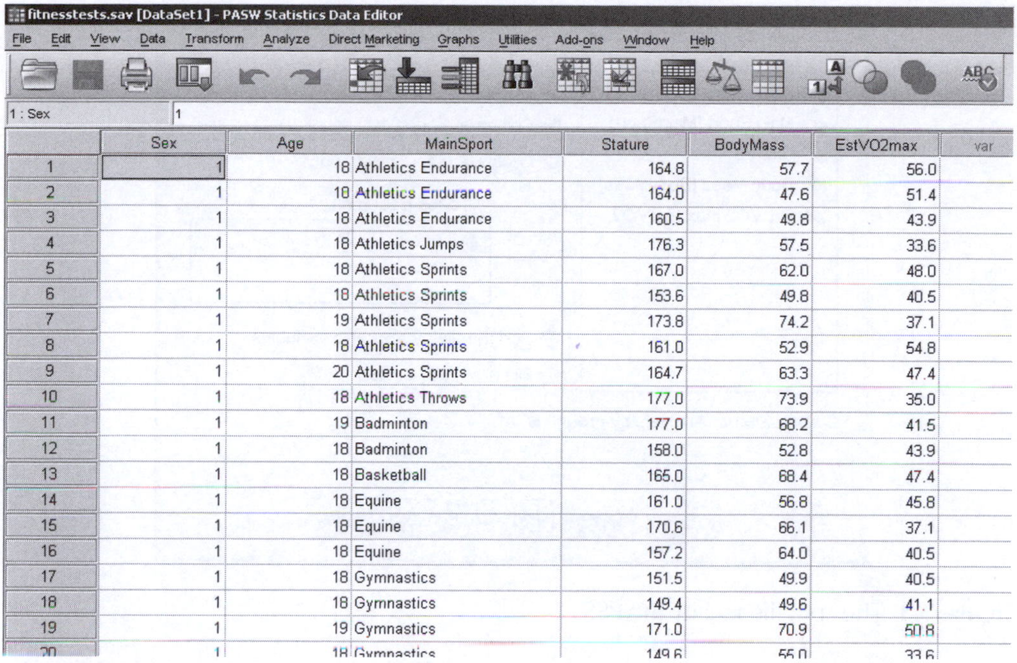

	Sex	Age	MainSport	Stature	BodyMass	EstVO2max	var
1	1	18	Athletics Endurance	164.8	57.7	56.0	
2	1	18	Athletics Endurance	164.0	47.6	51.4	
3	1	18	Athletics Endurance	160.5	49.8	43.9	
4	1	18	Athletics Jumps	176.3	57.5	33.6	
5	1	18	Athletics Sprints	167.0	62.0	48.0	
6	1	18	Athletics Sprints	153.6	49.8	40.5	
7	1	19	Athletics Sprints	173.8	74.2	37.1	
8	1	18	Athletics Sprints	161.0	52.9	54.8	
9	1	20	Athletics Sprints	164.7	63.3	47.4	
10	1	18	Athletics Throws	177.0	73.9	35.0	
11	1	19	Badminton	177.0	68.2	41.5	
12	1	18	Badminton	158.0	52.8	43.9	
13	1	18	Basketball	165.0	68.4	47.4	
14	1	18	Equine	161.0	56.8	45.8	
15	1	18	Equine	170.6	66.1	37.1	
16	1	18	Equine	157.2	64.0	40.5	
17	1	18	Gymnastics	151.5	49.9	40.5	
18	1	18	Gymnastics	149.4	49.6	41.1	
19	1	19	Gymnastics	171.0	70.9	50.8	
20	1	18	Gymnastics	149.6	55.0	33.6	

Figure 2.2 Datasheet viewer window in SPSS.

Logical partitioning of a data sheet

Figure 2.3 shows the split file facility that allows us to logically split the file using some variable, in this case gender. Once we have identified the partitioning variable, we can choose to present the output for the groups separately or within the same tables and charts to allow direct comparison. Any analysis done once the file has been split will be done for each gender separately. This saves us having to make two additional copies of the datasheet: one for the females and one for the males.

Computing new variables

The Transform menu contains a feature that allows us to produce new variables using existing variables. For example, Figure 2.4 shows how we can produce a variable BMI (Body Mass Index) using the existing variables body mass and stature. Note that in the expression entered in Figure 2.4, it was necessary to divide Stature by 100 because the datasheet represents stature in centimetres rather than metres. In forming such expressions, we should be aware that power ('**') has a higher priority then multiplication and division ('*' and '/') which in turn have a higher priority than addition and subtraction ('+' and '-'). If we wish to override these priorities, then we need to use parentheses ('(' and ')').

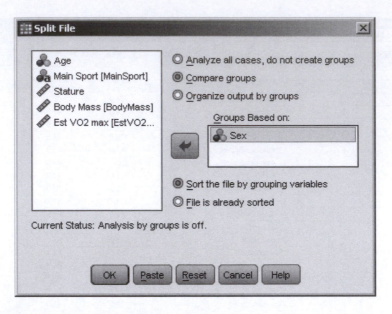

Figure 2.3 The split file facility in SPSS.

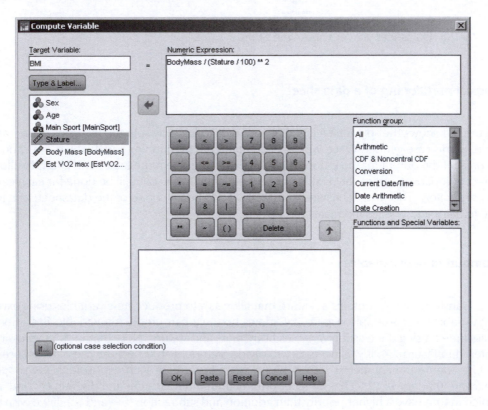

Figure 2.4 Computing a new variable in SPSS.

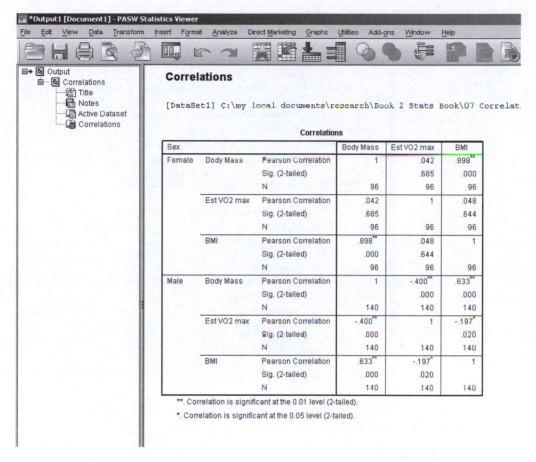

Figure 2.5 The SPSS output viewer.

Figure 2.5 shows the output viewer window of SPSS that contains all the results of any analyses that are done on the data. This particular example shows correlation results presented once we have split the file into separate logical partitions for females and males.

Notation

SPSS is used as an example statistical analysis software package in this book and, therefore, the notation used to describe communication between an operator and SPSS is covered in this chapter. SPSS provides a menu bar with menus for analysis, chart creation, data operations, data transformations, general file management and other facilities. The menu bar can be thought of as a first level of menu choices. Each menu item provides a secondary menu with a choice of actions some of which lead to a third level menu. For example, Figure 2.6 shows the use of the 'Analyse' menu to select 'Correlations' from the drop down menu which reveals a third level menu from which we might wish to select 'Bivariate'. In this book,

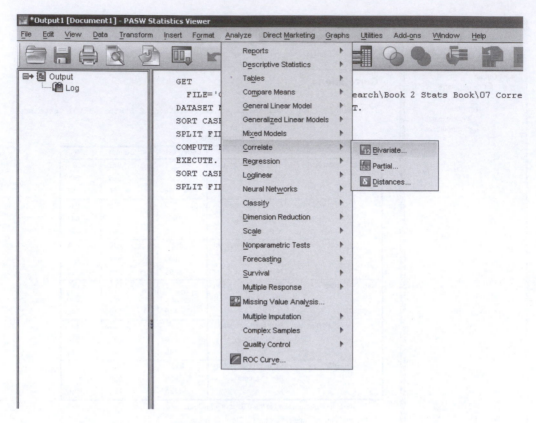

Figure 2.6 Menu navigation in SPSS.

menu navigation is represented in bold text with forward arrows (→) showing which choice to make next. So this example of selecting bivariate correlations is shown as follows:

Analyse → Correlate → Bivariate

The book illustrates statistical analysis procedures using data provided in SPSS datasheets. These datasheets are provided on an associated internet site (see the back cover of the book for details) and this book refers to the file names in normal text followed by the file type extension. For example, we have an SPSS datasheet 08-fitnesstests.SAV which is used as an example in Chapter 8 and for exercises in some of the other chapters. There are also Microsoft Excel spreadsheets used in other examples and exercises and these filenames will be seen with the .XLS extension after them. Microsoft PowerPoint presentations containing slides associated with different chapters use the .PPT extension to the file name while Word documents containing detailed solutions to exercises use the .DOC extension.

SPSS facilities for statistical analysis, chart building, data operations and data transformation all require the user to transfer variables of interest into areas of pop-up windows which Ntoumanis (2001) referred to as dialog boxes. An example of such a pop-up window is the

28
using this book

pop-up window for bivariate correlations shown in Figure 2.7. Users can communicate with SPSS through these pop-up windows and often the windows can stack up on top of each other as users select more detailed features. For example, if we were using the pop-up window in Figure 2.7, we could click on the **Options** button to activate a further pop-up window to advise SPSS of options we wish to include when the correlation analysis is performed. The Options pop-up window will have a **Continue** button allowing us to close the window once we have made our optional choices and wish to continue defining the correlation analysis we wish to perform. As you can already see in the text of this paragraph, bold is used to represent the on-screen buttons within a pop-up window such as **Options**, **Bootstrap**, **OK**, **Paste**, **Reset**, **Cancel** and **Help**. Like Ntoumanis's (2001) *Step-by-Step Guide to SPSS for Sport and Exercise Studies*, the current book uses italics to represent areas of a pop-up window where variables need to be transferred to. The variables themselves are included in single quotes. Therefore, what has been done in Figure 2.7 is described as transferring 'Stature', 'Body Mass' and 'Est $\dot{V}O_2$ max' to the *Variables* area. When selecting or deselecting items from check boxes (such as Pearson's r, Kendall's τ and Spearman's ρ in Figure 2.7) there is no special notation used because these have been straightforward enough for the author to describe without a special notation. Similarly, choices such as one-tailed or two-tailed analyses are described in the book without the use of a special notation.

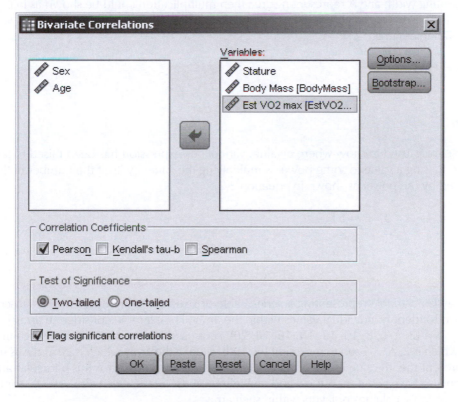

Figure 2.7 Bivariate correlations pop-up window.

EQUATION NOTATION

Equations and expressions

This book contains equations in most chapters and these are numbered in parentheses to the right of the equation. An equation typically uses an equal sign (=) to equate a variable or expression on the left-hand side of the equal sign to the expression on the right hand side of the equal sign. Sometimes, there may be more than one equal sign where three or more different expressions are being shown as being equal. There are occasions in the book where the numbered equations include stand-alone expressions which are not equated to any other expression or variable. For example, equations 5.28, 5.29 and 5.30 in Chapter 5 are expressions for the conditional probability of the serving player winning a game within a tennis match given that deuce has been reached.

Multiplication

There are some equations that use a '.' to represent multiplication rather than '×'. This is to avoid confusion where readers might think '×' is the name of a variable. There are some basic equations where '×' has been used. So if we have variables H representing height, W representing width and A representing area then multiplication could be shown as in equation 2.1. The use of the '.' also prevents situations where readers might think 'HW' was a single variable name.

$$A = H.W \tag{2.1}$$

Power

Superscript is used to show where a value, variable or expression has been raised to some power. Raising a value to some power is multiplying the value by itself the number of times specified by the power as shown in equation 2.2.

$$3^4 = 3 \times 3 \times 3 \times 3 = 81 \tag{2.2}$$

Arrays

An array is a variable representing a series, table or structure of numbers where subscripts are used to identify individual values within the array. For example consider an array of 10 values, $A = (2, 4, 6, 8, 10, 12, 14, 16, 18, 20)$. A_1 is used to represent the first value in the array which is 2, A_{10} is used to represent the 10th value in the array which is 20 and A_i is used to represent the ith value in the array where i is between 1 and 10 for this particular array. We can also have two-dimensional and multidimensional arrays which also use subscripts to represent rows, columns or layers within such arrays.

There are occasions where subscripts are used for other purposes within variables. For example, in Chapter 19 we use MS_R and MS_{C+E} to represent two different mean square values. In this situation, MS is not an array.

Precedence of arithmetic operators

Some equations use a combination of powers, multiplication, division, addition and subtraction. These are not simply applied from left to right within the equation. The powers are evaluated first because they have a higher priority than multiplication, division, addition and subtraction. The powers are evaluated from left to right within the equation. Second, multiplication and division are evaluated from left to right and finally addition and subtraction is done. Consider equation 2.3 that involves powers, addition, subtraction, multiplication and division. First, the two powers are evaluated as shown in equation 2.4 and then the divisions and multiplications are evaluated from left to right as shown in equation 2.5. This means that where we have $6 \times 3 / 2$, we first multiply the 6 and the 3 and then divide the resulting product by 2. Finally, we apply the additions and subtractions from left to right resulting in a value of y of 20.

$$y = 31 + 2^4 / 4 + 6 \times 3 / 2 - 5^2 + 1 \tag{2.3}$$

$$y = 31 + 16 / 4 + 6 \times 3 / 2 - 25 + 1 \tag{2.4}$$

$$y = 31 + 4 + 9 - 25 + 1 \tag{2.5}$$

Magnitude

If we have a value, x, the magnitude of x is denoted $| x |$. This is the value without the sign and is always positive. For example $| -4 | = 4$ and $| 4 | = 4$.

Set notation

When discussing probability, it is necessary to talk about sets of experimental outcomes. In reasoning about these, we need to apply some set operators within expressions. A set is expressed using '{ }' brackets to show the elements or members of a set. For example, equations 2.6 and 2.7 show the set of international rugby union teams that have won the World Cup, W, and the set of international teams that participate in the European Six Nations tournament, S.

$$W = \{Australia, England, New Zealand, South Africa\} \tag{2.6}$$

$$S = \{England, France, Ireland, Italy, Scotland, Wales\} \tag{2.7}$$

Scotland is an element of S which is denoted Scotland $\in$ S. A set containing Ireland and Wales is a subset of S which is denoted {Ireland, Wales} $\subseteq$ S. The symbol '$\cup$' is used to represent the union of two sets which is the set containing any member of one or both of the two sets being unioned. For example, the union of W and S is shown in equation 2.8.

$$W \cup S = \{Australia, England, France, Ireland, Italy,$$
$$New\ Zealand, Scotland, South\ Africa, Wales\} \qquad (2.8)$$

The symbol '∩' is used to represent the intersection of two sets which is a set that only contains members that are members of both of the sets. For example, equation 2.9 shows the intersection of W and S.

$$W \cap S = \{England\} \qquad (2.9)$$

Sums

One of the most commonly used types of equation in this book is the sum of some series of values or expressions. Consider the notation used to express a sum in equation 2.10. The symbol 'Σ' represents a sum. The only purpose of the variable i is to distinguish the terms being summed; the scope of i is solely within the sum. The 'i=1' subscripting the 'Σ' symbol and the 'n' superscripting the 'Σ' symbol mean that we are summing the terms A_i for all values of i from 1 to n.

$$\sum_{i=1}^{n} A_i = A_1 + A_2 + A_3 + A_4 + \ldots + A_n \qquad (2.10)$$

There are some occasions where the sum uses elements of a set rather than a numerical index. For example, in equation 2.11 we may have a set of numbers, SET, and this expression is the sum of each number, e, that is an element of SET.

$$\sum_{e \in SET} e \qquad (2.11)$$

Two uses of the bar accent

When dealing with a numerical variable, x, we use $\bar{x}$ to represent the mean of a sample of participants for the variable x. However, when dealing with sets, the complement of a set, S, is expressed as $\bar{S}$ and represents those known elements that are not members of the set S.

SUMMARY

This chapter has provided guidance on how users should use the book as a reference text to support their needs. This chapter has also introduced the way statistical procedures will be covered in the book including the notation to be used, the typical chapter structure and the use of the resources available on the website associated with this book.

CHAPTER 3

DESCRIPTIVE STATISTICS

INTRODUCTION

Two main types of statistics are descriptive statistics and inferential statistics. These are often used in a complementary way within quantitative research studies. As the name suggests, 'descriptive' statistics describe samples using sample statistics such as averages or frequencies. Descriptive statistics are used within a reductionist research approach whereby we gather data for a sample of individual participants from a population of interest and summarize the sample using descriptive statistics. If a large enough random sample is used, then researchers may generalize the findings for the sample to be relevant to the whole population from which the sample was drawn. Descriptive statistics are concerned with facts such as 'how many', 'how few', 'how high', 'how low'. The 'how many' and 'how few' type questions are answered using frequency profiles. How many of the sample are males? How many are females? The 'how high' and 'how low' type questions are answered using averages such as the mean or median value for the sample. When considering how high or low some average value is, we are often also interested in how consistent the sample is about this average. Do all of the participants have values close to the sample average? Do the participants have a wide range of values either side of the average? Descriptive statistics, such as the standard deviation or the inter-quartile range, are used to describe how consistent a sample is about the mean and the median respectively.

Descriptive statistics can also be used to compare samples. For example, we may wish to compare female athletes and male athletes in terms of some variables of interest. What percentage of female athletes prefer team sports? What percentage of male athletes prefer team sports? Is the percentage of female athletes who prefer team sports similar or different from that of male athletes? How tall are female athletes? How tall are male athletes? Is the height of female athletes similar or different from that of male athletes?

Inferential statistics support descriptive statistics by indicating the significance of any differences observed between samples being compared. A significant difference is one where the probably that the difference observed between the samples is not due to random chance. So if we see that a greater percentage of male athletes prefer team sports than female athletes, is this a significantly greater percentage? If we see that male athletes are taller than female athletes, are they significantly taller? The significance of differences between samples is covered by a large set of tests that are covered in Chapters 7, 10, 11, 12, 13, 14 and 15, with significance also coming into Chapters 8, 9, 16 and 20. Many students make the mistake of thinking that inferential statistics are more important than descriptive statistics. They usually make this mistake simply because the inferential statistical tests are more complex, more difficult to select and more difficult to apply than descriptive statistics. Dissertation students often view inferential statistics as a 'must do' task of a research project, believing that they will fail the dissertation if inferential statistics are not done. There are two very important messages that this author tries to convey to students embarking on research projects; these are discussed in the next two paragraphs.

The descriptive statistics are far more important than inferential statistics. Ideally we would like to see the descriptive statistics and the supporting inferential statistics, but if there was a choice of one or the other this author would choose descriptive statistics on every occasion. If a research report tells us that the heights of male and female athletes are significantly different, we are left asking a number of questions. Which group is taller? Are both male and female athletes short but with one group significantly taller than the other? Are both male and female athletes tall but with one group significantly taller than the other? Is one group tall and the other short with the difference in height being significant? Even if the report states the direction of the difference (for example, male athletes are significantly taller than female athletes), we are still left wondering if both groups are tall, both groups are short or if the males are tall and the females are short. It is the descriptive statistics that answer these questions. It is the descriptive statistics that allow us to compare the values within the study to known norms or relevant values reported in previous research. This author regards the inferential statistics as 'the icing on the cake', but we need the cake!

The second message that the author tries to convey to dissertation students is that the results section of a dissertation is typically allocated 15 per cent of the marks in a dissertation marking scheme. A results section containing descriptive statistics but no inferential statistics will usually get a higher mark than a results section that provides inferential statistics but fails to report the all important descriptive statistics. A dissertation that reports the descriptive statistics well but does not include inferential statistics could achieve a mark of 8 or 9 out of 15. Even if the results were not written up well enough to be given a pass mark (say a mark of 4 or 5 out of 15 was awarded), the student can pass the dissertation with 35 or 36 marks out of the 85 per cent allocated for other sections such as the literature review, description of methods and the discussion.

The remainder of this chapter will discuss the descriptive statistics used to summarize variables measured on different scales. The different descriptive statistics available, how they are produced in SPSS and how they should be reported are all covered.

DESCRIPTIVE STATISTICS FOR NOMINAL VARIABLES

Frequency profiles and the mode

Nominal variables, as mentioned in Chapter 1, are used to classify participants using named categories. For example, the nominal variable gender is used to classify participants as female or male. The main descriptive statistics used to summarize samples in terms of nominal variables are:

- Frequency distributions. These show the number of participants in each named category of the nominal variable, for example the frequency of females and the frequency of males. Frequencies can also be expressed as percentages of cases to aid interpretation due to the common use of percentages.
- Cross-tabulation of frequencies allows two nominal variables to be considered together. For example, it may not be sufficient to know the gender and sports preference frequencies for a sample. We may also want to look at the variables in combination. How many female participants prefer team sports? How many female participants prefer individual sports? How many male participants prefer team sports? How many male participants prefer individual sports?
- The mode is the most commonly occurring named category of a variable. For example, if a study has more female than male participants then the modal gender will be female.

Example: European soccer cities

Conway and O'Donoghue (2001) gathered data on 92 European cities that had a population of over 500,000 or which were capital cities or which had a soccer team that had won a European trophy (European Cup, European Cup Winners Cup and UEFA Cup). The purpose of the study was to compare those cities that had at least one team that had won a European trophy with those cities that did not. The cities with at least one team that had won a European trophy were termed 'trophy-winning cities'. The variables collected included a nominal variable indicating whether the country that the city was in hosted one of Europe's major soccer leagues. The major soccer leagues were the English FA Premier League, Serie A in Italy, the German Bundesliga and La Liga in Spain. The number of trophy-winning cities and non-trophy-winning cities was of interest. This was determined for the overall sample of cities as well as for the cities within the four nations hosting major soccer leagues and those nations that didn't. The appropriate descriptive statistics used to provide this information were frequency distributions, cross-tabulated frequency distributions and the mode.

SPSS

This example uses the file 03-soccercities.SAV. The first point to make about this data file is that the scale of measurements of each variable has been set up. This is very important in SPSS as the scale of measurement of the variable dictates how the variable can be used within various analyses. There is a tab at the bottom of the SPSS data viewer window that allows us to view and edit properties of variables as shown in Figure 3.1 (Variable view).

To obtain a frequency profile for any nominal variable we use **Analyse → Descriptive Statistics → Frequencies** which activates the pop-up window shown in Figure 3.2. We transfer the variable of interest, 'major' into the *Variable(s)* area and then click on **OK** to see the frequency profile. Table 3.1 shows that the default output for frequency profiles includes an equivalent percentage breakdown. This shows that 47 of the 92 cities (51.1 per cent) were trophy-winning cities making trophy-winning city the modal type of city within the sample.

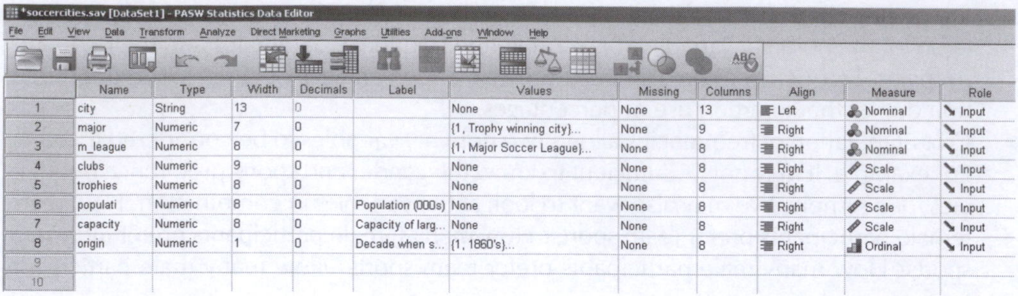

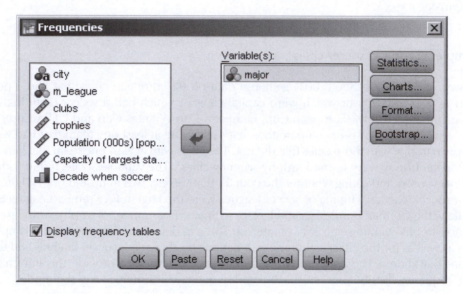

Figure 3.1 Variable view.

Figure 3.2 Frequencies pop-up window.

Table 3.1 SPSS output for frequency distributions

				Major		
		Frequency	Percent	Valid Percent	Cumulative Percent
Valid	Trophy winning city	47	51.1	51.1	51.1
	Other City	45	48.9	48.9	100.0
	Total	92	100.0	100.0	

Cross-tabulation of frequencies allows us to consider the combined frequencies of two variables. We use **Analyse → Descriptive Statistics → Crosstabs** to set up a cross-tabulation of frequencies. This activates the pop-up window shown in Figure 3.3. One variable is transferred into the *Row(s)* area and the other is transferred into the *Column(s)* area. This author

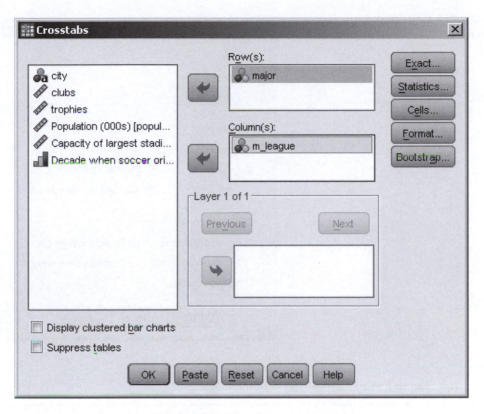

Figure 3.3 Cross-tabulation pop-up window.

typically uses columns to represent the variable with the lower number of values. This makes better use of an A4 page used to present results.

The **Options** button activates the pop-up window shown in Figure 3.4 which allows us to include percentage breakdowns within the output. These percentages could be for individual columns, rows or the whole cross-tabulation. In this example, we are interested in the percentage of cities that are trophy-winning cities within the two different types of country. Therefore, we choose to have percentage breakdowns of the columns. Table 3.2 shows the output produced by SPSS. This shows that 28 out of 32 cities (87.5 per cent) within England/Wales, Germany, Italy and Spain were trophy-winning cities compared with 19 out of 60 (31.7 per cent) of cities in other countries. Therefore, the modal type of city in England/Wales, Germany, Italy and Spain is a trophy-winning city but it is not a trophy-winning city in other countries.

Reporting results

In truth, the initial frequency analysis could have been done with the cross-tabulation of frequencies as the frequency profile for each variable individually is shown in the row and

Figure 3.4 Cell options for cross-tabulation.

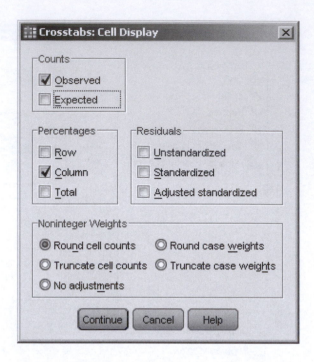

Table 3.2 Output for cross-tabulation of frequencies

				m_league	
major * m_league Cross-tabulation					
			Major Soccer League	m_league Other League	Total
major	Trophy winning city	Count	28	19	47
		% within m_league	87.5%	31.7%	51.1%
	Other City	Count	4	41	45
		% within m_league	12.5%	68.3%	48.9%
Total		Count	32	60	92
		% within m_league	100.0%	100.0%	100.0%

column totals in Table 3.2. We may have done the frequency profiles for the individual variables for exploratory purposes, but a single summary table would be used as shown in Table 3.3. The associated text would be as follows:

> Table 3.3 shows the frequency of different types of city with respect to soccer success in Europe. The modal type of city was trophy-winning city in England/Wales, Germany, Italy and Spain but not in other countries.

There are other forms of presentation used with frequencies; for example, column charts and pie charts could be used show the frequency distribution of one nominal variable while

descriptive statistics

Table 3.3 Distribution of European trophy winning cities from nations hosting Europe's four major soccer leagues and other nations

Soccer success	Nations hosting Europe's major soccer leagues	Other nations	Total
Trophy winning	28 (87.5%)	19 (31.7%)	47 (51.1%)
Not trophy winning	4 (12.5%)	41 (68.3%)	45 (48.9%)
Total	32 (100.0%)	60 (100.0%)	92 (100.0%)

a clustered column chart could be used to present the cross-tabulation of a pair of nominal variables. Sometimes we can think of ways of presenting results that have greater visual impact. For example, Conway and O'Donoghue (2001) used a map to compare the proportion of trophy-winning cities in the countries hosting Europe's major soccer leagues with other countries. This is shown in Figure 3.5 where we can see that there is a greater proportion of dark dots representing trophy-winning cities in England/Wales, Italy, Spain and Germany than there is in other countries. The use of maps to display results of statistical analysis has also been proposed by Graham (2006: 64–5).

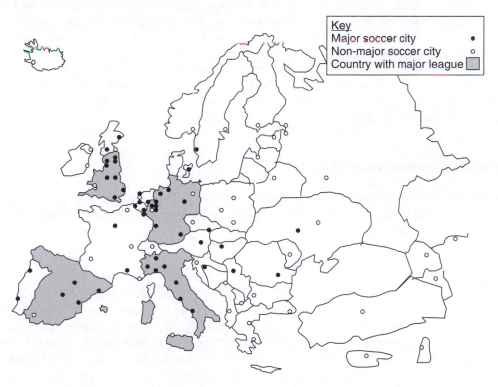

Figure 3.5 Using a map to present descriptive statistics (Conway and O'Donoghue, 2001).

DESCRIPTIVE STATISTICS FOR ORDINAL VARIABLES

Median, minimum, maximum, lower and upper quartiles

Ordinal variables are nominal variables but with the additional property that the named categories are ordered. Given that ordinal variables are discrete with a finite number of values, it is possible to describe them using the mode, frequency profiles and cross-tabulations of frequencies in the same way that nominal variables are described. There are other descriptive statistics that can be used with ordinal variables. These are the median, minimum and maximum. It is also possible to report the lower and upper quartile values. The median value is the one in the middle when all cases are arranged in ascending order of our ordinal variable of interest. However, if we have an even number of values, then there will be two values in the middle. If these two values are actually the same, then this value is the median. If, however, they are different values, we cannot take the mean of the two values because that assumes an interval between different points on the measurement scale and ordinal variables do not have such an interval. For example, if we had an ordinal variable with five possible values ('strongly disagree', 'disagree', 'undecided', 'agree', 'strongly agree') and we determined that the median was somewhere between 'undecided' and 'agree', we would not be able to determine the mean of these two values. So in this situation we simply state that the median is somewhere between 'undecided' and 'agree'. The same process is applied to determining the lower quartile, upper quartile and any other percentile we may be looking for. If the percentile location is at an actual data value from the sample, we use that value. If, however, we have data values on either side of the percentile point and these two values are the same, we use that single value. If we have data values on either side of the percentile point but these two values are not the same, we state that the percentile is somewhere in between these two values. The location of lower and upper quartiles within an ordered sample will be discussed more in the next part of this chapter which deals with interval and ratio scale variables. Chapter 4 on standard scores also discusses percentile points and quantiles further.

Example: European soccer cities

Describing ordinal variables is also illustrated using the example of European soccer cities (Conway and O'Donoghue, 2001). This time the variable of interest is the decade when soccer was introduced to the city ('origin').

SPSS

When we use **Analyse → Descriptive Statistics → Frequencies** we can use the **Statistics** button to request the median and quartiles as shown in Figure 3.6. In this example, we again use the file 03-soccercities.SAV and the variable is 'Origin'. This provides the output shown in Tables 3.4 and 3.5. The median, lower and upper quartiles are reported as 4, 2 and 5 respectively in Table 3.4. These are numeric codes for which data values have been defined but it is the codes that appear in this first table of the output. The frequency distribution in

40

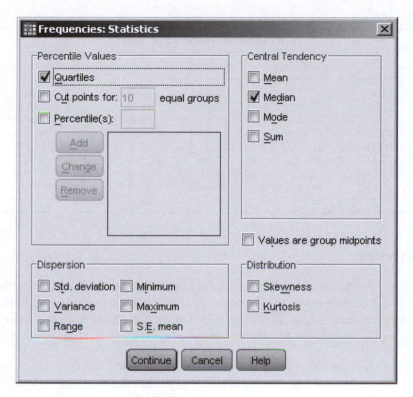

Figure 3.6 Statistics for ordinal variables.

Table 3.4 SPSS output showing quartiles

Statistics		
Decade when soccer originated		
N	Valid	92
	Missing	0
Median		4.00
Percentiles	25	2.00
	50	4.00
	75	5.00

Table 3.5 does use the value labels that were defined when the data file was created. This allows us to determine that 2, 4 and 5 represent the 1870s, 1890s and 1900s respectively. We can also see that the modal decade of origin of soccer was the 1870s with 25 of the 92 cities having soccer introduced in that decade. Note that the mode, median, lower and upper quartiles are values on this ordinal scale of measurement. Therefore the mode is the 1870s and not 25; 25 is the frequency of the mode. Similarly lower quartile, median and upper quartile are the 1870s, 1890s and 1900s.

Table 3.5 SPSS output describing ordinal variables

		Frequency	Percent	Valid Percent	Cumulative Percent
	Decade when soccer originated				
Valid	1860s	7	7.6	7.6	7.6
	1870s	25	27.2	27.2	34.8
	1880s	6	6.5	6.5	41.3
	1890s	12	13.0	13.0	54.3
	1900s	21	22.8	22.8	77.2
	1910s	3	3.3	3.3	80.4
	1920s	2	2.2	2.2	82.6
	1930s	16	17.4	17.4	100.0
	Total	92	100.0	100.0	

Reporting results

The frequency profile shown in Table 3.5 could be presented as a summary table such as Table 3.6 or in a column graph. In either case, symbols could be used to highlight the modal, median, lower and upper quartile decades. The following text could be used to support the tabular or column chart summary of the results:

> Table 3.6 shows the frequency of cities where soccer originated in different decades. The median decade of origin was the 1890s but the modal decade of origin was the 1870s with soccer being introduced into 27.2 per cent of the cities in this decade.

DESCRIPTIVE STATISTICS FOR INTERVAL AND RATIO VARIABLES

Mean, standard deviation, median, range and inter-quartile range

Frequency profiles are not feasible for interval and ratio scale variables that are measured on a continuous scale. Frequency profiles require values to be grouped into subranges. This would also permit the modal subrange to be identified. It is more common for interval and ratio scale variables to be summarized using measures of location, other than the mode, as well as measures of dispersion (Anderson *et al.*, 1994: 60–80).

Table 3.6 Decade of origin of soccer in European cities

Decade	1860s	1870s ^ $	1880s	1890s &	1900s #	1910s	1920s	1930s
Frequency	7	25	6	12	21	3	2	16

^ Lower quartile
& Median
# Upper quartile
$ Mode

The difference between interval and ratio scale variables is that a value on a ratio scale can meaningfully be divided by another value on that scale. This is not the case for interval scale variables because zero does not represent a complete absence of the concept being measured. However, this difference does not affect the use of descriptive statistics with interval or ratio scale variables. Therefore, packages such as SPSS often combine the interval and ratio scales (in SPSS the combined scale of measurement is simply called scale). There is division involved in the calculation of descriptive statistics such as the mean, but remember this type of division does not divide one value on the measurement scale (for example, 6m) by another (3m) to obtain a ratio between the two values (for example, 2). Instead, a sum of values measured on the scale (for example, 50m) is divided by a frequency of values (for example, 10) to give a mean value that is on the measurement scale (for example, 5m).

The measures of location that can be used are the mean or the median. The mean is the preferred measure of location because the value of the mean is computed using all the other values. The mean is an appropriate average where values are distributed symmetrically; that is, the number and spread of values above the mean is similar to the number and spread of values below the mean. Asymmetrical distributions are referred to as skewed distributions. Where data are skewed, the median is sometimes used instead. An example of this might be prize money for the top 1,000 tennis players in the world where 10 per cent of participants have values that are so high that they raise the mean to a value greater than 90 per cent of the values. In situations like this, the median would be a more suitable average than the mean.

Measures of dispersion include the variance, standard deviation, range and inter-quartile range. The standard deviation is the preferred measure of dispersion for interval and ratio scale variables. This is because the standard deviation is calculated using all the values in the sample. The variance is the square of the standard deviation but is not used as a measure of dispersion as much as the standard deviation would be. This is because the variance does not use the same units of measurement as the variable. It would be a bit like using area (m^2) as an indication of variability of distance (m). The coefficient of variation expresses the standard deviation, SD, as a percentage of the mean as shown in equation 3.1. This is a good measure of dispersion if variability increases as values increase. That is, the standard deviation would be a given percentage of the mean for different subranges of a variable.

$$CV = 100 \, SD \, / \, mean \tag{3.1}$$

Where variables are skewed, the inter-quartile range can be used. The inter-quartile range is the difference between the values for the lower and upper quartiles. The range is the difference between the largest and smallest values in the sample. The inter-quartile range is preferred to the range because the values at the lower and upper quartiles are located at those places within the ordered samples due to their relation with all the other values in the sample. The range, however, can be much more influenced by the minimum and maximum values. These could take many different values and still be outside the range of the remaining values.

Consider the 10 values shown in Table 3.7 which range from 8 to 73. The ranks of these data values are also shown. There is an even number of values and, therefore, the median value is the mean of the 5th and 6th values. This gives a median of 44. There is no 5.5th value but

Table 3.7 Arbitrary ranked data

Rank	1st	2nd	3rd	4th	5th	6th	7th	8th	9th	10th
Value	8	19	29	36	42	46	48	52	60	73

the 5.5th point was chosen because the size of the sample is 10 with ranks starting at the point 1 and ending at the point 10. This makes range of 9 (10–1) value points with 5.5 being the point in the middle, half way between 1 and 10.

To determine the lower quartile and upper quartile, a similar process can be used. The lower quartile would be located at the data point 1 + (10–1)/4 = 3.25 while the upper quartile would be located at the data point 1 + 3(10–1)/4 = 7.75. Just as there is no actual median data value in the sample at the data point 5.5, there are no values at the points 3.25 or 7.75 either. Therefore, we have to use the values at either side of these data points to compute the upper and lower quartile. One way is to simply use the mean so the lower quartile would be (29 + 36)/2 = 32.5 while the upper quartile would be (48 + 52)/2 = 50. This method does not use interpolation and gives Tukey's Hinges. However, the 3.25th data point is closer to the 3rd data point than the 4th data point and 7.75 is closer to the 8th data point than to the 7th data point. An alternative method is to use interpolation. The lower quartile will be a value that is the 3rd value (29) plus 0.25 of the difference between the 3rd and 4th values (36–29) × 0.25 = 1.75. This gives a lower quartile value of 29 + 1.75 = 30.75. Similarly, the upper quartile is a value that is the 7th value (48) plus 0.75 of the difference between the 7th and 8th values (52–48) × 0.75 = 3. This gives an upper quartile value of 48 + 3 = 51.

Example: European soccer cities

The example of European soccer cities (Conway and O'Donoghue, 2001) will be continued using two ratio scale variables: the population of the city and the capacity of the largest soccer stadium in the city. We wish to determine the mean and standard deviations for these variables and, if they are skewed, we also wish to determine the median and inter-quartile range for each variable. This can be done for the sample of cities as a whole as well as for trophy-winning cities and non-trophy-winning cities individually to allow a comparison.

SPSS

This example uses the file 03-soccercities.SAV and the specific variables capacity and population. To determine the mean and standard deviation of interval or ratio scale variables in SPSS, we use **Analyse → Descriptive Statistics → Descriptives** transferring the variables of interest into the *Variable(s)* area of the Descriptives pop-up window as shown in Figure 3.7. The **Options** button activates a further pop-up window which is shown in Figure 3.8 which allows us to tailor the output produced by SPSS.

When we click on **OK**, the output shown in Table 3.8 is produced. As we can see the populations of the cities is severely skewed with a greater standard deviation than the mean.

44

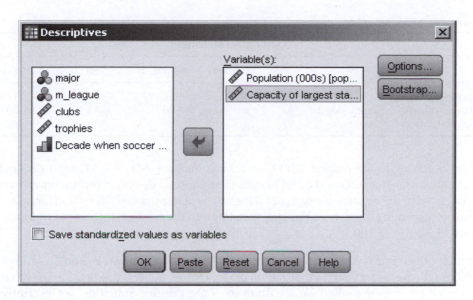

Figure 3.7 Descriptive pop-up window.

Figure 3.8 Options within descriptive statistics.

Table 3.8 SPSS output for descriptive statistics

Descriptive Statistics					
	N	Minimum	Maximum	Mean	Std. Deviation
Population (000s)	92	30	8967	1344.95	1668.114
Capacity of largest stadium (000s)	92	3	100	45.51	25.097
Valid N (listwise)	92				

When reporting to the nearest 1,000 we have a mean $\pm$ SD = 1.345 $\pm$ 1.668 million inhabitants. This is obviously skewed because the standard deviation represents an average deviation from the mean and we do not have values of less than 0.000 million for any cities. When representing capacity of the largest stadium to the nearest 100, we have a mean $\pm$ SD of 45,500 $\pm$ 25,100.

We cannot be certain whether the capacity of the largest stadium is skewed or symmetrical based on the mean and standard deviation alone. However, SPSS can provide us with infor-mation about skewness when we use **Analyse** → **Descriptive Statistics** → **Descriptives** if we ask for **Options** as shown in Figure 3.8 but tick that we wish to see information about the skewness of the distribution. Table 8.9 shows the SPSS output for descriptive statistics including statistics for skewness. When we divide the statistic for skewness by the standard error for skewness, we obtain the value of z_{Skew} which should be between –1.96 and +1.96 for the variable to be considered sufficiently symmetrical (Ntoumanis, 2001: 45). A value of less than –1.96 indicates that the variable is negatively skewed while a value greater than +1.96 indicates that the variable is positively skewed. Population is positively skewed (z_{Skew} = 11.74) while capacity of the largest stadium can be considered symmetrical (z_{Skew} = –1.3).

A more visual way of assessing the symmetry or skewness of variables is by using box plots which can be produced by the Explore facility of SPSS. We use **Analyse** → **Descriptive Statistics** → **Explore** and transfer our variables on interest into the *Dependent List* as shown in Figure 3.9. The **Statistics** button activates the pop-up window shown in Figure 3.10 that allows us to request percentiles which include the median, lower and upper quartiles. Once we click on **Continue** and then on **OK**, SPSS explores the variables providing the output including the percentiles shown in Table 3.10 and the box plots shown in Figures 3.11 and 3.12.

Table 3.9 Descriptive statistics including statistics about the skewness of the distribution

Descriptive Statistics							
	N	Mean	Std. Deviation	Skewness	Kurtosis		
	Statistic	Statistic	Statistic	Statistic	Std. Error	Statistic	Std. Error
Population (000s)	92	1344.95	1668.114	2.948	.251	9.406	.498
Capacity of largest stadium (000s)	92	45.51	25.097	.358	.251	–.619	.498
Valid N (listwise)	92						

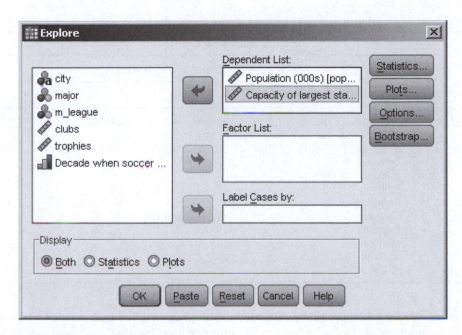

Figure 3.9 Exploring variables in SPSS.

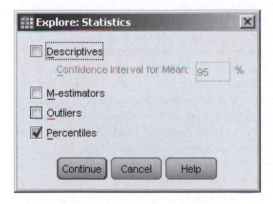

Figure 3.10 Requesting percentiles when exploring variables.

The 25th, 50th and 75th percentiles of the first part of Table 3.10 are the lower quartile, median and upper quartile respectively. These values for population are 0.448 million, 0.841 million and 1.500 million respectively. Table 3.8 shows that the minimum and maximum values are 0.030 million and 8.967 million respectively. The capacity of the largest stadium of a city has a median of 41,000 with lower and upper quartiles of 29,500 and 62,800 respectively. The minimum and maximum values for capacity of the largest stadium are 3,000 and 100,000 respectively. Recall that we are expressing population to the nearest 1,000 and capacity of the largest stadium to the nearest 100. It is very important to select an appropriate number of decimal places. Very often when students express values to 4, 6 or even 8 decimal places, it is because they have not thought about what the variables actually

Table 3.10 Percentiles produced by the Explore facility in SPSS

					Percentiles			
					Percentiles			
		5	10	25	50	75	90	95
Weighted Average (Definition 1)	Population (000s)	162.95	260.60	448.00	841.00	1500.00	2968.20	5792.05
	Capacity of largest stadium (000s)	6.95	10.00	28.50	41.00	62.75	80.00	93.75
Tukey's Hinges	Population (000s)			448.00	841.00	1485.00		
	Capacity of largest stadium (000s)			29.00	41.00	62.50		

represent and how they should be presented to aid decision making. Presenting results using different numbers of decimal places to the number used by SPSS can demonstrate some understanding of the variables.

Figures 3.11 and 3.12 show the box plots produced for the two variables. The box represents the inter-quartile range of a variable while the 'whiskers' represent the range once outliers and extreme values have been removed. Outliers, shown as 'o', occur where a value is more than 1.5 inter-quartile ranges outside the inter-quartile range. Extreme values, shown as '*', occur where a value is more than three inter-quartile ranges outside the inter-quartile range. As we can see in Figure 3.11, population is seriously skewed as it is not symmetrical and contains outliers and extreme values on one side of the distribution but not on the other. Figure 3.12 shows that capacity is more symmetrical but could still be interpreted as positively skewed based on this visual inspection. Therefore, in both cases the median, lower quartile and upper quartile will be used rather than the mean and standard deviation. This author prefers the use of the lower and upper quartiles rather than the inter-quartile range because an inter-quartile range of 33,300 for capacity of the city's largest stadium could be a range from 5,000 to 38,300 which with a median of 41,000 would suggest negative skewness, it could be 30,000 to 63,300 which would indicate positive skewness or it could be between 24,300 and 57,600 which would be interpreted as a symmetrical distribution. Basically, the median and inter-quartile range on their own do not provide sufficient information about the distribution of a variable.

To compare the descriptive statistics for major (trophy-winning cities) with other cities (non-trophy-winning cities), we use **Analyse → Compare Means → Means** transferring 'major' into the *Independent list* and the two ratio scale variables of interest into the *Dependent list* as shown in Figure 3.13. The default output includes the mean and standard deviation but not the median, lower or upper quartile. Therefore, we use the **Options** button to activate the pop-up window shown in Figure 3.14. This allows the median to be included but not

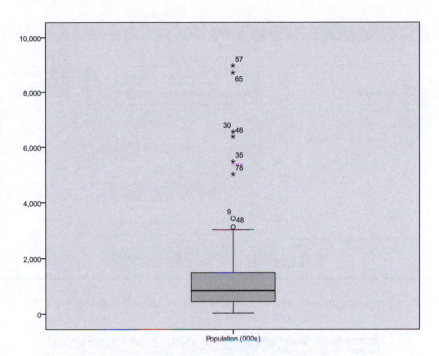

Figure 3.11 Box and whiskers plot for population.

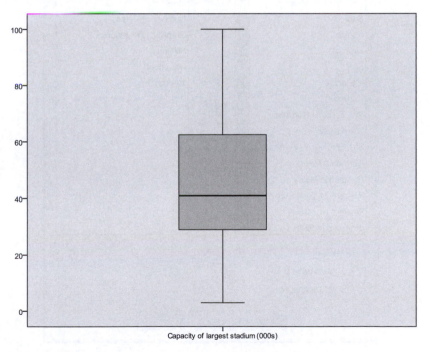

Figure 3.12 Box and whiskers plot for capacity of the largest stadium in the city.

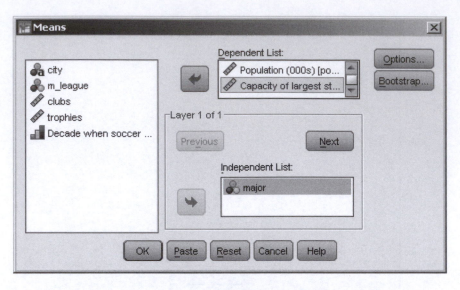

Figure 3.13 Comparing means for groups identified by an independent variable.

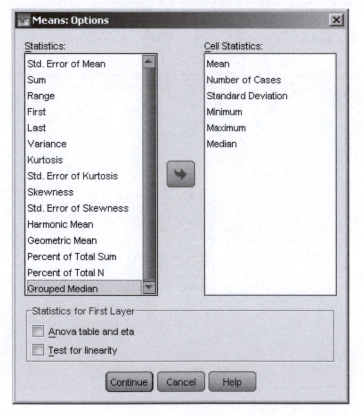

Figure 3.14 Options for comparing means.

descriptive statistics

the lower and upper quartiles. If we really wanted to see the lower and upper quartiles, we could logically split the file on the variable 'major' and then use Explore asking for percentiles. The Explore output would be presented for both values of 'major' (trophy-winning city or non-trophy-winning city) and would include the lower and upper quartiles.

The output for compare means is shown in Table 3.11 which reveals that trophy-winning cities have a lower median population but greater capacity for their largest stadium than non-trophy-winning cities. The population result is particularly interesting because it shows that even when using the exact same data, we can draw a different conclusion if using the mean and median. The values for some very populous cities have raised the mean more for the trophy-winning cities than for the other cities. If we sort the data file on population (**Data → Sort** see Figure 3.15) we find that only 27 of the 92 cities have populations greater than the mean.

As mentioned earlier, an alternative way of comparing interval or ratio scale variables between the two types of city is to use the Explore facility (**Analyse → Descriptive Statistics → Explore**) with 'major' transferred into the *Factor list* as shown in Figure 3.16. We would use the **Statistics** pop-up to ask for percentiles to be added to the output. The output would be similar to that shown in Figures 3.11 and 3.12 except with multiple box plots as shown in Figures 3.17.

The use of the **Statistics** button allows us to specify that we wish to include percentiles in the output. This can be done whether or not factors are included in the analysis. Table 3.12 shows the percentiles output when type of city ('major') is included as a factor.

Table 3.11 SPSS Output for compare means

Report			
	Major	*Population (000s)*	*Capacity of largest stadium (000s)*
Trophy winning city	Mean	1371.30	54.36
	N	47	47
	Std. Deviation	1726.008	23.353
	Minimum	116	9
	Maximum	8707	100
	Median	763.00	52.00
Other City	Mean	1317.42	36.27
	N	45	45
	Std. Deviation	1624.468	23.695
	Minimum	30	3
	Maximum	8967	100
	Median	848.00	34.00
Total	Mean	1344.95	45.51
	N	92	92
	Std. Deviation	1668.114	25.097
	Minimum	30	3
	Maximum	8967	100
	Median	841.00	41.00

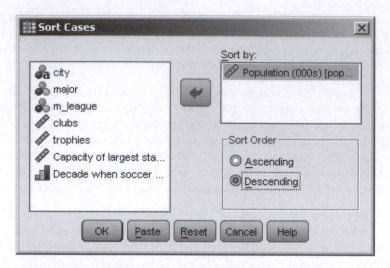

Figure 3.15 Sorting cases into descending order of population.

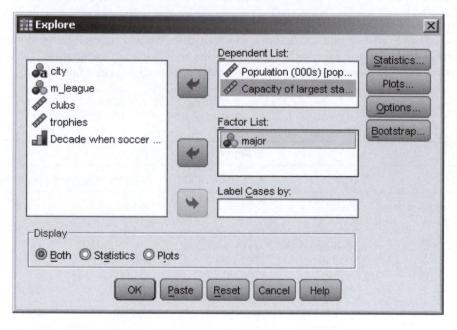

Figure 3.16 Explore with a factor.

Reporting results

In this situation where the variables appear to be skewed, we would choose to report medians as a measure of location and the lower and upper quartiles as an indication of dispersion. This could be done in tabular form as shown in Table 3.13. A table like this has an

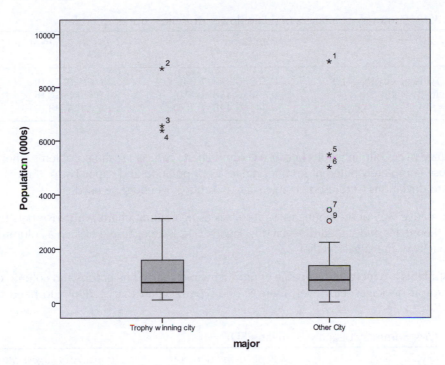

Figure 3.17 Clustered box and whisker plots for population when type of city (major) is used as a factor.

Table 3.12 Percentiles for variables when type of city is entered as a factor

			Percentiles						
		major				*Percentiles*			
			5	10	25	50	75	90	95
Weighted Average (Definition 1)	Population (000s)	Trophy winning city	167.4	247.4	397.0	763.0	1635.0	2889.6	6484.6
		Other City	154.2	267.8	458.0	848.0	1410.0	3189.8	5339.2
	Capacity of largest stadium (000s)	Trophy winning city	21.2	25.0	35.0	52.0	72.0	88.0	98.8
		Other City	3.0	6.8	20.0	34.0	50.0	76.0	80.0
Tukey's Hinges	Population (000s)	Trophy winning city			403.0	763.0	1593.0		
		Other City			458.0	848.0	1378.0		
	Capacity of largest stadium (000s)	Trophy winning city			35.5	52.0	71.5		
		Other City			20.0	34.0	50.0		

Table 3.13 Summary of analysis (median (lower quartile – upper quartile))

Type of city	Population ('000,000)	Capacity of largest city ('000)
Trophy winning city (n = 47)	0.763 (0.397–1.635)	52.0 (35.0–72.0)
Other city (n = 45)	0.848 (0.458–1.410)	34.0 (20.0–50.0)
All (n = 92)	0.841 (0.448–1.500)	41.0 (29.5–33.3)

advantage over column graphs because column charts can give scaling problems when variables use completely different number ranges as population and capacity do. If means and standard deviations were used, a table such as Table 3.14 could be used.

An alternative way of presenting these results is using column charts with error bars to represent the means and standard deviations. Figure 3.18 is an example of such a column chart for capacity of the largest stadium.

Most students are more than capable of producing pie charts, line graphs and column charts in Microsoft packages. However, there is one type of chart that students do need to ask

Table 3.14 Summary of analysis (mean$\pm$SD)

Type of city	Population ('000,000)	Capacity of largest city ('000)
Trophy winning city (n = 47)	1.371$\pm$1.726	54.4$\pm$23.4
Other city (n = 45)	1.317$\pm$1.624	36.3$\pm$23.7
All (n = 92)	1.345$\pm$1.668	45.5$\pm$25.1

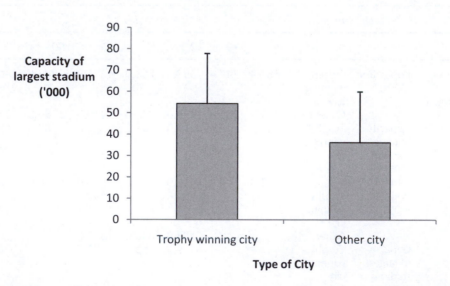

Figure 3.18 Capacity of largest stadium.

descriptive statistics

supervisors about in the author's experience and that is a column chart with error bars such as the one shown in Figure 3.18. Therefore, the way in which this type of chart is produced in Microsoft Excel is described in the file 03-error_bars.DOC that can be found on the internet site that supports this book.

SUMMARY

Descriptive statistics are the most important statistics used in research. They tell us what percent, how many, how high, how low, how consistent, how inconsistent, how short, how tall, how heavy, how light, how fast, how slow, and so on. These are the all important results to report as they are values in ranges that we often understand very well for commonly used variables. Furthermore, the descriptive results can be compared with values reported by previous research. The descriptive statistics used depend on the scale of measurement of the variables of interest. Categorical variables measured on nominal and ordinal scales can be represented by frequency profiles, percentage distribution of cases and the mode. Ordinal variables can also be described using the median as a measure of location with the minimum and maximum values and/or the lower and upper quartiles providing a measure of dispersion. Interval and ratio scale variables are continuous and can only be represented by frequency profiles if values are classified into subranges of the numerical scale of measurement. The mean and standard deviation are used to describe interval and ratio scale variables where their distributions are reasonably symmetrical. Where such variables are skewed, it is better to use the median as an average and the lower and upper quartiles to indicate variability.

EXERCISES

Exercise 3.1. Relative age in Grand Slam singles tennis

The file ex3.1-2009-grand-slam-tennis-senior.SAV contains details of female and male tennis players who participated in the singles events of Grand Slam tennis tournaments in 2009. The file contains variables for the gender, month, quarter year and half year of birth of each player.

a) How many male and female players were there? What does this tell us given that 128 different players participate in each singles event of each Grand Slam tournament?
b) Compare the quarter year of births of male and female players. What does this tell us?

Exercise 3.2. Burnout potential in student athletes

The file ex3.2-burnout.SAV contains data for 175 participants of four different experience levels (school, club, national and international). We are interested in the experience levels of the participants. Determine the frequency profile for experience level as well as the median, mode, minimum, maximum, lower-quartile and upper-quartile values for experience level. Do this for:

a) The sample as a whole.
b) The male and female participants individually.

Exercise 3.3. Fitness survey

The file ex3.3-fitnesstests.SAV contains 'stature', 'body mass' and 'estimated $\dot{V}O_2$ max' for a sample of 236 participants: 96 females and 140 males. Determine the mean and standard deviation for these three variables for the males and females separately.

PROJECT EXERCISE

Exercise 3.4. Descriptive statistics for height and body mass of your class

Gather data for height and body mass for your class of students and summarize the data using appropriate descriptive and inferential statistic for male and female students separately.

CHAPTER 4

STANDARDIZED SCORES

INTRODUCTION

In any area, it is essential to be able to interpret whether the values of given variables are low, average or high. If we do not understand what a value means, then we cannot make a decision based on the value and the variable is invalid. Some variables have become well understood through every day use; for example, in athletics it is understood that a 10,000m time of under 27 minutes is world class for a senior man while a time faster than 40 minutes could be considered good for a recreational runner. There are other variables that are not be so well understood requiring research to determine what low, average and high values are. Note that we are not using the terms poor, average and good because this requires relating the variable to some valid indicator of quality. There are some variables where the higher the value the better (for example, how far an athlete can throw a javelin), somewhere the lower the value the better (for example, the time it takes to run 100m) and others where an optimal value is better (for example, Body Mass Index).

In statistics, the word 'norm' is used to represent a standard or average value for some quantifiable aspect of human performance. Norms are typically used to represent average performance for different gender and age groups. The norms also give an indication of the spread of values for the population so that a user will know a bit more than whether they are above average or below average. These norms are evidenced by large studies that determine average values for the relevant populations. Hoffman's (2006) book *Norms for Fitness, Performance, and Health* as well as the ACSM's (2005) *Health-related Physical Fitness Assessment Manual* use three different ways of presenting norms for various human performance variables:

- Descriptive statistics – means and standard deviations.
- Percentile norms – these are values that divide the relevant population into ordered subgroups using the variable. These subgroups contain a given percentage of the population.
- Normative ranges – these are ranges of values for people classified by risk (low, moderate, high or very high) or performance rating (poor, fair, average, above average or excellent for example).

Percentiles are a particular type of quantile, as will be explained in this chapter. In this chapter, quantiles will be covered along with three other methods of standardizing scores: z-scores, T-scores and stanines. Each method gives an understanding of the average value and the spread of values for a given variable. These standards can be used to interpret individual values by relating them to the spread of values for the variable for the relevant population. The standardization of values for different variables also allows completely different variables to be compared. For example, we may wish to know whether an athlete needs to work on their trunk endurance or their flexibility. Determining standardized scores for the athlete for a test of trunk endurance and a test of flexibility allows these two different measures to be compared in real terms by relating the athlete's performance of each to the relevant population that the athlete is a member of. For each of the four methods, this chapter describes how to determine the norms for a variable and how a value can be interpreted using them.

NORMS

Norms are typically computed using samples rather than whole populations. For example, percentiles for maximal aerobic power in males and females aged 20s, 30s, 40s, 50s, 60s and 70s produced by ACSM (2010) used group sizes of 209 to 16,534. It should also be noted that there are different variables, tests and protocols used to represent aspects of fitness. For example, ACSM (2010) listed four different protocols for measuring maximal aerobic power: Balke treadmill (time), maximum $\dot{V}O_2$, 12 min run (miles) and 1.5 mile run (time). Values are ordered low to high or high to low depending on what indicates a better performance. For example, in Table 4.1, 1.5 mile run time is ordered from low values to high values because

Table 4.1 Norms for 1.5 mile run time for males (min:s): Source: ACSM (2010: 130–2)

%	Age					
	20–29 (n=2,606)	30–39 (n=13,158)	40–49 (n=16,534)	50–59 (n=9,102)	60–69 (n=2,682)	70–79 (n=467)
99	8:22	8:49	9:02	9:31	10:09	10:27
95 Superior	9:10	9:31	9:47	10:27	11:20	12:25
90	9:34	9:52	10:09	11:09	12:10	13:25
85	9:52	10:14	10:44	11:45	12:53	13:57
80 Excellent	10:08	10:38	11:09	12:08	13:25	14:52
75	10:34	10:59	111:32	12:37	13:58	15:38
70	10:49	11:09	11:52	12:53	14:33	16:22
65	11:09	11:34	11:58	13:25	14:55	16:46
60 Good	11:27	11:49	12:25	13:53	15:20	17:37
55	11:34	11:58	12:53	13:58	15:53	18:05
50	11:58	12:25	13:05	14:33	16:19	18:39
45	12:11	12:44	13:25	14:35	16:46	19:19
40 Fair	12:29	12:53	13:50	15:14	17:19	19:43
35	12:53	13:25	14:10	15:53	17:49	20:28
30	13:08	13:48	14:33	16:16	18:39	21:28
25	13:25	14:10	15:00	16:46	19:10	22:22
20 Poor	13:58	14:33	15:32	17:30	20:13	23:55
15	14:33	15:14	16:09	18:22	21:34	25:49
10	15:14	15:56	17:04	19:24	23:27	27:55
5	16:46	17:30	18:39	21:40	25:58	30:34
1 Very Poor	20:55	20:55	22:22	27:08	31:59	33:30

standardized scores

shorter times are better performances than longer times. Table 4.1 also shows that the ACSM classified particular percentiles as being associated with superior, excellent, good, fair, poor and very poor performance.

QUANTILES

Purpose

Quantiles are used to split an ordered set of data into equally sized partitions. A very commonly used quantile is the 2-quantile which splits the set of data into two equally sized partitions; the 2-quantile is also called the median. Common types of quantile are:

- The 2-quantile or median
- The 3-quantiles or tertiles
- The 4-quantiles or quartiles
- The 5-quantiles or quintiles
- The 10-quantiles or deciles
- The 20-quantiles or vigintiles
- The 100-quantiles or percentiles
- The 1000-quantiles or permilles.

In general, Q-quantiles split a set of data into Q partitions such that there are Q – 1 quantiles such that the probability of a randomly chosen case having a value less than the kth Q-quantile is k/Q, where k has a value from 1 to Q – 1. Let us consider percentile as a type of quantile. For a given variable and a given population, a percentile is the value such that the given per cent of the population have values that are below that value. For example, the 50th percentile for the height of boy aged 13 years (13 years to under 14 years) is the height that 50 per cent of 13-year-old boys would have heights below. Percentiles are commonly used in fitness (Hoffman, 2006), health (Hulens *et al.*, 2001) and developmental studies (Ulijaszek *et al.*, 1998) allowing values to be evaluated with respect to the relevant population.

There is some disagreement in previous statistics textbooks as to whether the quantiles represent the values that divide the population into partitions or whether the quantiles are the actual partitions. Vincent (1999: 36) shows a diagram that uses the term 'quartile' to represent a 25 per cent band of the population with the first quartile being the 25 per cent of the population with the lowest values and the fourth quartile being the 25 per cent of the population with the highest values. The diagram used by Vincent (1999) does the same thing with deciles. Other textbooks use the term 'quartile' to represent a cut off point between two 25 per cent bands of the population (Anderson *et al.*, 1994: 67–8; Groebner *et al.*, 2005: 88). In this book, we use the term 'quartile' to represent a value used at the border of two 25 per cent bands of the population. There are three quartiles: the lower quartile (Q1), the median (Q2) and the upper quartile (Q3). These are the values such that 25 per cent, 50 per cent and 75 per cent of the relevant population respectively would have lower values. We only need three values to divide the population into four groups, each containing 25 per cent of the population. It is not always possible to state the lowest or highest possible value because records keep being broken in sport. Therefore, we do not use a 0th percentile or a 100th percentile. This approach of using quantiles to represent points in an ordered set of data to divide the data into partitions is used in the current book for all types of quantiles. Consider the ordered series of N = 16 values shown in Table 4.2.

The values range from 2 to 29 and are placed at the locations 1 to 16. The quartiles do not occur exactly at the 4th, 8th and 12th value in the series even though 16 is divisible by 4. This is because the locations commence at 1 rather than 0. The median occurs somewhere between locations 8 and 9. The lower and upper quartiles occur somewhere between the locations 4 and 5 and 12 and 13 respectively. Different methods have been reported for calculating the location of a percentile. Groebner et al. (2005: 89) used equation 4.1 to determine the location, i, of a percentile, p, within an ordered series of n values.

$$i = \frac{p}{100}(n+1)$$
(4.1)

Therefore, Q1, Q2 and Q3 would be located at positions 4.25 (=25 × (16+1)/100), 8.5 (=50 × (16+1)/100) and 12.75 (=75 × (16+1)/100). Remember that these are just the locations of Q1, Q2 and Q3 rather than the actual values for these quartiles. Therefore, Q1 would be the value one quarter of the way between the 4th and 5th values (4 and 8) and would therefore be 5. Q2 would be half way between the 8th and 9th values, but these values are the same and so the median is 10. Q3 would be three quarters of the way between the 12th and 13th values (16 and 18) and would therefore be 17.5. If the number of values in the series, n, was an odd number, then the median, Q2, would be at the location of a particular value in the series because equation (4.1) would give a positive integer result for the position.

The method described by Groebner et al. (2005: 89) could generate locations between 0 and 17 in our example if a full range of percentiles from 0 per cent to 100 per cent were used. Although this author does not recommend the use of the 0th and 100th percentile, the 5th and 95th percentiles (1st and 19th vigintiles) still return locations less than 1 and greater than 16 respectively. Given that we do not have any values positioned outside the locations 1 to 16, it would not be possible to do an interpolation to compute such a value. In saying this, the author would not recommend using vigintiles based on only 16 values.

The Microsoft Excel package uses a slightly different method of determining percentiles within its PERCENTILE function. This can be found in the 'Quartiles' sheet of the 04-arbitrary. XLS spreadsheet file. Equation 4.2 is the equation used by Microsoft Excel to determine percentiles and in our example it gives locations ranging from 1 to 16. This still gives the location 8.5 for the median, but Q1 and Q3 would be taken from the locations 4.75 and 12.25 respectively. Therefore, Q1 would be the value three-quarters of the way between the 4th and 5th values (4 and 8) and would therefore be 7. Q3 would be one-quarter of the way between the 12th and 13th values (16 and 18) and would therefore be 16.5.

$$i = 1 + \frac{p}{100}(n-1)$$
(4.2)

Table 4.2 Sixteen arbitrary ordered values and their locations

Location, i	1	2	3	4	5	6	7	8	9	10	11	12	13	14	15	16
Value, x_i	2	2	3	4	8	8	9	10	10	13	16	16	18	20	24	29

60

standardized scores

For the remainder of this book, we will use equation 4.2 in the calculation of the locations of medians, quartiles or any other percentiles. Equation 4.3 is a more general equation to determine the location, i, of the kth Q-quantile within an ordered population of n values.

$$i = 1 + \frac{k}{q}(n-1)$$

(4.3)

Indeed, this method has already been used to calculate the median, lower and upper quartiles in Chapter 3. The median can be used as a measure of location while the inter-quartile range (IQR = Q3 − Q1) can be used as a measure of dispersion. The median, lower and upper quartiles can be displayed graphically using a box and whiskers plot as shown in Figure 4.1 (see page 65). The box represents the inter-quartile range with the line in the middle of the box showing the median. The whiskers represent a range of values 1.5 IQRs outside the IQR; that is the lower whisker is Q1 − 1.5 IQR while the upper whisker is Q3 + 1.5 IQR. Any value outside this range is marked as an 'outlier' while any values that are more than three IQRs outside the IQR are marked as 'extreme values'. Box and whisker plots are good for showing whether distributions of variables are symmetrical or skewed. In a symmetrical distribution, the mean and median are similar and also coincide with the modal subrange of values. A skewed distribution is one where the spread of values above and below the median are different. A negatively skewed distribution is where a small number of very low values cause more than 50 per cent of the values to be above the mean. A positively skewed distribution is where a small number of very high values cause more than 50 per cent of the values to be below the mean. Skewed distributions will be illustrated further in Chapter 5 on probability distributions.

Example: English National Superleague Netball performance

Indicators of netball performance include possession statistics (O'Donoghue *et al.*, 2008). Data from 118 team performances were used to develop a set of decile norms for the following performance indicators:

- Percentage of centre passes resulting in goals
- The number of turnovers made
- The percentage of turnovers resulting in goals
- The percentage of shots that are scored
- The number of goals scored.

This data is found in the file 04-netballperformance.XLS on the 'Data' worksheet. The 'Deciles' worksheet shows how the nine deciles are determined for each indicator using the PERCENTILE function. This function has two arguments, the first of which is an array of cells containing the values from which we are producing a percentile and the second argument is the particular percentile we wish to produce. For example, PERCENTILE (array of cells, 10%) will produce the 10th percentile which is the 1st decile of the values in the array of cells. The resulting deciles for the five performance indicators are shown in Table 4.3.

Table 4.3 Decile norms for English National Super League netball performance

	Performance Indicator				
Decile	% C.Pass to goal	No of turnovers	% Turnovers to goal	% Shots scored	Goals
1	30.0	38.7	25.4	53.0	24.7
2	37.0	41.0	31.0	59.5	32.4
3	39.6	44.0	37.2	64.0	34.1
4	42.9	48.0	40.7	66.7	38.0
5	47.4	49.0	43.8	68.9	41.0
6	50.2	51.0	47.7	71.2	43.2
7	53.1	54.0	50.6	73.3	47.0
8	55.8	57.0	53.6	76.7	52.6
9	59.4	64.0	58.2	79.8	59.3

These decile norms can be used in two ways. First, they can be used to assess a team's performance given a performance indicator value. Imagine a team scores from 52 per cent of their centre passes and 38 per cent of their turnovers. The value of 52 per cent of centre passes resulting in goals is in the seventh 10 per cent band of Superleague performances such that 60 per cent of performances have lower values and 30 per cent of performances have higher values than this 10 per cent of teams. The value of 38 per cent of turnovers resulting in goals is in the fourth 10 per cent band of Superleague performances such that 30 per cent of performances have lower values and 60 per cent of performances have higher values than this 10 per cent of teams. This shows that in real terms, using deciles to standardize the performance indicators, the team needs to work on their use of possessions where they takes the ball in a turnover more than it needs to work on centre pass possessions. These deciles can also be used to evaluate an opponent's possession statistics against the team. This gives an indication of the team's defence and how capable they are of preventing opposition turnovers from happening, opposition shots from being scored and opposition possessions being capitalized on. Here, lower 10 per cent inter-decile bands reflect better performances by the defence.

The second way in which these norms can be used is to determine the performance indicator values that are equivalent to a particular quantile. For example, a coach may decide to use targets at the 65th percentile for each of the performance indicators for the team, while wanting to restrict opponent performance indicator values to below the 45th percentile for each performance indicator. If the coach does not possess a copy of the 04-netballperformance.XLS spreadsheet and only has Table 4.3 available, then it is necessary to use linear scaling to determine the 65th and 45th percentiles for each variable. The 65th percentile would be estimated as being half-way between the 60th and 70th percentiles while the 45th percentile would be estimated as being half way between the 40th and 50th percentiles. Table 4.4 shows the resulting targets.

Sports performance is not a stable characteristic of a performer such as anthropometric and fitness characteristics are. One week a team may play an opponent ranked below them and the next week the team may play a team ranked above them. The quality of the opposition will have an effect on the performance of the team. Therefore, it is better to have different norms for different performance situations (O'Donoghue et al., 2008). The 'Deciles for

Table 4.4 Targets for a team wanting performances above the 65th percentile for its own possessions and restricting their opponents' possessions to below the 45th percentile

	Performance Indicator				
Percentile	% C.Pass to goal	No of turnovers	% Turnovers to goal	% Shots scored	Goals
Target for team performance (lower limit)	51.7	52.5	49.2	72.3	45.1
Target for opponent performance (upper limit)	45.2	48.5	42.2	67.8	39.5

levels' sheet of the file 04-netballperformances.XLS shows how decile norms are produced for four different types of performance:

- TvT: where a netball team in the top half of the English National Superleague plays a team that is also in the top half
- TvB: where a netball team in the top half of the English National Superleague plays a team that is in the bottom half
- BvT: where a netball team in the bottom half of the English National Superleague plays a team that is in the top half
- BvB: where a netball team in the bottom half of the English National Superleague plays a team that is also in the bottom half.

Table 4.5 shows the norms that are used in these different situations allowing more appropriate interpretations of performances taking the quality of the opposition into account. Consider a team in the bottom half of the league scoring from 35 per cent of their centre passes against top half opposition one week and 42 per cent of their centre passes against bottom half opposition a week later. Table 4.5 allows a coach to recognize that the 35 per cent of centre passes scored against top half opposition is in the 10 per cent of performances that 60 per cent of fellow bottom half teams would be below and 30 per cent of bottom half teams would be above. Table 4.5 also allows the coach to recognize that the 42 per cent of centre passes scored against a fellow bottom half team is in the 10 per cent of performances that 20 per cent of fellow bottom half teams would be below and 70 per cent of bottom half teams would be above. This allows the coach to recognize that the team's performance is not as good in 'real terms' in the second match.

SPSS

This example uses the same data from the 118 performances as used in the Microsoft Excel example. This time the data are stored in an SPSS data sheet 04-netball_performance_stats. SAV. Percentiles can be produced in SPSS using the Explore facility which is accessed using **Analyse → Descriptive Statistics → Explore**. The variable 'Type' is transferred into the *Factor List* and 'C.Pass%' is transferred into the *Dependent List*. Click on **Statistics** and place a tick in percentiles. Click on **Continue** to close down the Statistics pop-up window and on **OK** to close down the Explore pop-up window and execute the analysis. The percentiles given are 5 per cent, 10 per cent, 25 per cent, 50 per cent, 75 per cent, 90 per cent and

Table 4.5 Deciles for the different types of performance (O'Donoghue et al., 2008)

Performance indicator	Decile								
	1	2	3	4	5	6	7	8	9
TvT (n = 26)									
% Centre Passes to Goal	34.6	40.0	48.9	50.0	52.6	53.3	55.5	59.3	63.9
Number of Turnovers	34.0	36.0	38.0	40.0	40.5	43.0	45.0	49.0	54.0
% Turnover to goal	30.3	37.3	40.8	44.1	46.2	51.1	55.1	59.1	61.8
% Shots scored	63.6	68.1	69.2	71.2	71.8	73.3	75.4	76.8	80.0
Number of Goals	31.0	37.0	38.5	40.0	42.5	44.0	46.5	49.0	52.5
TvB (n = 25)									
% Centre Passes to Goal	39.8	46.1	47.9	51.7	54.3	56.1	56.7	59.3	62.3
Number of Turnovers	52.8	56.6	61.0	62.0	64.0	64.4	67.0	69.4	72.8
% Turnover to goal	42.1	47.2	49.3	50.0	50.8	53.6	54.1	57.0	59.7
% Shots scored	65.7	67.6	70.5	72.8	75.9	76.6	77.5	78.8	84.2
Number of Goals	43.0	46.4	52.2	53.6	57.0	58.4	60.0	61.2	68.2
BvT (n = 25)									
% Centre Passes to Goal	21.4	24.9	26.2	29.0	31.0	34.9	38.2	39.8	42.3
Number of Turnovers	40.0	41.0	44.0	48.6	49.0	50.4	51.8	52.0	54.0
% Turnover to goal	13.3	18.0	23.2	24.6	27.5	28.4	30.6	35.0	41.5
% Shots scored	44.0	48.6	52.5	54.2	58.0	58.8	61.2	65.0	68.5
Number of Goals	16.8	20.8	23.0	24.0	25.0	28.4	32.8	34.0	34.6
BvB (n = 42)									
% Centre Passes to Goal	38.1	41.2	42.5	45.2	46.9	50.0	51.1	53.6	55.8
Number of Turnovers	41.1	43.0	44.0	46.4	48.5	50.0	51.0	52.8	55.0
% Turnover to goal	33.3	36.5	38.3	40.6	43.2	46.6	49.6	51.7	53.3
% Shots scored	55.7	60.8	64.5	66.7	68.0	69.9	71.4	73.7	79.4
Number of Goals	33.1	34.4	37.3	40.0	41.5	42.0	44.7	47.4	51.0

95 per cent for each type of match. Figure 4.1 is part of the output and was mentioned earlier when discussing quartiles. Table 4.6 shows the percentiles output. In the four box and whiskers plots, we have two outliers in total.

Z-SCORES

Purpose

Because different characteristics have different means and standard deviations, it is sometimes difficult to determine whether the value for a particular case is low, high or average. Even if we know the values of two variables are above average, we may be uncertain as to which one is higher in real terms. The purpose of the z-score is to standardize a value so that it represents the number of standard deviations the value is above the mean. Therefore, a z-score of 0 shows that the value is the mean, +1 shows one standard deviation above the mean and –1.5 shows that the value is 1.5 standard deviations below the mean. A z score

standardized scores

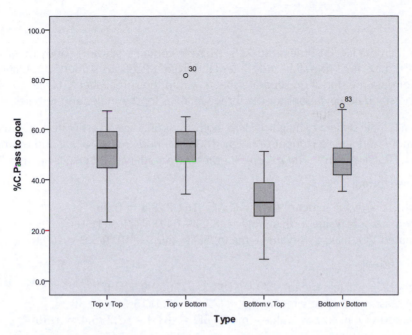

Figure 4.1 Box and whiskers plot for the percentage of centre passes leading to goal in netball matches between different qualities of team

Table 4.6 Percentiles produced by Explore

			Percentiles						
		Type	*Percentiles*						
			5	*10*	*25*	*50*	*75*	*90*	*95*
Weighted	% C.Pass	Top v Top	25.855	31.790	43.525	52.650	59.150	64.190	66.350
Average	to goal	Top v Bottom	35.350	38.460	46.750	54.300	59.150	63.500	76.620
(Definition 1)		Bottom v Top	8.600	16.220	25.300	31.000	39.200	44.780	50.120
		Bottom v Bottom	36.320	37.890	41.900	46.950	52.750	56.290	66.280
Tukey's	% C.Pass	Top v Top			44.700	52.650	59.100		
Hinges	to goal	Top v Bottom			47.200	54.300	59.100		
		Bottom v Top			25.600	31.000	38.800		
		Bottom v Bottom			41.900	46.950	52.500		

for a value $\times$ where the mean and standard deviation are μ and σ respectively is computed using equation 4.4.

$$z = \frac{x - \mu}{\sigma} \tag{4.4}$$

Interpretations of z-scores can be made in terms of numbers of standard deviations removed from the mean or by relating to percentage of the population assuming a normal distribution. The normal distribution will be discussed in Chapter 6 where z-scores will be revisited.

Example

Consider the data file '04-fitnesstests.XLS'. This file contains 'stature', 'body mass' and 'estimated $\dot{V}O_2$ max' for 140 males and 97 females from a fictitious student population. These two sets of data are stored on separate sheets within the spreadsheet. Table 4.7 shows the means and standard deviations for the three variables for the males and females.

Consider a female student of height 173cm, body mass 85.1kg and estimated $\dot{V}O_2$ max of 40.5 mL.kg⁻¹.min⁻¹ and a male student of height 180.3cm, body mass 72.9kg and estimated $\dot{V}O_2$ max of 61.1 mL.kg⁻¹.min⁻¹. The z-scores for these two students' test results are as follows:

The female student:

> Height: z = (value − mean) / SD = (173 − 164.7)/8.4 = 0.98
> Body mass: z = (value − mean)/SD = (85.1 − 64.0)/14.1 = 1.50
> Estimated $\dot{V}O_2$ max: z = (value − mean)/SD = (40.5 − 40.3)/5.9 = 0.03

The male student:

> Height: z = (value − mean) / SD = (180.3 − 178.4)/9.1 = 0.21
> Body mass: z = (value − mean)/SD = (72.9 − 77.4)/13.6 = −0.33
> Estimated $\dot{V}O_2$ max: z = (value − mean)/SD = (61.1 − 52.7)/5.4 = 1.56

Although the female student is shorter than the male student, she has a higher z-score than the male student for stature meaning that she is taller relative to her gender than the male student is relative to his gender.

T-SCORES

Purpose

There are some who find z-scores difficult to use (Hoffman, 2006: 17) due to 0 representing the mean and negative values being used. Therefore, z-scores for basketball performance indicators have been mapped onto a 0 to 4 scale to aid interpretation (Swalgin, 1998; Swalgin and Knjaz, 2006, 2007). A more commonly used method of avoiding negative numbers in standardized scores is the use of T-scores (Thomas and Nelson, 1996: 235). T-scores use 50 to represent the mean value for the variable of interest with 10 being used to represent an interval equivalent to 1 standard deviation. Therefore, 40 represents a value that is 1 standard deviation below the mean while 60 represents a value that is 1 standard deviation above the mean. T-scores have been used to interpret physical and performance tests done by athletes (Brown et al., 2008). Table 4.8 shows the relationship between raw scores,

Table 4.7 Descriptive statistics for fitness tests performed by a fictitious student sample (mean+SD)

Gender	Stature (cm)	Body mass (kg)	Est. $\dot{V}O_2$ max (mL.kg⁻¹.min⁻¹)
Female	164.7+8.4	64.0+14.1	40.3+5.9
Male	178.4+9.1	77.4+13.6	52.7+5.4

Table 4.8 Raw scores, z-scores and T-scores for a variable with a mean, μ, and a standard deviation, σ

Raw score	μ-4σ	μ-3σ	μ-2σ	μ-σ	μ	μ+σ	μ+2σ	μ+3σ	μ+4σ
z	−4	−3	−2	−1	0	1	2	3	4
T	10	20	30	40	50	60	70	80	90

z-scores and T-scores for a variable with a mean of μ and a standard deviation of σ. This shows that T = 10 z + 50. Equation 4.5 is the actual equation for a T score for a value $\times$ given a mean μ and a standard deviation σ.

$$T = 50 + 10\,(x - \mu) / \sigma \tag{4.5}$$

Example: Marathon running performance

For the purpose of this book, the author has analysed the finishing times of 22,524 male athletes who completed the 2011 London Marathon (http://results-2011.virginlondonmarathon.com/2011/, accessed 1 July 2011). On completion of this analysis, the author revisited the website for the London Marathon and found that an additional male finisher has been added to the results (accessed 28 July 2011). The author has decided not to redo the analysis for the sake of this additional finisher, instead using this as an illustration of a limitation of using the internet as a source of data. The frequencies of athletes finishing within each minute were used to estimate that the mean finishing time for the 22,524 male athletes was 4 hours 27 minutes and 7s with a standard deviation of 56 minutes and 0s. Table 4.9 shows T-scores, equivalent finishing times and the number of athletes who finished the race in those times or quicker. Nobody has broken 2 hours for the marathon at the time of writing and nobody has broken 50 minutes for the half marathon at the time of writing. The finishing times equivalent to T-scores of 10 and 20 are clearly outside the range of the variable and have resulted from the impact of the slowest times on the standard deviation. Basically, marathon finishing time is a positively skewed variable. Indeed, there were 30 athletes who ran slower than the time of 8 hours 11 minutes 10s which is standardized to a T-score of 90. The problem with T-scores of 10 and 20 in this example would also be a problem for z-scores of −5 and −4. In examples like this, it would be better to use quantiles.

Table 4.9 T-Scores for different men's marathon finishing times

T	Time (hour:min:s)	Cumulative frequency
10	0:43:06	0
20	1:39:06	0
30	2:35:07	95
40	3:31:07	3,601
50	4:27:07	11,907
60	5:23:08	19,112
70	6:19:08	21,770
80	7:15:09	22,346
90	8:11:10	22,495

If we were to use T-scores to assess performances, the author could check his own personal best of 2:21:54 against this data. Using equation 4.5 gives a T-score of 27.6 because 2:21:54 is 2:05:13 less than the mean which is 2.24 standard deviations below the mean resulting in a T-score of 22.4 less than 50. Readers should be advised that the author's personal best was set in 1992 on the Dublin course rather than London. With the men's world record being three minutes faster than when the author ran his own personal best, the validity of this particular T-score is obviously questionable.

STANINES

Purpose

The stanine scale comes from the words 'standard nine' and divides a distribution of values into nine partitions using z-scores (Vincent, 1999: 73–5). This is done using eight specific z-scores: $-1.75, -1.25, -0.75, -0.25, 0.25, 0.75, 1.25$ and 1.75. This does not necessarily divide the values into nine equally sized partitions as there may be more values between one pair of z-scores than there is between others. Table 4.10 shows the stanine that a value will be classified into based on its corresponding z-score. Unlike z-scores, stanines are not standardized versions of exact raw scores. Instead, each stanine represents a range of raw scores.

Example: Male finishing times in the 2011 London Marathon

The times of 22,524 males who completed the 2011 London marathon can be grouped into stanines because we know the mean is 4 hours 27 mins 7 s and the standard deviation is 56 mins 0 s. Table 4.11 shows the range of finishing times for each stanine. Using this method, the author's time of 2 hours 21 minutes and 54s falls in the 1st stanine.

SUMMARY

Different numerical variables often have different units and use different ranges of numbers. This makes direct comparisons between different variables difficult. There are four main ways of transforming raw values into standardized values: quantiles, z-scores, T-scores

Table 4.10 Stanines and their z-score ranges

Stanine	Range of z-scores
1st Stanine	$z < -1.75$
2nd Stanine	$-1.75 \leq z < -1.25$
3rd Stanine	$-1.25 \leq z < -0.75$
4th Stanine	$-0.75 \leq z < -0.25$
5th Stanine	$-0.25 \leq z < 0.25$
6th Stanine	$0.25 \leq z < 0.75$
7th Stanine	$0.75 \leq z < 1.25$
8th Stanine	$1.25 \leq z < 1.75$
9th Stanine	$1.75 \leq z$

standardized scores

Table 4.11 Stanines for male finishing times in the 2011 London Marathon

Stanine	Range of times	Frequency
1st Stanine	Quicker than 2:49:07	417
2nd Stanine	2:49:07 to 3:17:06	1,835
3rd Stanine	3:17:07 to 3:45:06	2,920
4th Stanine	3:45:07 to 4:13:06	4,403
5th Stanine	4:13:07 to 4:41:06	4,594
6th Stanine	4:41:07 to 5:09:07	3,643
7th Stanine	5:09:08 to 5:37:07	2,359
8th Stanine	5:37:08 to 6:05:07	1,249
9th Stanine	Slower than 6:05:07	1,104

and stanines. Quantiles divide the population into roughly equal sized partitions. The main quantiles used in sport and exercise research are percentiles, deciles and quartiles. Z-scores map values onto the number of standard deviations the values are above the mean for a given variable. T-scores are similar to z-scores except a value equivalent to the population mean is mapped onto 50 rather than 0 and each standard deviation the value is above the mean adds 10 to the T-score rather than 1 which is added to the equivalent z-score. Stanines are a further method that uses the mean and standard deviation to partition values into nine groups. All four of these standardization techniques allow different variables to be compared. Furthermore, standardized scores can be used in the opposite way. That is, desired standardized scores can be used to determine the raw scores that are equivalent to them. This can be effective for setting reasonable performance targets.

EXERCISES

Exercise 4.1. Deciles to interpret netball performance

Consider Table 4.5 showing norms for netball performance indicators in matches between teams of different qualities. A team in the top half of the league plays two matches: the first against a team in the bottom half of the league and the second against a team in the top half. Table 4.12 summarizes the performances of the team and their opponents in these two matches. The team's statistics are indicators relating to their use of the ball when attacking. The team wish to maximize these values. The opponents' statistics reflect how the team performed in defence; the team would wish to minimize the values achieved by the opponents.

a) Using Table 4.5 earlier in this chapter, has the team's attacking performance improved from the first match to the second match?
b) Using Table 4.5 and considering the opponents' statistics, has the team's defensive performance improved from the first match to the second?

Exercise 4.2. Quartiles for netball performance

Use of the PERCENTILE function in Microsoft Excel to determine quartile norms for the Top v Top, Top v Bottom, Bottom v Top and Bottom v Bottom types of netball performance in the 04-netballperformances. XLS spreadsheet.

Table 4.12 Performances of a team and its opponents in two netball matches

Performance Indicator	First Match		Second Match	
	Team (T)	Opponents (B)	Team (T)	Opponents (T)
% Centre Passes to Goal	48	39	47	42
Number of Turnovers	64	54	51	49
% Turnover to goal	53	26	45	45
% Shots scored	77	70	81	78
Number of Goals	55	31	43	40

Exercise 4.3. Decile norms for fitness test performances

The spreadsheet '04-fitnesstests.XLS' contains stature, body mass and estimated $\dot{V}O_2$ max values for 140 fictitious males and 97 fictitious females. Use the percentile function to determine the deciles for males and females for all three variables.

Exercise 4.4. Using z-scores to compare running event performances

Consider Table 4.13 which shows the mean and standard deviation for different athletic events by boys aged 16 to 18 in a school. Table 4.13 also shows in the right-hand column the personal bests of a particular athlete you are coaching. What event does the athlete's personal best help him beat the most opponents in?

Table 4.13 Running event performances

Event	School performances (s)		Athlete you are coaching	
	Mean	SD	Personal best (s)	z?
200m	26.5	1.5	28.0	
400m	62.7	5.0	62.5	
800m	136.2	10.8	130.8	

PROJECT EXERCISE

Exercise 4.5. Fitness assessment

With the aid of qualified personnel, perform a series of fitness tests covered in the ACSM's (2010) *Health Related Fitness Assessment Manual* and use the published norms to compare different fitness test scores you have achieved.

standardized scores

CHAPTER 5

PROBABILITY

INTRODUCTION

Probability is widely used in many business and sports contexts and decision makers need to be comfortable with probability (Groebner *et al.*, 2005: 128). Probability is used every time we perform an inferential statistical test. The 'p' values that we see in quantitative research papers are probabilities. Therefore, the purpose of this chapter is to explain what probability is, how probabilities can be determined and the use of probability in research studies.

Probability is defined as the chance that a particular event will occur (Groebner *et al.*, 2005: 128). There are many uncertainties in most areas of industry, business and life in general. Sport, exercise, leisure and physical education are no exceptions. What is the chance of a mechanical failure preventing a racing vehicle from completing the course in a motor sport event? What is the chance of each judge awarding above a certain figure for a gymnast or a figure skater? How likely is it that a goalkeeper will choose to move the right way when facing a penalty in soccer? Probability places a numerical value for the chance of an event occurring. This numerical value is between zero and one. A probability of zero means that the event will certainly not occur. A probability of one means that the event will certainly occur. Numerical values greater than zero and less than one represent the chance of the event occurring where there is uncertainty.

PROBABILITY IN RESEARCH

In order to fully understand inferential statistics that are covered in Chapters 7 to 15, it is necessary to understand probability. Consider the following sentence from the abstract of a published research paper on bone mass, bone mineral density and muscle mass in professional golfers:

> The only effect of professional golf participation on regional body composition was a 9% increase in muscle mass of the dominant arm ($p < 0.05$).
>
> (Dorado *et al.*, 2002)

Most investigations involve samples and there is a probability that sample statistics do not reflect population parameters perfectly. The '$p < 0.05$' in parenthesis at the end of the sentence is a recognition that there is a small chance (less than one in 20) that the result for muscle mass of the dominant arm found in the study might not be the true result for the population. Census studies of full populations are rarely done and, therefore, we rarely know the extent of sampling error that has occurred. Therefore, we can only report an estimate of the probability that a sampling error has happened. In general in statistics, the p value is the probability that a difference or relationship observed in a sample is due to sampling error rather than there being a real effect in the relevant population (Fallowfield et al., 2005: 97).

Probability is also important in understanding the distribution of variables. A variable, such as height, will have a range of values and different numbers of people will have values within different subranges of height. A probability distribution is used to represent the probability of a person chosen at random being within a given subrange of the possible values of a variable. What is the chance that this person is less than 165cm? What is the chance that this person has a height between 170cm and 180cm? There may be more people with heights in some 10cm subranges than others. Most statistical tests use the probability distributions of variables or critical test statistics in the calculation of the p value illustrated using Dorado et al.'s (2002) investigation. Probability distributions will be covered in Chapter 6 and hypothesis testing will be covered in Chapter 7. At this stage, they are mentioned so that readers will understand why probability is covered in this book.

Probability is not only important in the process of research but also in some of the topics investigated in sports science. For example, in the study of talent development in sport, expert performers make better use of situational probability (Singer and Janelle, 1999). Probabilistic models have also been used in the study of tennis strategy (Gale, 1971).

EXPERIMENTS

Terminology

Anderson et al. (1991: 112–13) used the terms 'experiments', 'experimental outcomes', 'sample space' and 'events' to describe probability. Groebner et al. (2005: 129) used the same terms except that they used the term 'elementary events' instead of 'experimental outcomes'. Like many words in the English language, the word 'experiment' is used to represent different concepts. Indeed, the word 'experiment' is used to represent two different things in this book. Therefore, to avoid confusion in this chapter or later on, it is worth acknowledging the two different uses of the word. The main use of the word 'experiment' in sport and exercise research is a research process to control and manipulate a set of participants to deliberately test if some treatment or other intervention makes a difference to some response variable(s) of interest. This has already been covered in Chapter 1 and will be revisited in Chapter 7.

In the current chapter, the word 'experiment' is used differently to be consistent with how the word is used in the well-developed area of probability. When discussing probability, an 'experiment' is a process that results in an outcome from a defined set of possible outcomes. Every time the experiment is performed, there must be one and only one experimental outcome. The set of all possible experimental outcomes for a given experiment is termed 'the sample space'. In Table 5.1, the '{ }' brackets are used to form this set (the sample space) from the possible experimental outcomes.

Consider an experiment which is playing a game of soccer. The experimental outcomes are *win*, *draw* and *lose*. Therefore, the sample space is the set of these experimental outcomes shown in equation 5.1. The example of a soccer game shown in Table 5.1 uses a different set of experimental outcomes involving venue. This was quite intentional by the author to make the point that outcomes of experiments can be classified in different ways.

Sample space, S = {win, draw, lose}   (5.1)

The possible experimental outcomes in the sample space must be complete and mutually exclusive (Groebner *et al.*, 2005: 132). Mutually exclusive means that there is no outcome that can be interpreted as a win and a draw, for example. In any experiment, there can be one and only one outcome: the match will be won, drawn or lost. Completeness means that there will be no outcome to the match that is outside the sample space: the match will be won, drawn or lost.

The final row of Table 5.1 should be of interest to those already familiar with scientific research. It shows that the two uses of the word 'experiment' described in this chapter are not so inconsistent. An experimental research project is one type of quantitative research project. Quantitative research projects are designed to answer quite specific research questions. For example, an experimental study may be undertaken to determine whether a 12-week interval training programme will have an effect on estimated $\dot{V}O_2$ max. Hypotheses are the possible answers to the research question. The null hypothesis is that there will be no effect, while the alternative hypothesis is that there will be an effect. This forms the sample space of the experiment (using the word 'experiment' as it is used in probability). The sample space consists of just two experimental outcomes: the null hypothesis is accepted or the null hypothesis is rejected.

Multistep experiments

Groebner *et al.* (2005: 129) listed three steps involved in preparing to assign probabilities: defining the experiment, defining experimental outcomes and defining the sample space.

Table 5.1 Experiments and their sets of experimental outcomes

Experiment	Sample Space
Take a service in tennis	{In, Fault, Let}
Role a 6 sided die	{1, 2, 3, 4, 5, 6}
Rolling a pair of 6 sided dice	{2, 3, 4, 5, 6, 7, 8, 9, 10, 11, 12}
A game of soccer	{Home win, draw, away win}
Toss a coin	{Heads, Tails}
A quantitative research study	{the null hypothesis is rejected, the null hypothesis is accepted}

This is especially true where experiments involve complex outcomes or multiple steps. Multistep experiments are experiments that are composed of lower level experiments; for example, a tennis match is composed of sets which in turn are composed of games which in turn are composed of points. Therefore, we could have a hierarchical structure of experiments to reason about. This may involve determining the different number of combinations of outcomes that can occur in a sequence of experiments. A simpler example is when a football team plays two matches, each of which has three outcomes. The total number of experimental outcomes will be $3 \times 3 = 9$. The sample space for each match has three experimental outcomes but the sample space for the two matches together will have nine outcomes, each of which is a structure expressed in '( )' parentheses as shown in equation 5.2. This can also be represented by a tree diagram (Graham, 2006: 112; Wood, 2003: 70–7) such as Figure 5.1.

$$S = \{(Win, Win), (Win, Draw), (Win, Lose),$$
$$(Draw, Win), (Draw, Draw), (Draw, Lose),$$
$$(Lose, Win), (Lose, Draw), (Lose, Lose)\} \tag{5.2}$$

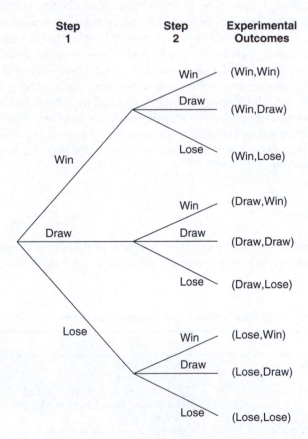

Figure 5.1 Tree diagram to represent a multistep experiment

Multistep experiments typically require some temporal ordering of steps in the experiment. Consider the example of rolling two six-sided dice which was included in Table 5.1. There are 11 outcomes in the sample space {2, 3, 4, 5, 6, 7, 8, 9, 10, 11, 12} rather than 36 outcomes (given by 6 × 6). This is because there is more than one way of achieving each experimental outcome except 2 and 12. This would not be considered as a multistep experiment because the dice are rolled simultaneously and the total shown by the two die is used as a single outcome of the experiment.

Probability in multistep experiments

In a multiple step experiment, we may be interested in the probability that a team will win its next three games or that a tennis player will win the next four points on his or her service game. The probability of gaining the desired outcome is the product of the individual step probabilities. For example, if a tennis player's probability of winning a point on serve is 0.7, then the probability of the player winning the first four points on serve is $0.7 \times 0.7 \times 0.7 \times 0.7 = 0.7^4 = 0.24$.

ASSIGNING PROBABILITIES

The classical method

Once we have determined the possible outcomes of the experiment, we can assign probabilities to these. For an experiment's sample space to be complete, the sum of the probabilities of the experimental outcomes must be one. Therefore, when the experiment is performed, one and only one of the outcomes will occur. Anderson et al. (1991: 116–18), Levine and Stephan (2005: 67–70) and Groebner et al. (2005: 134) described three approaches to assigning probabilities to experimental outcomes. These are the classical approach, using relative frequencies and the subjective approach.

The classical method assumes that each of the n experimental outcomes has an equal chance of occurring. Therefore, a probability of 1/n is assigned to each. For example, in the coin tossing experiment shown in Table 5.1, we might expect the probability of the outcome being heads to be 0.5 and the probability of the outcome being tails to be 0.5. Similarly, with a roll of a six-sided die, we would expect each of the six outcomes to have the same probability of occurring if it is a fair die (1/6 = 0.166667).

A variation of the classical method is used when we have an example such as rolling two six-sided dice. If we assume that the chance of each die showing any number is equal, we can use this information to determine the probability of each experimental outcome of multi-dice experiments. Table 5.2 shows the number of different ways that each experimental outcome can occur in this example and how this information is used to assign probabilities to the experimental outcomes. There are 36 combinations of values that result from two dice being thrown. Therefore, 36 is divided into the number of ways of achieving an experimental outcome to determine the probability of that outcome using the classical approach.

Table 5.2 Assigning probabilities for the outcomes of rolling two six-sided dice

Experimental outcome	Combinations of 2 six sided dice yielding outcome	Number of ways of achieving outcome	Probability
2	(1,1)	1	1/36 = 0.028
3	(1,2),(2,1)	2	2/36 = 0.056
4	(1,3),(2,2),(3,1)	3	3/36 = 0.083
5	(1,4),(2,3),(3,2),(4,1)	4	4/36 = 0.111
6	(1,5),(2,4),(3,3),(4,2),(5,1)	5	5/36 = 0.139
7	(1,6),(2,5),(3,4),(4,3),(5,2),(6,1)	6	6/36 = 0.167
8	(2,6),(3,5),(4,4),(5,3),(6,2)	5	5/36 = 0.139
9	(3,6),(4,5),(5,4),(6,3)	4	4/36 = 0.111
10	(4,6),(5,5),(6,4)	3	3/36 = 0.083
11	(5,6),(6,5)	2	2/36 = 0.056
12	(6,6)	1	1/36 = 0.028
Any		36	36/36 = 1.000

The relative frequency method

For other experiments, the probabilities of experimental outcomes of individual steps or overall experiments might not be equal. For example, in a game of soccer, we might expect a greater chance of a home win than an away win. Relative frequencies can be used when data are available to estimate the proportion of occasions that the experiment will produce each experimental outcome. This method assumes that the same distribution of experimental outcomes will occur if the experiment is repeated on a large enough number of occasions. In the 2009–10 season, there were a total of 193 home wins, 96 draws and 91 away wins in the 380 matches played in the English FA Premier League. We could use these to determine crude probabilities for match outcomes in the 2010–11 season:

- The probability assigned to a home win, P(Home Win) = 193/380 = 0.508.
- The probability assigned to a draw, P(Draw) = 96/380 = 0.253.
- The probability assigned to an away win, P(Away Win) = 91/380 = 0.239.

In the 2010–11 season, the frequency of home wins, draws and away wins were 179, 111 and 90 respectively. The same broad pattern of more home wins than draws than away wins was observed, but with noticeably fewer home wins and more draws than in 2009–10.

The subjective method

The subjective method of assigning probabilities to experimental outcomes involves using experience, expertise and intuition. This method is used where equal probabilities cannot be assumed and where no relevant data are available to allow the relative frequency method to be used. A probability is chosen to represent our belief about the chances of the outcome occurring. Groebner *et al.* (2005: 137) used an example of bidding for a contract and the perceived probabilities of being awarded the contract with different mark up percentages and knowledge of competitors who might also be bidding.

COUNTING RULES

Complex sample spaces

The primary purpose of this section is to explain how to determine the number of possible experimental outcomes for the purposes of assigning probabilities. The examples we have used so far have been very simple and it has been relatively easy to assign probabilities. Consider a more complex example. A football fan wins a prize of a ticket and all-expenses paid trip to an English FA Premier League soccer match in the coming season. The fan really wants to watch the team he supports, Fulham, playing against Manchester United, Liverpool, Arsenal, Chelsea or Manchester City. To work out the probability of the randomly chosen match being one that the man really wants to go to, we need to work out how many matches there are in the English FA Premier League and assume that each has an equal chance of being chosen. There are many situations such as this where we need to determine the total number of possibilities. We will consider two such situations which are combinations and permutations.

Counting for combinations

The number of combinations of n particular items from a total of N different available items is given by equation 5.3.

$$C_n^N = \binom{N}{n} = \frac{N!}{n!(N-n)!} \tag{5.3}$$

Note that N! is N factorial which is the product of all the positive integers from one to N (equations 5.4 and 5.5).

$$N! = 1 \times 2 \times 3 \times 4 \times 5 \times \ldots \times N \tag{5.4}$$

$$0! = 1 \tag{5.5}$$

Consider the Six Nations international rugby union tournament. The number of matches required to ensure that each team plays against each other team is the number of combinations of two teams that can be formed from the complete set of six teams. This is because each game involves two teams. Using equation 5.6 with N being six and n being 2 tells us that there are 15 matches needed to ensure this is a round robin tournament.

$$C_2^6 = \binom{6}{2} = \frac{6!}{2!(6-2)!} = \frac{720}{2 \times 24} = \frac{720}{48} = 15 \tag{5.6}$$

Similarly, if we consider a pool of four teams within the first stage of the FIFA World Cup for soccer, equation 5.7 can be used to show that we need six matches to ensure each team in the pool plays each other team.

$$C_2^4 = \binom{4}{2} = \frac{4!}{2!(4-2)!} = \frac{24}{2 \times 2} = \frac{24}{4} = 6 \tag{5.7}$$

Being able to determine the total number of combinations of items can be useful in determining the number of experimental outcomes to assign probabilities to. The authors own favourite example is the UK National Lottery where entrants choose six numbers from 49 available numbers. Six numbers are drawn, and if the entrant's six chosen numbers match those that are drawn then the user wins the jackpot (or at least a share of the jackpot). The number of possible combinations of six numbers out of 49 available is given in equation 5.8.

$$C_6^{49} = \binom{49}{6} = \frac{49!}{6!(49-6)!} = \frac{49 \times 48 \times 47 \times 46 \times 45 \times 44 \times (43 \times ... \times 1)}{(6 \times 5 \times 4 \times 3 \times 2 \times 1) \times (43 \times ... \times 1)} \tag{5.8}$$

$$C_6^{49} = \binom{49}{6} = \frac{49 \times 48 \times 47 \times 46 \times 45 \times 44}{(6 \times 5 \times 4 \times 3 \times 2 \times 1)} \tag{5.9}$$

This can be reduced to equation 5.9 by dividing the nominator and denominator by 43! to give 13,982,816 different combinations of six numbers out of 49. If we assume that each number has an equal chance of being drawn, then we can assume that each possible combination of six numbers has an equal chance of matching the numbers on the six balls to be drawn when the lottery is played. Effectively, if we have one lottery ticket with six chosen numbers, we have one chance in 13,982,816 of winning the jackpot. If more than 14 million lottery tickets are bought by the public in a week, the chances are that someone will win, but will it be you? (The author is of course referring to a marketing slogan that has been used by the National Lottery 'It could be you!') To understand how low the chances of winning are, let us assume that someone buys 100 lottery tickets (with different combinations of six numbers on) each week to improve their chances each week to about one in 139,828. There are 52 weeks in a year which makes the chance of winning in a given year one in 2,689 approximately. The entrant might be lucky enough to win on the first year, or unlucky enough to wait 2,689 years after they started to play the lottery; there is actually no guarantee that they will win in the first 2,689 years at all. If a lot of people tried this, the average person would win half way between year one and year 2,689. That is, we can expect to wait about 1,345 years to win the UK National Lottery if we buy 100 tickets per week. Readers should not be surprised that this author has never entered the UK National Lottery. If readers are still not convinced about how unlikely it is that they will win the UK National Lottery, Wood (2003: 2) referred to a report concluding that one is 750 times more likely to be killed by an asteroid striking Earth than winning the UK National Lottery.

Counting for permutations

The difference between a combination and a permutation is that in a permutation the order of the n items from N possible items is important. For example, in our example of the Six

Nations rugby union tournament, the permutation (Wales, Ireland) is different to the permutation (Ireland, Wales).

Equation 5.10 gives the number of possible permutations of n items form a total of N items.

$$P_n^N = n!\binom{N}{n} = \frac{N!}{(N-n)!} \tag{5.10}$$

Therefore, if the Six Nations rugby tournament was to be revised so that each pair of teams had to play each other twice (each team having a home and away match against the other), the number of matches required would be 30 as given in equation 5.11.

$$P_2^6 = 2!\binom{6}{2} = \frac{6!}{(6-2)!} = \frac{6\times5\times4\times3\times2\times1}{4\times3\times2\times1} = \frac{720}{24} = 30 \tag{5.11}$$

This can also be applied to determine the number of matches in the English FA Premier League where there are 20 teams who each must play the other 19 teams twice: once at home and once away from home. We can see from 5.12 that there are 380 matches required.

$$P_2^{20} = 2!\binom{20}{2} = \frac{20!}{(20-2)!} = \frac{20\times19\times(18\times17\times...\times1)}{(18\times17\times...\times1)} = 20\times19 = 380 \tag{5.12}$$

If you are interested in what would happen if winning the jackpot in the UK National Lottery required the six numbers to be entered in the same order they were drawn, you can use equation 5.12 with n being six and N being 49. The number of permutations of six numbers out of 49 is 10,068,347,520. If the UK National Lottery was won this way and we bought 100 tickets per week, we would expect to win after 968,110 years; not quite 'never in a million years' but close to it.

EVENTS AND THEIR PROBABILITIES

Events

An event is a subset of the sample space. Where an event consists of a single experimental outcome, it is called a simple event. Other events can be composed of more than one experimental outcome. We will use rolling a six-sided die as an example. There are six experimental outcomes within the sample space. An event, E, could be achieving a result of three or more when rolling the die. The probability of the event, E, occurring is the sum of the experimental outcomes that make up the event as shown in equation 5.14 where e is used to represent the individual experimental outcomes of E. Therefore, the probability of the event E, P(E) = 1/6 + 1/6 + 1/6 + 1/6 = 4/6 = 0.6667.

$$E = \{3,4,5,6\} \tag{5.13}$$

$$P(E) = \sum_{e \in ET} P(e) \tag{5.14}$$

Rules for computing event probabilities

Consider Table 5.3 which represents 108 points played between two tennis players A and B. There are eight different experimental outcomes which have been labelled 'a' to 'h'. The frequencies of the eight experimental outcomes are used to determine probabilities for point outcomes in a future match between the two players.

A Venn diagram can be used to show the experimental outcomes within each event (Graham, 2006: 113). Figure 5.2 shows the sets of experimental outcomes for each player serving and each player winning the point.

The probability of an event is the sum of the probabilities of the experimental outcomes that are included in that event. For example, P(Player A Serves) = P({a,b,e,f}) = P(a) + P(b) + P(e) + P(f). Equation 5.15 computes the probability for this event and equations 5.16 and 5.17 compute the probabilities for two other events.

Table 5.3 Frequency of points of different classes (probabilities assigned in parentheses and experimental outcome codes are shown in square brackets)

Point Outcome	Server: Player A		Server: Player B	
	1st Serve in	2nd Serve required	1st Serve in	2nd Serve required
Player A wins point	30 (0.278) [a]	12 (0.111) [b]	8 (0.074) [c]	10 (0.093) [d]
Player B wins point	6 (0.056) [e]	6 (0.056) [f]	25 (0.231) [g]	11 (0.102) [h]

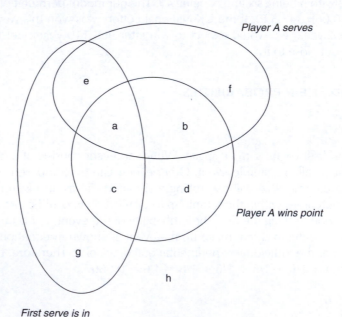

Figure 5.2 Venn diagram showing experimental outcomes within each event

$$P(\text{Player A serves}) = .278 + .111 + .056 + .056 = 0.5 \tag{5.15}$$

$$P(\text{Player A wins point}) = .278 + .111 + .074 + .093 = 0.556 \tag{5.16}$$

$$P(\text{2nd serve required}) = .111 + .056 + .102 + .093 = 0.362 \tag{5.17}$$

Events and set operations

Events like those shown in equations 5.15 to 5.17 and in the Venn diagram in Figure 5.2 can be combined. For example, we may wish to know the probability of player A serving and winning the point or the probability of player A serving or winning the point. Because these events are sets, set operations such as union and intersection are used to combine them.

The complement of an event E is denoted and its probability $P(E) = 1 - P(\bar{E})$. For example, the probability of not requiring a second serve is given by P($\overline{\text{2nd serve required}}$) = 1 – P(2nd serve required) = $1 - 0.361 = 0.639$

Given two events E_1 and E_2, the union ($\cap$) of E_1 and E_2 is the event containing those experimental outcomes belonging to E_1, E_2 or both. For example P(Player A serves Player A wins point) = P({a,b,c,d,e,g}) = 0.667. We cannot simply add the probabilities of the two events in this union because some experimental outcomes would be counted twice. This uses the addition rule described by Groebner et al. (2005: 140) and shown in equation 5.18. If the events E_1 and E_2 are mutually exclusive, with no intersection, then their probabilities could be added to give the probability of the union of the two events.

$$P(E_1 \cap E_2) = P(E_1) + P(E_2) - P(E_1 \cap E_2) \tag{5.18}$$

The intersection ($\cap$) of E_1 and E_2 is the event containing those experimental outcomes belonging to both E_1 and E_2. For example, P(Player A serves $\cap$ Player A wins point) = P({a,b}) = 0.389.

Mutually exclusive or disjoint events are events with no intersection. In other words, two disjoint events do not have any common experimental outcomes. For two disjoint events, E_1 and E_2, the probabilities for union and intersection are shown in equations 5.19 and 5.20 respectively.

$$P(E_1 \cap E_2) = 0 \tag{5.19}$$

$$P(E_1 \cup E_2) = P(E_1) + P(E_2) \tag{5.20}$$

Conditional probability

Often the probability of an event is influenced by whether or not a related event has occurred. Consider the details for the tennis match between Player A and Player B described in Table

5.3 and Figure 5.2. The probability of the first service being in given that Player B is serving may be of interest. This is the concept of conditional probability. In this case, the probability depends on the probability that Player B served (experimental outcomes c, d, g and h) and the probability that Player B served and the first serve was in (experimental outcomes c and g). In general, the probability of an event E_1 given that an event E_2 is known to have occurred is denoted $P(E_1 \mid E_2)$, which is calculated using equation (5.21). Think of the '$\mid$' symbol as being pronounced 'given that'.

$$P(E_1 \mid E_2) = \frac{P(E_1 \cap E_2)}{P(E_2)} \tag{5.21}$$

Therefore, P(First serve was in | Player B served) = 0.333 / 0.5 = 0.667.

Dependent and independent events

Conditional probability can tell us whether two events, E_1 and E_2, are dependent or independent. The two events are independent if the relation in equation 5.22 holds.

$$P(E_1 \mid E_2) = P(E_1) \text{ or } P(E_2 \mid E_1) = P(E_2) \tag{5.22}$$

For example, if the probability that first serve was in was the same no matter who was serving, then P(First serve was in) and P(Player B served) would be independent.

Consider the results of a soccer team over a 40-match season that are shown in Table 5.4. We wish to determine from these frequencies retrospective probabilities to study whether winning is dependent on either the venue or that the opponents wear red kit. Half of the matches are played at home so P(Home) = 20/40 = 0.5. The team won 12 matches, so P(Win) = 12 / 40 = 0.3. The probability of the match being played at home and being won is P(Home $\cap$ Win) = 9 / 40 = 0.225. Substituting our events into equation (5.22), we can check equation if P(Win | Home) = P(Win). From equation (5.21) we can determine the conditional probability P(Win | Home) = 0.225 / 0.5 = 0.45 which exceeds the probability of 0.3 for P(Win) and, therefore, there is a dependence between the two events.

Considering the whether the opposing team wears red shirts, we can see that there are 10 matches where the team plays opposition in red and so P(Opponents wear red) = 0.25.

Table 5.4 Outcomes of matches for a fictitious soccer team

Match Condition	Outcome			
	Win	Draw	Lose	Total
Home	9	5	6	20
Away	3	7	10	20
Opponents wear red	3	3	4	10
Opponents do not wear red	9	9	12	30
Total	12	12	16	40

The probability of the team winning a match when the opponents wear red is P(Opponents wear red ∩ Win) = 3 / 40 = 0.075. As before, the probability of winning in general P(Win) = 12 / 40 = 0.3. From equation (5.21) we determine that the conditional probability P(Win | Opponents wears red) = 0.075 / 0.25 = 0.3 which is equal to the probability of 0.3 for P(Win) and, therefore, the two events are independent.

PROBABILISTIC MODELLING

The first thing to say about this section is that the information is not necessary for the under-standing of probability in the statistical testing chapters of this book. This is a specialist use of probability and so some readers may decide not to read this section at all. In this section, we illustrate probabilistic modelling using the example of winning a game of tennis based on the probability of winning a point (Morris, 1977; Croucher, 1986). Let us call the probability of the serving player winning a point p and the probability of the server losing a point q = 1 − p. We further assume that tennis points are independent identically distributed random vari-ables (Klaassen and Magnus, 2001; Newton and Aslam, 2006). When we use a tree diagram to show the different ways in which the serving player can win or lose the game, we find that we do not get a perfect tree and that the game could be infinitely long. In Figure 5.3, the arrows running from top left to bottom right represent points being won by the serving player (probability p) and the arrows running bottom left to top right represent points being won by the receiving player (probability q). Let us first consider the number of ways that the server could win the game without the score reaching deuce.

There is only one way of winning the game without losing a point and the probability of this happening is $p \times p \times p \times p = p^4$ as shown in equation 5.23.

$$pppp = p^4 \tag{5.23}$$

There are four ways of winning the game losing a single point, and the probability of doing so is given by equation 4.24.

$$qpppp + pqppp + ppqpp + pppqp = 4p^4q. \tag{5.24}$$

You can trace these four different paths on Figure 5.3. Note that pppppq is not included because the server would have won the game after winning the first four points in this case. In general, the serving player always wins the last point of the game when holding serve. There are 10 ways of winning the game losing two points and the probability of doing so is shown in equation 5.25.

$$qqpppp + qpqppp + qppqpp + qpppqp + pqqppp + pqpqpp +$$

$$pqppqp + ppqqpp + ppqpqp + pppqqp = 10p^4q^2. \tag{5.25}$$

The probability of the serving player winning the game without any deuce points being played is the sum of equations 5.23 to 5.25 which is shown in equation 5.26.

$$p^4 + 4p^4q + 10p^4q^2 = p^4(1 + 4q + 10q^2) \tag{5.26}$$

The next step in determining the probability of the serving player winning the game is to determine the probability of the first deuce point being reached. When the first deuce is reached, the players will have played six points with the serving player having won three and lost three. There are 20 different ways of reaching the first deuce and the probability of doing so is given in equation 5.27.

$$qqqppp + qqpqpp + qqppqp + qqpppq + qpqqpp + qpqpqp +$$

$$qpqppq + qppqqp + qppqpq + qpppqq + pqqqpp + pqqpqp +$$

$$pqqppq + pqpqqp + pqpqpq + pqppqq + ppqqqp + ppqqpq +$$

$$ppqpqq + pppqqq = 20p^3q^3. \tag{5.27}$$

The final and most difficult step is to determine the conditional probability of the serving player winning the game given that the first deuce has been reached. There could be an infinite number of deuces in the game and the conditional probability of the serving player winning the game from this score is given by equation 5.28.

$$pp + (pq + qp)pp +$$

$$(pq + qp)(pq + qp)pp +$$

$$(pq + qp)(pq + qp)(pq + qp)pp +$$

$$(pq + qp)(pq + qp)(pq + qp)(pq + qp)pp + \ldots \ldots \tag{5.28}$$

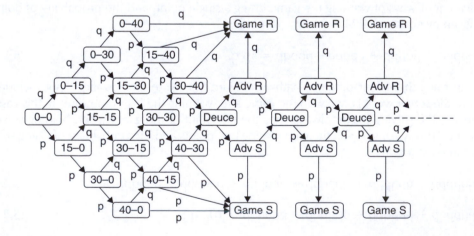

Figure 5.3 A probabilistic model for winning a game in tennis ('Game R' is game won by receiving player, 'Game S' is game won by serving player)

As we can see from equation 5.28, there are two ways for the score to return to deuce having been deuce: the server wins a point and then loses a point or the server loses a point and then wins a point. The above conditional probability simplifies to equation 5.29.

$$pp + (2pq)pp + (2pq)^2pp + (2pq)^3pp + (2pq)^4pp + \ldots \ldots \tag{5.29}$$

This is the sum of a geometric progression (equation 5.30); note the first term, pp, can be thought of as $(2pq)^0pp$ because anything to the power of zero is one.

$$p^2 \sum_{i=0}^{\infty} (2pq)^i \tag{5.30}$$

In general, the sum of an infinite geometric progression (where 'a' is not equal to zero and 'r' is less than one) is given by equation 5.31.

$$a \sum_{i=0}^{\infty} r^i = \frac{a}{1-r} \tag{5.31}$$

Therefore, the conditional probability of the server winning the game given that the first deuce has been reached is given by equation 5.32.

$$p^2 \sum_{i=0}^{\infty} (2pq)^i = \frac{p^2}{1-2pq} \tag{5.32}$$

The probability of the serving player winning the game, P(HOLD), is given by the probability of winning the game without the game going to deuce plus the product of the probability of reaching the first deuce and the conditional probability of winning given that the first deuce has been reached. This is given in equation 5.33.

$$P(HOLD) = p^4(1 + 4q + 10q^2) + 20\frac{p^5q^3}{1-2pq} \tag{5.33}$$

SUMMARY

There is uncertainty in many areas, including sports performance. Probability represents this uncertainty on a scale of zero to one with a probability of zero meaning an event is certain not to happen, a probability of one means that the event will certainly occur and a probability of 0.5 means that the event is equally likely to occur or not occur. In order to reason about probability, events should be understood in terms of the experimental outcomes that form the sample space for the experiment of interest. In multistep experiments, tree diagrams can be used to identify experimental outcomes. There are classic, relative frequency and subjective methods of assigning probabilities to experimental outcomes. When using relative frequency methods of assigning probabilities, it is sometimes necessary to consider combinations and permutations of experimental outcomes that can occur at different steps of experiments. Events are expressed as subsets of the sample space and the probabilities of events can be determined based on the probabilities of the experimental outcomes included in the events.

EXERCISES

Exercise 5.1. Probability of an experimental outcome

The draw is being made for the third round of the FA Cup. There will be 32 matches involving a total of 64 teams. Eighteen matches have been drawn and Liverpool have just been drawn as the home side for the nineteenth match. There are still nine Premier League teams left in the hat that have not been drawn yet. What is the probability of Liverpool's opponents being a Premier League side?

Exercise 5.2. Throwing two six-sided dice

Consider the probabilities for the 11 experimental outcomes shown in Table 5.2. Determine the probabilities of the following events:

a)   The total for the two die is seven.
b)   The total for the two dice is an even number.
c)   The total for the two dice is 10 or greater.
d)   The two dice show the same value between two and six.

Exercise 5.3. Retrospective probability in tennis

Consider the frequencies for experimental outcomes shown in Table 5.3. Do the following:

a)   Determine the probability of a second serve being required.
b)   Determine the conditional probability of a point being won by the server given that a second serve is required. Is winning the point independent of whether a second serve is required?
c)   Determine the conditional probability of player A winning a point given that player A is serving. Is player A winning a point independent of whether player A serves?
d)   Determine the conditional probability of a second serve being required if player B is serving. Is a second serve being required independent of whether player B is serving?

Exercise 5.4. Probabilistic modelling of a tennis point

Let p1 and p2 represent the probability of the first and second serves being in respectively in a tennis match. Let q1 and q2 be the conditional probabilities that the serving player will win the point given that the first serve is in and the second serve is in respectively. Determine an equation for P(Server wins the point) in terms of p1, p2, q1 and q2. Hint, using a tree diagram when thinking about this really helps.

Exercise 5.5. Combinations and permutations

A basketball governing body is considering the following three tournament structures for a national competition involving 24 teams:

a)   Four groups of six teams where each team plays each other team in their group once at a neutral venue. The group winners and runners up progress to four quarter-finals finals, the winners of which progress to two semi-finals. The two semi-final winners play in the final.

probability

b) Eight groups of three teams where each team plays each other team in their group twice: once at home and once away. The group winners progress to four quarter-finals finals, the winners of which progress to two semi-finals. The two semi-final winners play in the final.

c) Three regional leagues of eight teams where each team plays each other team in their region once at a neutral venue. The three regional winners contest a round robin tournament where they play each other once each.

How many matches are involved in these tournament structures?

PROJECT EXERCISE

Exercise 5.6. Probability of upsets in sport

Examine world ranking data and tournament results in a sport of your choice to determine the probability of an upset where lower ranked players or teams defeat higher ranked opponents. Such ranking and result information exists for international soccer, international rugby union and Grand Slam singles tennis among other sports.

CHAPTER 6

DATA DISTRIBUTIONS

INTRODUCTION

In Chapter 5, we covered probabilities and the outcomes of 'experiments'. This chapter uses probability to discuss the distribution of values of numerical variables. Fallowfield *et al.* (2005: 97) recommended exploring data distributions before using inferential statistical procedures, warning against dismissing this activity as being too simplistic. The probability of a variable taking some valid value for a given case or element is one. That is, we are certain that each case will have a value for each variable of interest. Probability can be used to describe the likelihood of each potential value occurring. For example, the probability distribution for the number of sets in completed men's singles matches at Wimbledon might be 0.545 for three sets, 0.289 for four sets and 0.165 for five sets. This is based on the 121 completed matches in the 2011 championship where there were 66 three-set matches, 35 four-set matches and 20 five-sets matches. The number of sets in this example is a random variable because it is a variable that gives a numerical outcome (3, 4 or 5) to an 'experiment' (Anderson *et al.*, 1994: 158). The numerical values of the random variable together with the probabilities of each value occurring can be used to calculate the expected value and variance for such variables. There are discrete random variables that use a finite set of values, such as the number of sets in a men's singles tennis match at Wimbledon which has three possible values. There are also continuous random variables that could take any one of an infinite number of possible values. This chapter covers the different types of discrete and continuous probability distributions. Those distributions that are used in statistical testing are of particular interest and a pre-requisite to reading Chapter 7 of this book.

DISCRETE PROBABILITY DISTRIBUTIONS

The uniform discrete probability distribution

Consider an 'experiment' to throw a six-sided die; the result x is a discrete random variable that can take any one of six values (1, 2, 3, 4, 5, 6). Assuming that the chance of any value occurring is equal, we have a uniform probability distribution with each of the six values having a probability of 1/6 of occurring. We can use these values and their probabilities to determine the expected value E(x) and variance Var(x) of x using equations 6.1 and 6.2

respectively, where X is the set of possible values of x and f(x) is the function for the probability of a given value x resulting from the experiment.

$$E(x) = \Sigma_{x \in X} \, x \cdot f(x) \tag{6.1}$$

$$Var(x) = \Sigma_{x \in X} \, (x - E(x))^2 \cdot f(x) \tag{6.2}$$

Table 6.1 shows the calculation of E(x) and Var(x). The sums in the final row of Table 6.1 are the expected value and the variance. Examination of this table should help understand the equations 6.1 and 6.2. The dot symbol '.' is used to represent multiplication. The 'Σ' symbol is a sum, meaning that both equations 6.1 and 6.2 are sums: equation 6.1 sums the six 'x.f(x)' terms while equation 6.2 sums the six '(x − E(x))².f(x)' terms. There are six terms being summed in each case because there are six values of x in the set of possible values, X, and the sum uses each x that is an element of ($\in$) X. The standard deviation is the square root of the variance which in this case is 1.7078. The spreadsheet 06-discrete_probability_distributions.XLS has a sheet 'Six sided die' that shows this example.

The binomial distribution

The binomial distribution is used when there are a fixed number of events within an experiment, each with two outcomes and the probability of each outcome is the same for all events. Consider a netball player taking (n=) 10 shots from a location on the edge of the shooting circle where she has a probability of scoring, p = 0.8. We will assume that this probability satisfies the assumptions of stationarity and independence required by the binomial distribution. Independence means that the probability p is not influenced by the preceding shot(s), while stationarity means that the probability p is not influenced by the number of shots that have been scored at any point in the training session. The overall score is a random variable that takes a cardinal value between 0 and 10. There is only one way that the player can get a score of zero which is to miss all 10 shots. There are 10 different ways in which a player could get a score of one: by scoring any of the 10 shots and missing the other nine. This can be determined using equation 6.3 where x is the score, n is the number of events and C(x) is the number of different ways of obtaining that score and n! is the factorial of n which is $1 \times 2 \times 3 \times 4 \times \ldots \times n$.

Table 6.1 Calculation of E(x) and Var(x)

x	f(x)	x.f(x)	x − E(x)	(x − E(x))²	(x − E(x))².f(x)
1	.1667	.1667	−2.5	6.25	1.0417
2	.1667	.3333	−1.5	2.25	0.3750
3	.1667	.5000	−0.5	0.25	0.0417
4	.1667	.6667	0.5	0.25	0.0417
5	.1667	.8333	1.5	2.25	0.3750
6	.1667	1.0000	2.5	6.25	1.0417
		E(x)			Var(x)
Σ	1.0000	3.5000			2.9167

$$c(x) = \frac{n!}{x!(n-x)!}$$ (6.3)

Using equation 6.3, we find that there are $10! / (2!\ 8!)$ ways of getting a score of two. This evaluates to 45. You can check this on a calculator or in Microsoft Excel using the FACT function for factorial (!) or you can get out a pen and piece of paper and count up all the different ways there of getting a score of two out of 10 shots. The 'Binomial' sheet of the 06-discrete_probability_distributions.XLS file shows that there are 252 ways of scoring five shots out of 10 so don't try listing these with pen and paper. The probability of scoring a particular sequence of x shots out of n is given by equation 6.4. This is because the probability of each score is p and the probability of each miss is $1 - p$.

$$P(\text{a particular sequence of x scores and n-x misses}) = p^x.(1 - p)^{n-x}$$ (6.4)

However, this is just the probability of one particular way of scoring x goals and there are $C(x)$ different ways of scoring x goals. Therefore, the probability, $f(x)$, of a score of x is given by equation 6.5. This would give a probability of 0.20 for the player scoring exactly seven out of 10 shots.

$$f(x) = C(x).p^x.(1 - p)^{n-x}$$ (6.5)

Table 6.2 shows the calculation of the expected value and variance for x using equations 6.1 and 6.2. These equations are valid because the binomial distribution is a discrete probability distribution just like the discrete uniform probability distribution. The only difference is that the values of $f(x)$ are not the same for each value of x.

For much larger values of n, the method using equations 6.1 and 6.2 to determine $E(x)$ and $Var(x)$ respectively is not really practical. Therefore, alternative equations are used for the binomial distribution which achieve the same results as can be found on the 'Binomial' sheet of the 06-discrete_probability_distributions.XLS file. These equations are equations 6.6 and

Table 6.2 Calculation of $E(x)$ and $Var(x)$ for a Binomial distribution where $n = 10$ and $p = 0.8$

x	f(x)	x.f(x)	x – E(x)	(x – E(x))²	(x – E(x))².f(x)
0	0.0000	0.0000	−8	64	0.0000
1	0.0000	0.0000	−7	49	0.0002
2	0.0001	0.0001	−6	36	0.0027
3	0.0008	0.0024	−5	25	0.0197
4	0.0055	0.0220	−4	16	0.0881
5	0.0264	0.1321	−3	9	0.2378
6	0.0881	0.5285	−2	4	0.3523
7	0.2013	1.4093	−1	1	0.2013
8	0.3020	2.4159	0	0	0.0000
9	0.2684	2.4159	1	1	0.2684
10	0.1074	1.0737	2	4	0.4295
		E(x)			Var(x)
Σ	1.0000	8.0000			1.6000

6.7 respectively. These give values of 8.0 for the expected number of shots scored out of 10 with a variance of 1.6.

$$E(x) = n.p \tag{6.6}$$

$$Var(x) = n.p.(1 - p) \tag{6.7}$$

A binomial distribution with n events with a constant probability of p for one of the outcomes of each event can be approximated by a normal distribution with a mean of n.p and a standard deviation of $\sqrt{n. p. (1 - p)}$ provided that both n.p and n.(1 - p) are greater than or equal to five (Anderson *et al.*, 1994: 218). The normal probability distribution will be covered later in this chapter.

It is often impossible to apply the binomial probability distribution in sport due to the assumptions of the distribution. First, many events have more than two possible outcomes; for example win, draw and lose. Second, most sports contests do not involve a fixed number of trials. For example, tennis matches that are the best of five sets stop when one player has won three sets. This is also true of penalty shoot-outs in soccer which may require five, less than five or on some occasions more than five penalties for each team. Similarly, if we wanted to abstract a netball match to a sequence of goals, the total number of goals varies from match to match.

The Poisson distribution

The Poisson distribution differs from the binomial distribution in that there are not a fixed number of trials. Instead E(x) represents the expected number of occurrences of some event of interest in a period of observation. The value of the random variable x does not have a theoretical upper limit like the binomial distribution does. The Poisson distribution will be illustrated using the number of goals scored in a soccer match. From the English FA Premier League being founded in August 1992 to the end of the 2010–11 season, 19,523 goals have been scored in 7,466 matches. Therefore, we expect a mean of 2.615 goals to be scored in English FA Premier League football matches. It is possible that there may be 0, 1, 2, . . . goals in the match. It is difficult to set an upper limit due to some very high scoring matches in the history of club level and international soccer. For example, Arbroath once defeated Bon Accord 36–0 in a Scottish Cup match (12 September 1885) and Australia defeated American Samoa 31–0 in a World Cup qualifier (11 April 2001).

The Poisson distribution uses the probability function shown in equation 6.8 where e is a constant 2.71828 such that e^x is its own derivative; if we graphed e^x against x, the value of e^x and the gradient of the graph would be the same. E(x) is our expected mean of 2.615 goals per match.

$$f(x) = \frac{E(x).e^{-E(x)}}{x!} \tag{6.8}$$

So what is the probability of 11 goals being scored in a match? Well according to equation 6.8, f(11) = 0.00007. The reciprocal of this is 13,952, meaning we would expect a match of 11 goals once every 13,952 matches. Since the English FA Premier League being founded in August 1992 to the end of the 2010–11 season, there have been 7,466 matches and so we would have expected 0.5 matches of 11 goals so far, 2.3 matches of 10 goals and 8.6 matches of nine goals. These calculations were made using equation 6.8 and the 'Poisson' sheet of the 06-discrete_probability_distributions.XLS spreadsheet shows the calculations of f(x). Using the Poisson distribution gives a very close estimate of the number of matches of different numbers of goals. At the time of writing, there has already been one match of 11 goals (Portsmouth 7–4 Reading on 29 September 2007), two matches of 10 goals (Tottenham 6–4 Reading on 29 December 2007 and Tottenham 9–1 Wigan on 22 November 2009) and nine matches of nine goals. The probabilities f(x) are calculated for all values of x from 0 to 100 in the 'Poisson' sheet of the 06-discrete_probability_distributions.XLS file. This sheet uses equations 6.1 and 6.2 to check E(x) and Var(x) because the Poisson distribution, like the binomial distribution, is a special case of a discrete probability distribution. This confirms that Var(x) = E(x) which is the case in the Poisson distribution (Levine and Stephan, 2005: 84).

CONTINUOUS PROBABILITY DISTRIBUTIONS

The uniform continuous probability distribution

Continuous variables have an infinite number of potential values even where there are definite minimum and maximum values. Therefore, it is not appropriate to determine probabilities for particular values as we were able to in discrete probability distributions. Instead, the area underneath the graph for a continuous probability distribution is one meaning that we are certain that each case must have some value in the allowable range of values for the variable. Figure 6.1 shows an example of a uniform continuous probability distribution for an automatic athletics start simulator. The device emits the commands 'On your marks',

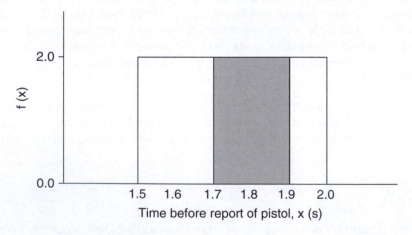

Figure 6.1 Uniform continuous probability distribution

'Get set' before the report of a starter's pistol is sounded. The machine takes a random time between 1.5 s and 2.0 s between the end of the 'Get set' command and beginning of the report of the starter's pistol. The time is random and uniform.

Because the area under the curve is one meaning and the probability function is uniform, there must be some minimum value and some maximum value otherwise the area would be infinite. Our range of x values is 0.5s (from 1.5s to 2.0s) and, therefore, f(x) is 2.0 for all 0.1s subranges of values in this range and 0.0 for all value outside this range. Consider the interval between 1.7s and 1.9s. The probability of the pistol reporting between these two times is 0.4 because the area of the curve between these two points is 0.2 × 2.0 (width × height). This method can be used to determine the probability of the time, x, being in any other sub-range of the possible values. We cannot compute the probability of x being exactly 1.7s because this is an instant or point in time with no duration. We must always specify some subrange of values of interest when using any continuous probability distribution if we wish to determine the probability of the value being in that subrange.

The normal probability distribution

The normal distribution is the most common parametric distribution that sport and exercise science data follow with variables such as height, mass, distance, temperature and speed following this distribution (Fallowfield *et al.*, 2005: 98). The normal probability distribution is a bell-shaped curve as shown in Figure 6.2. If we think of adults' heights, we probably know of more adults with heights in the range 1.6m to 1.7m and 1.7m to 1.8m than we know adults whose heights are between 1.5m and 1.6m or between 1.8m and 1.9m. This is typical of normally distributed variables where the mean value and median value are similar if not the same and fall within a modal subrange of values in the middle of the measurement scale. Like the uniform continuous probability distribution, the height of the curve shown in Figure 6.2 reflects the number of people with given heights. Figure 6.2 is assumed to be a distribution for an entire population with a mean of μ and standard deviation of σ for the variable of interest. Mathematically, the normal distribution is actually for an infinite range of numbers from $-\infty$ to $+\infty$ but the probabilities of very large and very small values will be so small that there may not be any cases in the population with such values. Consider the subrange of values shown in Figure 6.2 from three standard deviations below the mean ($\mu - 3\sigma$) to three standard deviations above the mean ($\mu + 3\sigma$). The probability of a value lying in this subrange is 0.997 in a true normal distribution. Therefore, in a normally distributed variable, 99.7 per cent of cases will be within three standard deviations of the mean. We can also see that 95.4 per cent of cases are within two standard deviations of the mean and 68.3 per cent of cases are within one standard deviation of the mean. Recalling from Chapter 4 that a z-score is the number of standard deviations a value is above the mean, we can see that the subrange of values from ($\mu - 3\sigma$) to ($\mu + 3\sigma$) maps onto the z-scores -3 to $+3$. There are many normal distributions which are located on the scale of real numbers at different means. All will use the full range of real numbers ($-\infty$ to $+\infty$) but the main part of the bell-shaped curve that accounts for 99.7 per cent of cases will be from three standard deviations below the mean to three standard deviations above the mean. This causes different variables to use different spreads of values around the mean to cover this 99.7 per cent of cases. Some may appear thin and tall, others may appear wide and short, but all are normally distributed but with different standard deviations. The perfect normal probability distribution with a mean of zero and a standard deviation of one is called the standard normal distribution and its values are z-scores. All other normal distributions can be mapped onto the standard normal curve by converting raw scores to z-scores using equation 4.4 from Chapter 4. The raw scores are also equivalent to T-scores as shown in Figure 6.2.

Although the curve is theoretically infinite, the area under the curve is one given that the probability function for the curve, f(x), gives very small values when x is outside three standard deviations of the mean. Just as we cannot determine the probability for a single value of x in a uniform continuous probability distribution, we cannot determine the probability for a single value of x in a normal probability distribution either. Once again, the value is an instantaneous value or single point on the scale and so there is no measurable area for that point because the width of the vertical line at that point is zero. Therefore, we need to determine the probability of a value being in some range of values. Using the standard normal distribution as an example (z-scores with a mean of zero and a standard deviation of one), some subranges and equivalent probabilities are listed in Table 6.3.

As has already been mentioned, the standard normal distribution curve is a theoretical distribution that uses equation 6.9. In reality, we will not have participants covering all values for some variable x. Even if we had data for all seven billion people on Earth, they would be represented by seven billion individual points on the x axis and would not cover

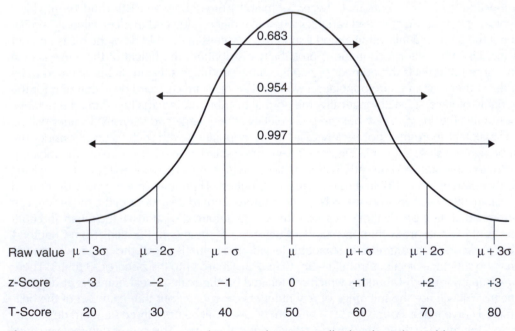

Raw value	$\mu - 3\sigma$	$\mu - 2\sigma$	$\mu - \sigma$	μ	$\mu + \sigma$	$\mu + 2\sigma$	$\mu + 3\sigma$
z-Score	−3	−2	−1	0	+1	+2	+3
T-Score	20	30	40	50	60	70	80

Figure 6.2 Raw scores, z-scores and T-scores for a normally distributed variable

Table 6.3 Probabilities for some subranges of the standard normal distribution

Subrange of z-scores	Probability of a z-value being in this subrange
Less than 0	0.5
Between −1 and +1	0.683
Greater than +1	0.1585 (= (1−.683)/2)
Less than −3	0.0015 (= (1−.997)/2)
Between 0 and 2	0.477 (= .954/2)
Between −2 and +1	0.8181 (=.954/2 + .683/2)

all the infinite number of values of x. This is worth bearing in mind when checking variable for normality.

$$f(z) = \frac{1}{\sqrt{2\pi}} \ e^{-(z^2/2)} \tag{6.9}$$

To illustrate this point, let us consider the finishing times of 22,524 male athletes completing the 2011 London marathon (http://results-2011.virginlondonmarathon.com/2011/, accessed 1 July 2011). Since this analysis was done by the author, the website has been updated to include 22,525 male finishers (accessed 28 July 2011). We could use small subranges such as one minute intervals to determine if the distribution of finishing times is close to normally distributed. Figure 6.3 is a histogram showing the number of athletes finishing in each minute. A value on the x axis such as 2:15 represents the one minute between 2:15:00 and 2:15:59 as finishing times were recorded to the nearest second. Expressing the variable to the nearest one minute does not make the variable a discrete variable, it is simply summarized using one minute intervals. The file 06-London_marathon_men_2011.XLS contains the frequency of the 22,524 athletes finishing in each one minute interval. There are some interesting observations that can be made in the data which are listed below:

- The median finisher completed the race in 4 hours 23 minutes which is not half way between the fastest time of 2 hours 4 minutes and 9 hours 40 minutes.
- The one minute intervals with the most finishers were 3:57:00 to 3:57:59 (n=208), 3:59:00 to 3:59:59 (n=203) and 4:20:00 to 4:20:59 (n=202).
- There are six athletes with times between 2:19:00 and 2:19:59 despite there only being six athletes with times between 2:14:00 and 2:18:59 and only 10 with times between 2:20:00 and 2:24:59.
- Similarly, there are spikes in the data at the one minute intervals 2:59:00 to 2:59:59 and 3:59:00 to 3:39:59 with athletes appearing to make an extra effort if they had a chance of finishing in under 3 hours or under 4 hours.
- There was a very high number of finishers (n=38) between 7:28:00 and 7:28:59 despite no other one minute interval in the 32 minutes before or 2 hours and 12 minutes after having any more than nine finishers.

The distribution of men's marathon times is positively skewed with a long 'tail' of high values. We have a median value of 4 hours 23 minutes and values of over 9 hours. For this distribution to be normally distributed, it would need to be symmetrical with some values of less than 0 hours 0 minutes. Of course such values are impossible for the marathon or any other running event. A more general example of a positively skewed distribution is shown in

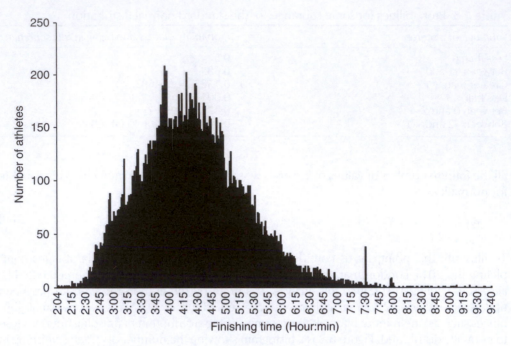

Figure 6.3 Distribution of men's finishing times in the 2011 London marathon

Figure 6.4(a). In positively skewed distributions, the mean is typically greater than the median with both coming after the modal subrange which forms the 'hump' of the curve (Vincent, 1999: 51). Negative skew is where there is a long tail below the middle of the curve with the mean being lower than the median with both being less than the modal subrange as shown in Figure 6.4(b). Figures 6.4(c) and 6.4(d) show a platykurtic and a leptokurtic distribution respectively. These are not skewed because they are symmetrical. However, the number of cases within one standard deviation of the mean is either too low or too high for the distribution to be considered normal. A platykurtic distribution is one where insufficient cases fall within one standard deviation of the mean and the distribution is, therefore, too flat in the middle to be normal. A leptokurtic distribution is one where too many cases fall within one standard deviation of the mean and the distribution is, therefore, too pointed in the middle to be normal.

Where a distribution is normally distributed, we can determine the probability of a randomly chosen participant falling within any subrange of values. We have already shown some examples of this in Table 6.3. The probability of a randomly chosen participant falling in a given subrange can be determined using Microsoft Excel. Microsoft Excel, provides the function NORMDIST(x, mean, standard deviation, Cumulative) to determine the area of a normal distribution curve of a given mean and standard deviation before a given value of x. The first argument of the function is the value of the variable we are interested in, x. The second and third arguments are the mean and standard deviation of the distribution for the variable. The final argument should be 1 to ensure that the probability given is for the area

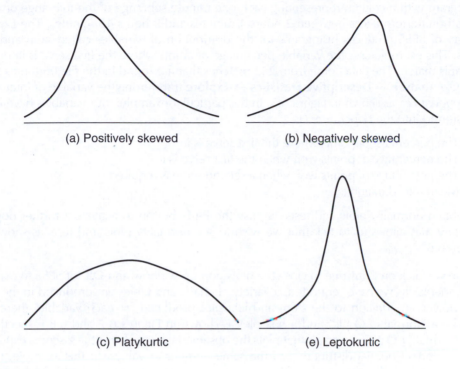

(a) Positively skewed

(b) Negatively skewed

(c) Platykurtic

(e) Leptokurtic

Figure 6.4 Non-normal distributions

preceding x. Therefore, NORMDIST(12, 10, 2, 1) would be evaluated to 0.8415 because 12 is one standard deviation above the mean of 10. Therefore, we have 0.5 for the lower half of the distribution of x and 0.683 / 2 for half of the area that is within one standard deviation of the mean. There is also a function NORMSDIST(z) which is evaluated to the area under the standard normal distribution curve that precedes a given value of z. Arguments for the mean and standard deviation are not required in this function because the standard normal distribution always has a mean of zero and a standard deviation of one.

Microsoft Excel also provides functions allowing us to determine the raw score where a given percentage of the proportion has a higher or lower score. For example, imagine 800m running times for a population of interest were distributed normally with a mean of 120s and a standard deviation of 6s and we wish to determine the 800m time for which only 25 per cent of the population would run faster. Remember that 800m running time is one of those variables where lower values are better. NORMINV(0.25, 120, 6) is evaluated to 115.95 so we would need to run 1 minute and 55.95s or faster to be in the fastest 25 per cent of 800m runners in this population.

The normality of variables can be checked using visual inspection or statistical criteria (Fallowfield *et al.*, 2005: 100). Visual checks include the use of histograms to represent the frequency of values in different subranges of a variable. Figure 6.3 is an example of such a histogram with a column representing each one minute subrange of the full range of male marathon performances between 2 hours 4 minutes and 9 hours 40 minutes. The Explore facility of SPSS produces histograms for the distribution of variables called stem and leaf plots. This example uses the variable percentage of points where the first serve is in during a tennis match. The data come from 252 matches that are found in the file 06-tennis.SAV. We use **Analyse → Descriptive Statistics → Explore** transferring the variable of interest to *Dependent List* as shown in Figure 6.5. In this particular example, four variables have been transferred into the *Dependent List*:

- The percentage of points where the first serve is in
- The percentage of points won when the first serve is in
- The percentage of points won when a second serve is required
- Mean rally duration (s).

To obtain normality plots and tests, we use the **Plots** button to activate a further pop-up window that allows us to tick that we wish to see normality plots and tests as shown in Figure 6.6.

When we click on **Continue** to close the plots pop-up window and **OK** for SPSS to explore the variables as we have requested, a variety of charts and tables are produced in the output viewer. In addition to the stem and leaf plot produced for each variable, there will also be a Normal Q-Q plot and a box plot as shown in Figures 6.7 and 6.8 respectively. The Normal Q-Q plot is a graph that plots the observed values against the z-scores expected from a perfect normal distribution of the same number of values. In this example, there are 252 observed values and so the 252 expected z-scores are determined based on

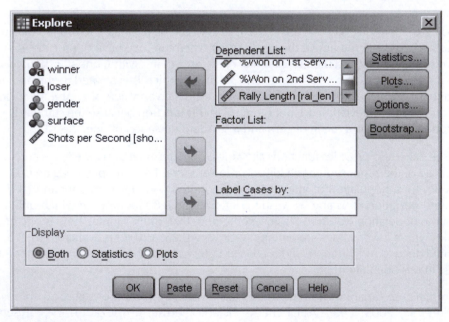

Figure 6.5 The Explore facility in SPSS

98

data distributions

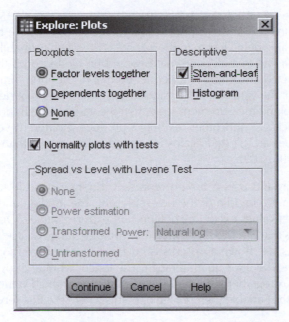

Figure 6.6 Selecting normality plots and tests when exploring variables

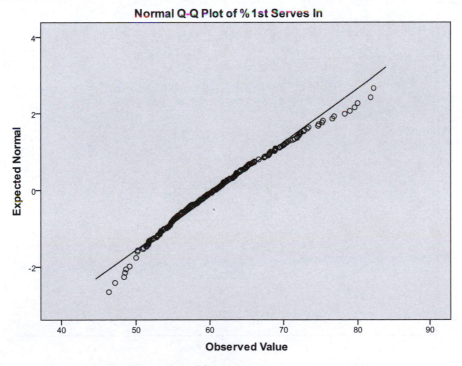

Figure 6.7 Normal Q-Q plot produced by SPSS

cumulative probabilities of 0.5 / 252, 1.5 / 252, 2.5 / 252, . . ., 250.5 / 252 and 251.5 / 252. The specific z-score for one of these probabilities is the z-score where the probability of a randomly chosen z-value being less than the z-score is the given probability. The line on the Normal Q-Q plot is a line that all coordinates would be plotted on if the observed values were exactly as expected according to a perfect normal distribution of the same mean and standard deviation for the variable. If the coordinates of the Q-Q plot deviate from this line in some subrange(s) of the measurement scale then the variable might not be normal. Some statistics software packages and other textbooks produce a different version of Q-Q plot where expected values are plotted against observed values rather than expected z-scores (Fallowfield *et al.*, 2005: 102).

The box plot in Figure 6.8 uses a box to represent the inter-quartile range of a variable. The horizontal line in the middle of the box represents the median and the whiskers represent the full range of values excluding outliers and extreme values. Outliers are shown as circles and are values that are more than 1.5 inter-quartile ranges outside the inter-quartile range. Extreme values are shown as asterisks and are values that are more than three inter-quartile ranges outside the inter-quartile range. The longer whisker above the box than the whisker below the box together with there being outliers with high values but not with low values indicates that the variable (percentage of points where the first serve is in) is positively

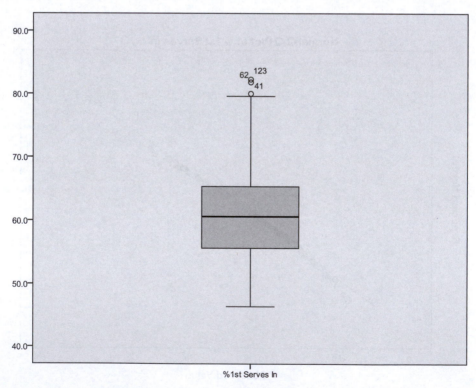

Figure 6.8 Box and whiskers plot for the percentage of points where the first serve was in

100
data distributions

skewed. Some statistics software packages and textbooks show box plots running from left to right with the variable's scale of measurement on the horizontal axis rather than the vertical axis (Newell *et al.*, 2010: 129).

The visual inspections described above can be useful, but sometimes we may have a box plot or a Normal Q-Q plot that could be judged as sufficiently normal by some and not by others. Therefore, this author prefers the use of definite criteria for deciding whether a variable is normally distributed or not. There are several quantitative tests of normality including the Kolmogorov–Smirnov test, the Shapiro–Wilk test, the Anderson–Darling test and the D'Agostino–Pearson Omnibus test (Newell *et al.*, 2010: 127). When we request normality plot and tests when exploring variables in SPSS, two tests will be performed: the Kolmogorov–Smirnov test and the Shapiro–Wilk test. The Kolmogorov–Smirnov test should be used when we have at least 50 cases otherwise the Shapiro–Wilk test should be used. Therefore, in this example the Kolmogorov–Smirnov test is used and the results are shown in Table 6.4. Where the p value (called Sig. in SPSS) is greater than 0.05, the data are deemed to be sufficiently normal. Therefore in our example, two of our four variables are normally distributed: the percentage of points where the first serve was in and the percentage of points won when the first serve was in. The meaning of p (or Sig.) will be discussed in Chapter 7, but for the moment all we need to know is that the Kolmogorov–Smirnov test and the Shapiro–Wilk test provide definite criteria for determining whether a variable is sufficiently normal.

A simpler way to check for normality is to test for skew and kurtosis separately (Fallowfield *et al.*, 2005: 103–9). The SPSS output provided by the Explore facility includes a table of descriptive statistics (Table 6.5) which shows results for skew and kurtosis. The statistic is divided by the standard error in each case to return a value z_{Skew} in the case of skewness and z_{Kurt} in the case of kurtosis. If z_{Skew} is less than -1.96, then the data are negatively skewed, if z_{Skew} is greater than $+1.96$, then the data are positively skewed, otherwise the data are symmetrical. If z_{Kurt} is less than -1.96, then the data are platykurtic, if z_{Kurt} is greater than $+1.96$, then the data are leptokurtic, otherwise the data mesokurtic. The absolute value of 1.96 used in these decisions was recommended by Ntoumanis (2001: 45). Other statisticians classify distributions with z_{Skew} and z_{Kurt} being between -2 and $+2$ as being normally distributed (Fallowfield *et al*, 2005: 103). Where a variable is mesokurtic and symmetrical, the variable is considered to be normal enough. The output for skew and kurtosis for the percentage of points where the first serve is in is shown in Table 6.5. It is unfortunate that

Table 6.4 SPSS output showing normality test results

| | Tests of Normality | | | | | |
| | Kolmogorov–Smirnova | | | Shapiro–Wilk | | |
	Statistic	df	Sig.	Statistic	Df	Sig.
% 1st Serves In	.045	252	.200*	.982	252	.003
% Won on 1st Serves	.047	252	.200*	.993	252	.294
% Won on 2nd Serves	.069	252	.005	.936	252	.000
Rally Length	.090	252	.000	.937	252	.000

a Lilliefors Significance Correction
* This is a lower bound of the true significance.

Table 6.5 SPSS output from the Explore facility including skewness and kurtosis

			Statistic	Std. Error
		Descriptives		
%1st Serves In	Mean		61.094	.4503
	95% Confidence Interval for Mean	Lower Bound	60.207	
		Upper Bound	61.981	
	5% Trimmed Mean		60.844	
	Median		60.575	
	Variance		51.102	
	Std. Deviation		7.1486	
	Minimum		46.4	
	Maximum		82.3	
	Range		35.9	
	Interquartile Range		9.8	
	Skewness		.473	.153
	Kurtosis		−.055	.306

z_{Skew} and z_{Kurt} are not included in the SPSS output, but they are easy enough to determine from the statistics and standard errors reported. In this example, z_{Skew} = .473 / .153 = 3.092 while z_{Kurt} = −.055 / .306 = −0.180. Therefore, the variable is positively skewed according to this test which is inconsistent with the conclusion drawn from the same data using a Kolmogorov–Smirnov test.

It is possible to obtain z_{Skew} and z_{Kurt} using **Analyse → Descriptive Statistics → Descriptives** and using **Options** to analyse the distribution of the selected variable for skewness and kurtosis. Figures 6.9 and 6.10 show the pop-up windows used to set up this analysis while Table 6.6 shows the output produced.

There are statistical tests that can only be used validly if dependent variables are normally distributed. If we wish to use such a test, but the variable is not normally distributed, there may be ways in which the variable can be transformed into a normally distributed variable (Nevill, 2000; Tabachnick and Fidell, 2007: 86–8). For example, if a variable is positively skewed, it is possible that the natural logarithm of the variable may be normally distributed. Consider the variable rally length which is positively skewed as shown by the skewness statistics in Table 6.7 (z_{Skew} = 1.059 / 0.153 = 6.922) and the box plot in Figure 6.11.

We can produce a logarithmically transformed version of rally length using **Transform → Compute** as shown in Figure 6.12. Select the Ln function, transfer 'rally length' into this function as an argument in the *Numerical Expression* area and specify a name for the new variable; for example, 'ln_ral_len'. When we click on **OK**, this new variable is produced in the SPSS datasheet to the right of the existing variables. Table 6.8 shows that this logarithmically transformed variable is sufficiently normal (z_{Skew} = −0.055 / 0.153 = −0.359 and z_{Kurt} = −0.128 / 0.306 = −0.418) as does the Kolmogorov–Smirnov test in Table 6.9 ($p > 0.05$).

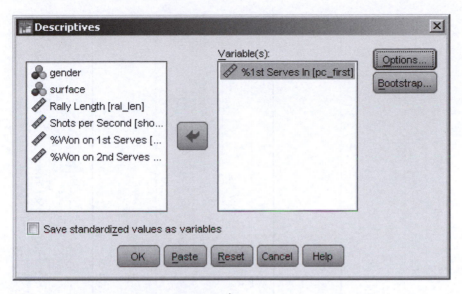

Figure 6.9 Descriptive statistics pop-up window

Figure 6.10 Selecting analysis of distribution in terms of skewness and kurtosis

Table 6.6 Descriptive statistics including results for skew and kurtosis

	N	Minimum	Maximum	Mean	Std. Deviation	Skewness		Kurtosis	
	Statistic	Statistic	Statistic	Statistic	Statistic	Statistic	Std. Error	Statistic	Std. Error
%1st Serves In	252	46.4	82.3	61.094	7.1486	.473	.153	-.055	.306
Valid N (listwise)	252								

Table 6.7 Descriptive statistics for rally length (s) in tennis

	N	Minimum	Maximum	Mean	Std. Deviation	Skewness		Kurtosis	
	Statistic	Statistic	Statistic	Statistic	Statistic	Statistic	Std. Error	Statistic	Std. Error
Rally Length	252	2.5	15.7	6.154	2.2559	1.059	.153	1.807	.306
Valid N (listwise)	252								

Table 6.8 Descriptive statistics for a logarithmically transformed version of rally length

	N	Minimum	Maximum	Mean	Std. Deviation	Skewness		Kurtosis	
	Statistic	Statistic	Statistic	Statistic	Statistic	Statistic	Std. Error	Statistic	Std. Error
ln_ral_len	252	.93	2.75	1.7533	.35937	-.055	.153	-.128	.306
Valid N (listwise)	252								

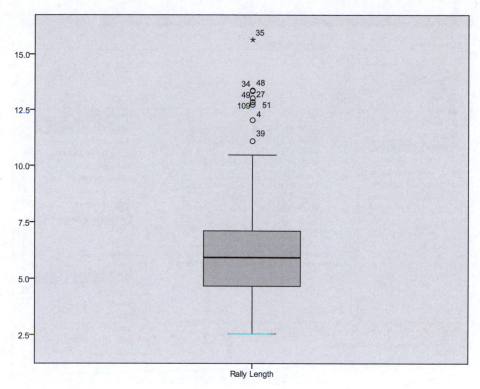

Figure 6.11 Box and whiskers plot showing positive skewness

Table 6.9 Normality results for the logarithmically transformed version of rally length

| | Tests of Normality | | | | | |
| | Kolmogorov–Smirnov[a] | | | Shapiro–Wilk | | |
	Statistic	df	Sig.	Statistic	df	Sig.
ln_ral_len	.056	252	.054	.989	252	.051

a Lilliefors Significance Correction

t distributions

Some of the statistical tests that are covered in this book use t distributions rather than the normal distribution. The t distributions are appropriate where sample sizes are less than 30 (Anderson et al., 1994: 278–84). They are similar to the normal distribution but are flatter and wider as shown in Figure 6.13. The different t distributions are distinguished by their number of degrees of freedom which depend on sample sizes. As the number of degrees of freedom approaches 30, the peak of the t distribution approaches that of the normal

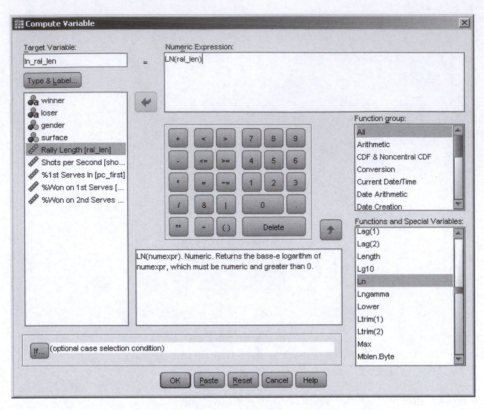

Figure 6.12 The Transform-Compute facility in SPSS

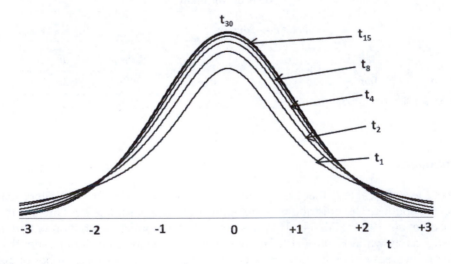

Figure 6.13 t distributions

data distributions

distribution. As the number of degrees of freedom approaches infinity (∞), the t distributions and the normal distribution almost match (Hinton, 2004: 66–7).

F distributions

Figure 6.14 shows five of the F distributions. The F distributions have two degree of freedom values as opposed to the t distributions which have one. The F distributions are related to the t distribution as $F_{1,DF} = (t_{DF})^2$ where DF is the degrees of freedom of a t distribution and the error degrees of freedom of an F distribution (Murphy et al., 2009: 34). This means that F distributions are always positive. F distributions are used in 'single tailed' testing because the purpose of the test is to see if the variance ratio in question is significantly greater than one. F distributions are used in tests of variance ratios such as analysis of variances tests and analysis of covariance (Hinton, 2004: 120–3) which will be covered in Chapters 11, 12 and 13.

Chi square distributions

Figure 6.15 shows some chi square (χ^2) distributions. Like the t distributions, there are a many chi square distributions that are distinguished by their degrees of freedom. Chi square distributions are used in the testing of observed and expected frequencies. Larger chi square values are expected when we have a larger number of observed frequencies and corresponding expected frequencies to compare. The number of degrees of freedom (DF) is a function the number of frequencies being analysed. Therefore, chi square distributions with a higher DF value will be located higher up the chi square scale shown on Figure 6.15. Chi square distributions are related to F distributions as $\chi^2_{DF} / DF = F_{DF,\infty}$ (Murphy et al., 2009: 34). The chi square distribution is used in chi square tests which will be covered in Chapter 15 as well as Kruskal–Wallis H tests and Friedman tests that will be covered in

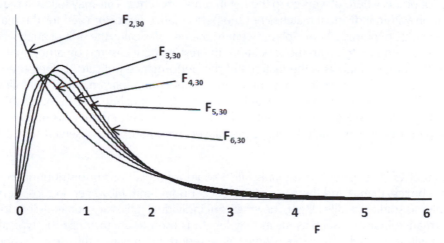

Figure 6.14 F distributions

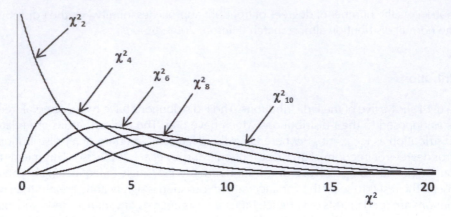

Figure 6.15 Chi square distributions

Chapter 14. When the number of degrees of freedom is 30 or more, chi square can be approximated by the normal distribution.

ISSUES WITH DISTRIBUTIONS

Some variables used in sports science research may be composite variables that are calculated in terms of other measured variables that are measured on different scales. If we consider the percentage of points in a tennis match where the first serve is played in, we divide the number of points where the first serve was played in by the total number of points expressing this fraction as a percentage. The binomial distribution seems best suited to analyse the percentage of points where the first serve is played in. One issue here is that the total number of points played is not the same for each match. A similar example is the percentage of points where players go to the net in tennis matches. This may follow a binomial distribution within individual matches but the performance indicator used for the match is the percentage of points where a player went to the net. Basically, the unit of analysis is not a point in a tennis match with the variable of interest being measured on a nominal scale. Instead, the unit of analysis is the match and the performance indicator is measured on a ratio scale. The performance indicator, percentage of net points played, might not follow the binomial distribution across a set of matches. Indeed O'Donoghue (2009) found this variable to be positively skewed across a set of different players but normally distributed across a set of performances by each of four tennis players who had been observed for 18 to 23 matches.

The number of shots played in a tennis rally should follow a Poisson distribution because it cannot be less than 0 and there is no theoretical upper limit. However, in most performance analysis studies of rally length, the lengths of individual rallies are not analysed. Instead a performance indicator such as mean rally length is used to characterize the typical rally duration within a match. Therefore, studies often report the 'mean of the means' where the first mean is the mean of mean values from different matches.

SUMMARY

Probability distributions represent the chance of a randomly chosen case or participant having a particular value or being in a particular range of values of the scale of measurement used. For discrete variables, probabilities can be determined for individual values. However, continuous variables have an infinite number of values making it impossible to determine a probability for a particular value when measured to an infinite number of decimal places. Therefore, the probability of a randomly chosen case or participant being in a given range of values is calculated instead. The binomial and Poisson distributions are applicable to discrete variables in sport. The main continuous probability distribution used in sport and exercise research is the normal probability distribution. Other distributions used are t, F and χ^2, as will be shown in later chapters of this book.

EXERCISES

Exercise 6.1. Probabilities and percentiles in a normal distribution

A population of athletes performs a 20m sprint test with a running start permitted recording the fastest 20m sprint out of three attempts timed by electronic speed gates. The times are normally distributed with a mean of 2.76s and a standard deviation of 0.12s. Use the NORMDIST function in Microsoft Excel to determine the probability that a randomly selected member of the population will have a 20m sprint time of:

a) Faster than 2.4s
b) Faster than 2.9s
c) Between 2.6 and 2.8s
d) Slower than 3.0s

Remember that the fourth argument of NORMDIST should be 'Cumulative' to return the probability of a randomly chosen sprinter having a time that is less than the specified value.

Exercise 6.2. Distribution of tennis performance variables in women's and men's singles

Using the 252 tennis matches in the file 06-tennis.SAV, determine whether the following five variables are normally distributed for women's singles and men's singles separately:

- Mean rally length
- Mean shots played per second
- % points where the first serve is in
- % points won when the first serve is in
- % points won when a second serve is required.

Use z_{Skew} and z_{Kurt} as well as Kolmogorov–Smirnov tests of normality. Are the results of these tests consistent? Where any of the variables are not normally distributed, will natural log or square root transformations produce versions of the variables that are normally distributed?

Exercise 6.3. English FA Premier League soccer

In the 380 matches of the English FA Premier League in the 2010–11 season, there have been 617 goals scored by the home team and 446 goal scored by the away team. Using the Poisson distribution, determine the probability that the home team will score 0, 1, 2, 3, 4+ goals. Do the same for the away team.

Exercise 6.4. Stanines

Table 4.10 of Chapter 4 shows the z-score ranges associated with the nine stanines of the stanine scale. If a variable is normally distributed, what is the probability of a randomly chosen member of the population being in each stanine?

Exercise 6.5. Chance of winning the toss more than other teams

a) In some sports, a coin is tossed to determine which team commences the match in possession or which side of the playing area they defend in the first period of an invasion game or whether they bat or field first in an innings sport. Imagine a sport where winning the toss gives a team an increased chance of winning the match based on historical records. The actual probability of winning the toss for any one match is 0.5. There are 20 teams in the league and each has to play the other 19 teams once during a season. Use the binomial distribution to determine the probability of a team winning the toss on 12 or more matches during the season.
b) Can the probability you have determined be used to determine how many teams are expected to win the toss more than 12 out of 19 times during the season? If so, how many teams?

Exercise 6.6. Athletic burnout questionnaire

A sports psychologist uses three variables scored from an athletic burnout questionnaire, (1) reduced sense of accomplishment RSA, (2) emotional and physical exhaustion EE and (3) devaluation DEV. Each of these variables takes a positive integer value between 5 and 25. The psychologist claims that despite the variable being an integer rather than a real number, it is still normally distributed. The sports psychologist reports the mean±standard deviation for the three variables to be: RSA 15.6±2.6, EE 13.7±2.1 and DEV 14.5±2.4. What is the percentage of cases that would be expected to have each value (5 to 25) for these variables to be normally distributed?

PROJECT EXERCISE

Exercise 6.7. British Indoor Rowing performance

The finishing times for 2,000m rowing on a row ergometer are found on the internet site of the British Indoor Rowing Championship (http://concept2.co.uk/birc/). Use an appropriate test to determine if the finishing times are normally distributed for each of the following categories:

■ Open heavyweight men
■ Open lightweight men
■ Open heavyweight women
■ Open lightweight women

data distributions

CHAPTER 7

HYPOTHESIS TESTING

INTRODUCTION

Quantitative research typically uses a reductive approach where concepts of interest are abstracted to categorical or numerical variables, data are collected from a sample drawn from the relevant population, the data are analysed and the results generalized to the population. This approach relies on an understanding of the distribution of the variables used and the probability of the sample being representative of the population of interest. Hence this book has covered probability and data distributions prior to the current chapter. Readers are, therefore, advised that Chapters 5 and 6 on probability and probability distributions respectively should be seen as prerequisites to reading the current chapter. Furthermore, reading Chapter 6 on probability distributions should be preceded by reading about z-scores in Chapter 4 on standardized scores.

A further characteristic of quantitative research is that it typically answers some specific research question. Once a research topic has been selected, a specific research question is formulated through a process of focusing the problem, defining variables and stating precise hypotheses (O'Donoghue, 2010: 78–101). There are usually two hypotheses: a null hypothesis, H_0, and an alternative hypothesis, H_A. The null hypothesis could be a statement that there is no difference between the groups or conditions being compared or that there is no relationship between variables being correlated. In these situations, the alternative hypothesis is a statement that there is a difference between the groups or conditions being compared or that there is a relationship between the variables being correlated. These hypotheses are mutually exclusive statements so that the research will conclude one or the other but not both. The quantitative research process can, therefore, be seen as specifying a precise research question using formal hypotheses and then gathering and analysing data to decide which hypothesis to accept and which to reject. This chapter covers sampling, sampling error and hypothesis testing in general, before describing what statistical procedures to apply in different situations. The selection of a statistical test is arguably the most important material covered in the book as once the researcher or student is capable of choosing the correct statistical procedures to answer their given research question, the application of the statistical procedure follows the 'cookbook' approach mentioned by Hinton (2004: xvii).

HYPOTHESES

The role of hypotheses in research studies

There is a considerable amount of work to be done in order to define a workable research question in the form of precise hypotheses that are expressed in terms of operationalized variables. A full coverage of criteria for selecting a research area, reviewing relevant literature, reasoning about the problem, selecting variables and defining those variables is beyond the scope of this book. However, students are encouraged to understand statistics in the context of quantitative research and are referred to other textbooks that describe the research process more fully (Gratton and Jones, 2004: 1–174; O'Donoghue, 2010: 29–102; Thomas and Nelson, 1996: 26–60). The purpose of this section of the current chapter is to describe the qualities of hypotheses, their use in quantitative research and their link to research design and the use of statistics.

Statistical procedures can be used well or they can be used poorly. One example of poor use of statistical procedures is where a research project has not been well defined and data have been gathered without a precise research question having been tied down. Poorly designed research such as this uses statistical procedures to compare samples and conditions or correlates variables to see if any significant patterns emerge.

There are different types of quantitative research including descriptive research (Thomas and Nelson, 1996: 314–26), experimental research (Thomas and Nelson, 1996: 344–63) and research synthesis (Thomas and Nelson, 1996: 292–304). A research project of any of these types should have clear aims and scope. The researcher should have surveyed sufficient previous research to identify a needed and workable research question. In good research, hypotheses form a simple logical framework for data gathering and analysis (Newell et al., 2010: 116). A hypothesis is an anticipated answer to the research question that can be tested using the data gathered. There is usually a pair of hypotheses: the null hypothesis, H_0, and the alternative hypotheses, H_A. Some textbooks denote the alternative hypotheses H_1 (Diamantopoulos and Schlegelmilch, 1997: 136; Fallowfield et al., 2005: 16) or H_a (Anderson et al., 1994: 288) rather than H_A. Some research questions are composed of different parts requiring a pair of hypotheses for each part; these might be denoted H_01, H_A1, H_02, H_A2, H_03, H_A3, etc. Each pair of hypotheses should be mutually exclusive and covering all possible outcomes for some sub-question of the research project. In quantitative research, we gather data and analyse that data to try to disprove the null hypothesis (Anderson et al., 1994: 289; Newell et al., 2010: 116). This means that the conclusion drawn will be based on the objective data that have been gathered and analysed rather than on the researcher's personal opinion. Hypotheses can be used to test the validity of a claim made by providers of services or products, to determine which of a choice of decisions is to be made or to provide evidence whether or not a change in practice is justified (Anderson et al., 1994: 290–1).

Hypotheses can be one tailed or two tailed (Anderson et al. 1994: 291–2). A one-tailed hypothesis is where the anticipated answer is in a particular direction. For example, our alternative hypothesis might be that adults are taller than children while the null hypothesis is that adults are not taller than children. If we were using two-tailed hypotheses, we would not be anticipating a particular direction to any difference that could be found. Therefore, the null hypothesis would be that there is no difference between the heights of adults and children while the alternative hypothesis is that there is a difference between their heights.

Qualities of hypotheses

O'Donoghue (2010: 92–101) gave some examples of how not to formulate hypotheses. Consideration of these examples gives an indication of the qualities that should be exhibited by hypotheses. Hypotheses should be observable and testable. They should be precise and unambiguous avoiding vaguely defined variables or, worse still, broad concepts. The null and alternative hypotheses should be mutually exclusive so that there is no possible outcome of the research project that could be interpreted as evidence for both. The null and alternative hypotheses should together completely cover all possible outcomes of the research project. The hypotheses should be formalized assumed outcomes whose testing is justified by a sound rationale based on a survey of previous research. This is more desirable than the unfocused testing of correlations between every pair of variables that have been measured. The hypotheses should be cautious and able to stand alone without assuming unproven preconditions. Methodological detail should be avoided when specifying hypotheses: the hypotheses should be concerned with what conclusions might be found rather than how the results will be produced. An alternative hypotheses should only specify a particular direction of difference (for example, less than or greater than some value) if there are sound theoretical reasons for assuming that any difference would be in one direction rather than the other. This author has never assumed a particular direction of any difference or relationship in any published research he has done. A particular direction of any difference should only be assumed where there is convincing evidence from previous research to justify this; for example, testing if the height of adults is greater than that of children. This also applies when testing the relationship between variables. Where there are multiple variables involved in a study, the hypotheses should be worded to ensure all possible combinations of outcomes are covered within the hypotheses. For example, O'Donoghue (2010: 100) used a performance analysis of tennis example which involved two categorical variables (gender and court surface) and a numerical dependent variable (mean rally duration):

H_0 None of gender, court surface or the interaction of gender and court surface have an influence on mean rally duration.

H_A One or more of gender, court surface and the interaction of gender and court surface have an influence on mean rally duration.

In this example, the alternative hypothesis includes the words 'one or more' to ensure that any differences found will mean that the null hypothesis will be rejected.

SAMPLING

Populations and samples

Quantitative research, whether descriptive or experimental, typically uses a sample of participants drawn from the population of interest. In a full census of a population, the population data would be summarized using descriptive statistics presented in tabular or graphical form. This reductive approach allows for concise communication of effective information to support decisions that need to be taken. Similarly, when studying a population using a sample drawn from that population, the sample is summarized using descriptive statistics. If a large enough random sample is drawn from the population, it is hoped that the sample would be representative of the population. In Chapter 1, the terms 'population parameter' and 'samples statistic' were introduced. If the sample is truly representative of the population, then the sample statistics should be good estimates of the population parameters.

Usually in research, we do not actually gather data for the whole population and, therefore, we never know how close the sample statistics are to the population parameters. However, as we will see in this chapter, if we have a good understanding of the distribution of the sample and we can assume that the wider population follows a similar distribution, there are mathematical theories that can be used to estimate sampling error. Sampling error is the difference between a sample statistic and its corresponding population parameter. The larger the sample size, the smaller the sampling error will probably be.

A practical exercise in sampling

In this section we will use a simple example of tossing a coin 10 times and counting up the number of times the result is 'heads'. The author synthesized a population of (N =) 100 people tossing a coin 10 times using the pseudo-random number generator in Microsoft Excel. We use N to represent the size of a population and n to represent the size of a sample. The 100 values obtained for the number 'heads' resulting from each person's 10 coin tosses are shown in Table 7.1. The population mean, μ, was 5.08 and the population standard deviation, σ, was 1.39.

To investigate the means of samples of different sizes, six sets of 20 samples were drawn from the population. There was a set of 20 samples of size n = 1, as well as sets of 20 samples of size n = 2, 3, 4, 5 and 6. Each sample has a mean, $\bar{x}$, and a standard deviation, s. Obviously, many of these samples overlapped and are not independent. This reflects what happens in research in the real world. If three students attending different universities were doing the same project but using different samples from the same population, they might not exclude

Table 7.1 Number of times a coin toss was 'heads' when the coin was tossed 10 times by 100 people

3	4	8	6	6	5	6	7	6	5
5	5	4	5	4	5	5	5	3	3
4	6	5	7	5	2	3	6	8	4
5	7	5	5	5	5	5	4	4	7
8	3	5	4	6	5	3	6	7	5
4	4	4	5	5	5	7	7	5	3
7	7	3	4	3	4	4	6	5	6
5	4	5	4	6	2	5	9	7	4
5	6	7	4	6	4	5	5	6	6
4	6	6	6	4	7	4	7	3	4

people who had participated in either or both of the other research studies; they might not ask about or even know about the other two studies that were being carried out.

The results of this exercise are shown in Table 7.2. There is a tendency for the mean sample mean, $\bar{x}$, to approach our population mean, μ, of 5.08 as the sample size increases. Table 7.2 shows the mean of the sample means, $\bar{\bar{x}}$, and the standard deviation of the sample means, $S_{\bar{x}}$. The exact match of 5.08 (to two decimal places) between μ and $\bar{\bar{x}}$ when we have a sample size of six does not always happen. However, when we do a study, we do not use multiple samples of a given size, we usually use a single sample drawn from the population. Therefore, we wish to know how much error there could be in any single sample. The minimum and maximum values for the sample means give the worst cases observed in this exercise. As we can see, the minimum value increases while the maximum value decreases as the sample size increases. When the sample size was one the maximum sampling error was 3.08 (5.08 – 2) but when the sample size was six, the maximum sampling error was 0.75 (either 5.08 – 4.33 or 5.83 – 5.08). This is not a general sampling pattern and readers should be aware that with a large enough population size and a very large number of samples there will be a small probability of a sample having a mean of 0 or a mean of 10.

The standard deviation of the sample means, $S_{\bar{x}}$, gives a better indication of the reduction of sampling error as sample size increases. This is because $S_{\bar{x}}$ is calculated using all sample means rather than just the minimum and maximum. The standard deviation represents an average spread of values about the mean. Therefore, the greater dispersion of sample means when sample sizes are small leads to a greater potential for sampling error when single samples are used. It is possible for a single sample of one participant to have a value five which is closer to the population mean of 5.08 than many of the means of samples of six participants. Indeed, only five of the samples of six participants had a mean of within 0.08 of the population mean of 5.08.

The coin tossing spreadsheet

If the previous section describing the coin tossing exercise makes perfect sense and you are completely satisfied that larger sample sizes will probably lead to lower sampling errors, then you can go straight onto the section on Central Limit Theorem. However, if like the author, you take a critical view of the work of others and ask questions such as 'How did they do that?', you should read this section. Being critical is to be encouraged in students where as blind acceptance is to be discouraged.

Table 7.2 Means of 20 samples of different sizes

Statistic	Samples Size					
	1 (n = 20)	2 (n = 20)	3 (n = 20)	4 (n = 20)	5 (n = 20)	6 (n = 20)
Mean	5.30	5.40	5.23	4.88	4.90	5.08
SD	1.53	0.90	1.19	0.73	0.57	0.46
Minimum	2.00	3.00	3.33	4.00	4.20	4.33
Maximum	7.00	7.00	7.33	6.50	6.20	5.83

The sheet 'Random-generation' of the spreadsheet 07-heads_or_tails.XLS was used to simulate the tossing of a coin ten times (rows 3 to 12 and 16 to 25) by a population of 100 participants (columns B to CW). Rows 3 to 12 simply used the RAND() function in Microsoft Excel to generate a random number from 0.0 to under 1.0. Because random numbers change every time any change is made to the spreadsheet, it was necessary to copy the values generated and paste these (using **Paste** → **Paste Special** → **Paste Values**) into the 'Population 100' sheet of the spreadsheet file. This was necessary so that all readers of this book would be using the same values as the author. The result is that we have 100 sets of ten random numbers in the cells B3 to CW12 of the 'Population 100' sheet which represent the coin-tossing task performed by the population. The cells B16 to CW25 categorize each random number in B3 to CW12 as representing a result of 'heads' or 'tails'. This is done using an IF function which assumes the probability of the result of any coin toss being 'heads' is 0.5. For example, the cell B16 would contain:

=IF(B3 >.5, 'HEADS', 'TAILS')

The cells B26 to CW26 contain the total number of 'heads' resulting from the 10 coin tosses of each member of the population. This is computed using the COUNTIF function in Microsoft Excel. For example, the result for the first participant in the cell B26 is evaluated using the following function:

=COUNTIF(B16:B25, '=HEADS')

Cells B41 and B42 contain the population mean, μ, of 5.08 and the population standard deviation, σ, of 1.39 respectively.

On the 'Samples' sheet of the 07-heads_or_tails.XLS spreadsheet file, these 100 values are found in column A. The purpose of this sheet is to draw 20 samples of a single participant as well as 20 samples of 2, 3, 4, 5 and 6 participants from our population. Random numbers had been used to generate the location of each sample member within the population, but the RAND() functions have had to be pasted over (with the resulting values) to prevent them from changing. This was necessary for all readers of the book and users of the spreadsheet to have the same data. The samples of different sizes are generated in different columns of the spreadsheet:

- Sample size 1: Columns D to F (coloured yellow)
- Sample size 2: Columns H to J (coloured light blue)
- Sample size 3: Columns L to N (coloured pink)
- Sample size 4: Columns P to R (coloured light green)
- Sample size 5: Columns T to V (coloured amber)
- Sample size 6: Columns X to Z (coloured grey).

The mean of each of the six sets of 20 samples, $\bar{x}$, is determined in the third of the three columns used before an overall analysis of the samples of different sizes is undertaken. For each sample size, the sample means are summarized using the mean of the sample means, $\bar{\bar{x}}$ (row 164), the standard deviation of the sample means, $S_{\bar{x}}$ (row 166), the minimum of the sample means (row 168) and the maximum of the sample means (row 170). It is these

116

values that have been used in Table 7.2. If you are still sceptical about all of this, as the author would be in your position, then you should take a new set of random numbers from the 'Random-generation' sheet, determine the population mean, take several samples of size two and several samples of size six and compare the sample means obtained. If you think the result is a fluke, repeat the exercise until you are happy to draw your own conclusions about the effect of sample size on sampling error.

CENTRAL LIMIT THEOREM

A practical exercise sampling from a normally distributed population

Chapter 6 focused much more on the normal probability distribution than it did on any of the others. The reasons for this are that the normal distribution is assumed by many of the inferential statistical tests used in research and there are many forms of sport and exercise data that are normally distributed (Fallowfield et al., 2005: 98). As has already been mentioned in the current chapter, research studies typically use single samples drawn from the relevant population and the true values of population parameters are unknown. However, if our data are normally distributed, then there are statistical theories that can be used with our sample statistics to reason about extent of sampling error. Central Limit Theorem can be used to estimate sampling error even though we are not aware of the population parameters. This will be explained in the next section of the current chapter. First though, the scene will be set using a larger exercise than the coin-tossing example used previously in this chapter.

The author has created a second spreadsheet to illustrate sampling error. The purpose of this exercise was to compare sampling error in samples of different sizes with what would be predicted from Central Limit Theorem. Specifically, 50 samples of size 10, 15, 25, 50 and 100 were created, with sampling errors calculated using the known mean for the full population of 10,000.

The spreadsheet file used in this exercise is called 07-Sampling10000.XLS and it consists of two sheets. The first sheet ('Population Rand') contains 10,000 normally distributed random numbers created using the following function:

= NORMINV(RAND(), 50, 5)

Microsoft Excel's NORMINV function produces a value from a normal distribution with a mean and standard deviation specified by the second and third arguments of the function. The first argument is the probability that a randomly selected value from this distribution will be less that the value returned by the function. For example, NORMINV(0.5, 50, 5) would be evaluated to 50 while NORMINV(0.16, 50, 5) would evaluate to about 45, which is one standard deviation below the mean. Remember that 68 per cent of values in a normal distribution are within one standard deviation of the mean, so approximately 16 per cent of values are lower than one standard deviation less than the mean and approximately 16 per cent of values are higher than one standard deviation above the mean. The use of RAND() as the first argument is to generate a random probability between zero and less than one. The 10,000 values are located in a 100 × 100 array of cells from A4 to CV103.

As with the coin-tossing example, the use of RAND() means that the random numbers will change every time a change is made to the spreadsheet. Therefore, the values produced which are used in this example have been copied into a second sheet called 'Population Fixed' using the same 100 x 100 cells. As we can see in the cells E1 and E2, although we tried to create a population of values with a mean of 50 and a standard deviation of 5, we have actually produced a population mean, μ, of 49.94 and a population standard deviation, σ, of 4.99. Figure 7.1 shows how the 100 × 100 array of cells is split up into 50 samples of 10, 50 samples of 15, 50 samples of 25, 50 samples of 50 and 50 samples of 100.

The mean of each sample, $\bar{x}$, within each set of samples is determined. The most important information for this section of the chapter is the mean, $\bar{\bar{x}}$, and standard deviations, $S_{\bar{x}}$, of sample means for samples of different sizes. This information is found in the cells BA115 to BC120 and is shown in Table 7.3. Once again, the standard deviations of sample means show that larger sample sizes have smaller sampling errors. There are three types of standard deviation within the synthetic data we are dealing with here. First, there is the standard deviation of the values for the population ($\sigma = 4.99$). Second, each sample of values will have a standard deviation, s. Third, because we are reasoning about the sample mean of samples of different sizes, there is the standard deviation of those sample means, $S_{\bar{x}}$. It is the third type of standard deviation, $S_{\bar{x}}$, that is shown in Table 7.3. The standard deviation of sample means is smaller than that of the population (Rowntree, 2004: 106). Similarly, there are three types of mean in the synthetic data: the population mean ($\mu = 49.94$), the mean of an individual sample, $\bar{x}$, and the mean of sample means of samples of a particular size, $\bar{\bar{x}}$. It is the third type of mean, $\bar{\bar{x}}$, that is shown in Table 7.3.

Central Limit Theorem

The sampling exercises described previously for tossing a coin 10 times and the arbitrary variable with a population mean, μ, of 49.94 and population standard deviation, σ, of 4.99

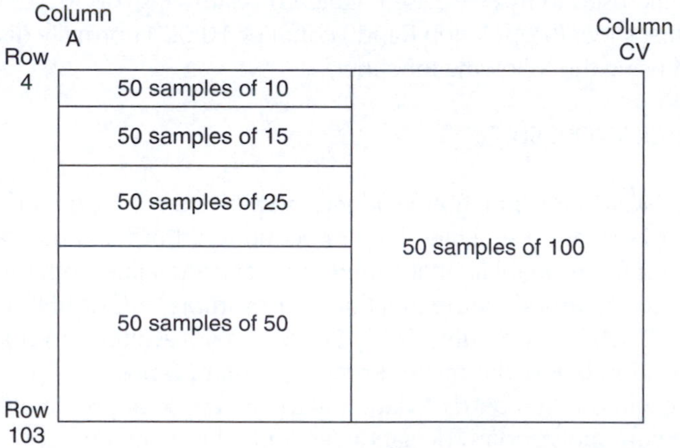

Figure 7.1 Spreadsheet design for a population of 10,000 divided into samples of different sizes.

hypothesis testing

Table 7.3 Means and standard deviations of sample means from samples of different sizes

Sample Size	Mean, $\bar{x}$	Standard Deviation, $S_{\bar{x}}$
10	50.28	1.52
15	49.97	1.09
25	49.89	0.83
50	50.07	0.72
100	49.85	0.49

from a sample of 10,000 were done to give an insight to the effect of sample size on sampling error. In reality, we perform studies where each population is represented by a single sample. Therefore, this type of multisample exercise is not something that researchers are required to do. What you need to know as researchers using statistical analysis techniques is how to estimate the amount of sampling error that there may be. Sampling error, e, for a population mean is given by the equation 7.1 where | x | is the magnitude of x (the value of x without a plus or minus sign in front of it).

$$e = |x - \mu| \qquad (7.1)$$

Central Limit Theorem provides a method for estimating sampling error and determining a range of error values such that we can be confident the true population mean is within this range of the sample mean. When drawing a large random sample from a population that has a mean, μ, and a standard deviation, σ, the sampling distribution of $\bar{x}$ is normally distributed with a mean of μ and a standard deviation of $\sigma/\sqrt{n}$ where n is the sample size. Large random samples are where n is at least 30. According to Central Limit Theorem, even if the distribution of the variable x is not normal, when the sample size is 30 or greater, the distribution of the sample mean $\bar{x}$ is normally distributed (Anderson et al., 1994: 252–3; Levine and Stephan, 2005: 104–6). Where the variable x is normally distributed, the distribution of the sample mean $\bar{x}$ is normally distributed even if n < 30.

Typically, the values of the population parameters, μ and σ, are unknown and, therefore, we need to use the sample statistics, $\bar{x}$ and s, as 'point estimates' (Diamantopoulos and Schlegelmilch, 1997: 116). These point estimates are used in the standard error of the mean (SEM = s/$\sqrt{n}$) and we substitute the values $\bar{x}$ and s into the distribution of the sample means getting the normal distribution with a mean $\pm$ standard deviation shown in equation 7.2.

$$\bar{x} \pm s/\sqrt{n} \qquad (7.2)$$

This distribution can be used to determine an interval estimate that we can have some level of confidence that the true population mean, μ, falls within. Mathematically, the normal distribution is infinite but we can use our knowledge of this probability distribution to determine a confidence interval that has lower limit and an upper limit. For example, if we wished to determine an interval such that we could have 95 per cent confidence that the true mean, μ, was within that interval, we would use the middle 0.95 of the distribution of the sample mean. We have already seen in Chapter 6 that 95.4 per cent of values of a normally distributed variable are within two standard deviations of the mean. The z-scores between which 95 per cent of values of a normally distributed variable are found are –1.96 and +1.96. Table 7.4 shows the absolute z-scores associated with different confidence levels that are typically used in sport and exercise research. The confidence coefficient $(1 - \alpha)$ expresses the confidence level as a proportion between zero and one rather than as a percentage. The area of a normal distribution curve that is outside the confidence interval is split in two: the area of $\alpha/2$ to the left of the curve and the area of $\alpha/2$ to the right of the curve. Therefore, the equivalent absolute z-score is $z_{\alpha/2}$ rather than z_α.

The value of α is the probability of a sampling error larger than the sampling error mentioned in the precision statement of the research study (Anderson *et al.*, 1994: 271). That is, α is the probability that the sampling error is outside the range of sampling errors that we have a given level of confidence level that the true mean, μ, is within (for example, 95 per cent confidence). Again, the author cannot emphasize enough that the population mean, μ, is unknown, so we do not know how much sampling error there actually is, therefore we can only give a range of values for which we have some level of confidence that the true mean is within. Having decided on an α value, the confidence interval is given by equation 7.3.

$$\bar{x} \pm z_{\alpha/2} \, s/\sqrt{n} \tag{7.3}$$

If we have a finite population of size N and the sample size n is larger than 5 per cent of that population, then the 'finite population correction' should be applied to the standard error of the mean (SEM). The finite population correction factor is shown in equation 7.4 and it is applied to the confidence interval of the mean as shown in equation 7.5.

$$\sqrt{\frac{N-n}{N-1}} \tag{7.4}$$

$$\pm \left(z_{\alpha/2} s/\sqrt{n}\right)\sqrt{\frac{N-n}{N-1}} \tag{7.5}$$

Let us return to our arbitrary population of 10,000 values with a mean, μ, of 49.94 and a standard deviation, σ, of 4.99. The largest sample drawn from this population had 100 members and so the finite population correction is not required. Table 7.5 compares the observed standard deviations of sample means with the theoretical sampling error (SEM) derived using Central Limit Theorem. The minor differences arise because values produced by Central Limit Theorem are theoretically based on all possible samples of the given sample size as opposed to the 50 samples used during this exercise. It is hoped that this exercise will give students some confidence in Central Limit Theory without the full mathematical proof being required.

Table 7.4 Confidence limits and associated z-scores

Confidence level	α	Absolute z-score ($z_{\alpha/2}$)
90%	0.1	1.64
95%	0.05	1.96
99%	0.01	2.58
99.9%	0.001	3.29

120

Table 7.5 Mean, $\bar{x}$, $\pm$ standard deviation, $S_{\bar{x}}$, of sample means from samples of different sizes

Sample Size	Observed (100 samples)	Theoretical (Central Limit Theorem)
10	50.28 ± 1.52	49.94 ± 1.58
15	49.97 ± 1.09	49.94 ± 1.29
25	49.89 ± 0.83	49.94 ± 1.00
50	50.07 ± 0.72	49.94 ± 0.71
100	49.85 ± 0.49	49.94 ± 0.50

Figure 7.2 shows the 95 per cent confidence intervals based on five of the samples of 100 members used in this arbitrary example: samples 3, 13, 23, 33 and 43 (in the 'Population Fixed' sheet of the 07-Sampling10000.XLS spreadsheet) where $\bar{x} \pm s$ are 49.71 ± 4.64, 49.53 ± 4.76, 50.49 ± 4.92, 50.60 ± 4.99 and 48.46 ± 4.29 respectively. These distributions are used to form the 95 per cent confidence intervals shown using horizontal lines in Figure 7.2. This was done using equation 6.3 with n = 100 and $z_{\alpha/2}$ = 1.96. Figure 7.2 refers to the population mean as being 'unknown' because it would be unknown to a researcher who has gathered data from a single sample. As we can see, the true unknown population mean falls outside the 95 per cent confidence interval generated by one of the five samples shown (Sample 43). In studies using samples drawn from populations of unknown parameters, we will often produce a confidence interval and conclude that we are 95 per cent confident that the true population parameter is within that interval. The true mean either will or will not be within the confidence interval, but we will not know this. A point that was made by Diamantopoulos and Schlegelmilch (1997: 120) is that the relationship between a confidence interval and the population parameter in question is not a probability. The population exists as does the population parameter (albeit unknown to us). Therefore, the popula-

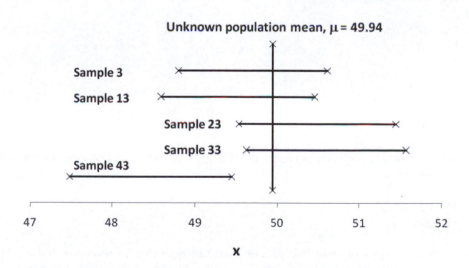

Figure 7.2 95 per cent confidence intervals determined from different samples.

tion parameter is either certainly within the confidence interval (a probability of one) or certainly not within the confidence interval (a probability of zero). Therefore, we express a level of confidence with confidence intervals rather than probability. According to Central Limit Theorem, we would expect 2.5 of our 50 samples of 100 to generate a confidence interval that did not include the true population mean. There were two such samples (Sample 11 and Sample 43) as readers will observe in the 'Population Fixed' sheet of the 07-Sampling10000.XLS spreadsheet.

HYPOTHESIS TESTING

Diamantopoulos and Schlegelmilch (1997: 135–47) described an approach to hypothesis testing that has five steps: formulating the hypotheses, specifying the significance level, selecting the statistical test to use, defining the region of the test statistic that indicates rejection of the null hypothesis and computing the value for the test statistic using the data collected and choose which hypothesis to accept and which to reject. The use of statistical analysis software packages available today allows the fourth and fifth stages to be altered. First, the test statistic and associated p (Sig.) value are often output without the necessity of showing the critical value of the test statistic. The decision as to which hypothesis to accept and which to reject can be made based on the p value produced and the significance level set.

In formulating the null and alternative hypotheses, the main issues to consider are the operationalized research question, the variables involved and whether the hypotheses should be one tailed or two tailed. A one-tailed null hypothesis assumes a particular direction of any difference (or correlation) that might be found. One-tailed hypotheses should only be used where there are sound theoretical reasons for assuming one direction of any difference over the other. If comparing an observed mean, μ, with a hypothesized mean, μ_0, one-tailed null and alternative hypotheses might be:

H_0: $\mu \leq \mu_0$
H_A: $\mu > \mu_0$

or

H_0: $\mu \geq \mu_0$
H_A: $\mu < \mu_0$

Where two-tailed hypotheses are used, the null and alternative hypotheses are more likely to be in the form:

H_0: $\mu = \mu_0$
H_A: $\mu <> \mu_0$

There are two types of error that can be made in quantitative research studies: Type I Errors and Type II Errors. Table 7.6 shows the situations where these different errors occur.

122

hypothesis testing

The probability of a Type I Error is α while the probability of a Type II Error is β. The choice of confidence level depends on the consequences of making a Type I Error compared to the consequences of making a Type II Error. There is typically a trade off between α and β because with a constant sample size one cannot be decreased without increasing the other. Two examples this author uses when introducing hypotheses are experiments on some medical treatment aimed at reducing some measurable symptom of a given illness. In the first experiment, there is a reduction in the symptoms according to our sample. There is a chance that a Type I Error may have been made and that there may be some undesirable side effects of the treatment. The second example is where an experiment does not find a significant reduction in measurable symptoms according to our sample. There is a chance that a Type II Error may have been made and that the treatment could actually benefit patients. The question is which of these errors is the more serious? Typically this leads to a discussion about the nature of the illness, the cost of introducing the treatment, the nature of potential side effects and the effectiveness and cost of current treatments being used for the illness. Both types of error are serious and the tolerable chances of making Type I and Type II Errors should be considered for each investigation individually. Very often Type I Errors are considered more serious than Type II Errors because they may lead to changes of policies, procedures and practices that are costly, time-consuming and ultimately ineffective for the population.

The probabilities in Table 7.6 are actually conditional probabilities. For example, α is the probability of rejecting the null hypothesis based on the sample when the null hypothesis is actually true according to the full population. Vincent (1999: 138–40) used a diagram similar to Figure 7.3 to illustrate Type I and Type Errors together with α and β. Figure 7.3 uses an example of an experiment where an intervention is made to see if it improves performance in a written physical education exam. An experimental group of 30 participants achieves a mean$\pm$SD score of 50.0$\pm$1.0 while the experimental group of 30 scores 50.4$\pm$1.0. The SEM for each group is $1.0 / \sqrt{30} = 0.185$. The researcher decides to use a one-tailed test assuming that any difference that might be found between the groups would be that the experimental group performed better. The researcher decides to do this one-tailed test at a 95 per cent confidence level. Therefore, the area of the distribution for the control group's mean which would lead to H_0 being rejected if the mean for the experimental group exceeded it is $50.0 + 1.0 z_{0.05} / \sqrt{30} = 50.0 + 1.0 \times 1.64 / \sqrt{30} = 50.3$. The area of the distribution of the mean for the experimental group that is less than this critical value of 50.3 is 0.292. This can be checked in Microsoft Excel using NORMDIST(50.3, 50.4, 0.183, 1).

Table 7.6 Hypothesis testing and potential errors (probabilities)

Decision made based on the sample	Population situation	
	H_0 true	H_0 false
H_0 not rejected	Correct (1−α)	Type II Error (β)
H_0 rejected	Type I Error (α)	Correct (1−β)

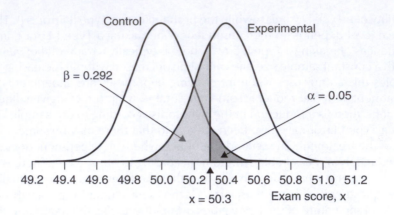

Figure 7.3 Estimated sampling distribution for the mean of the control and experimental group.

Some may look at Figure 7.3 and ask 'What about the true population distribution? There is nothing in this except what we know from the sample!' This was certainly the initial thought of the author on reading Vincent's (1999: 138–40) book. Vincent was very intentional in his use of his own diagram and the text supporting it. Figure 7.3 represents the decision that would be made by the researcher with the data they have available to them, and that data are exclusively from the sample. Vincent stated that the shaded area where α is marked covers those values of x where a Type I Error would be made if the null hypothesis was true (note the use of the word 'if' here). Similarly, Vincent described the other shaded area marked β as covering those values of x where a Type II Error would be made if the null hypothesis was false (again note the use of the word 'if' here). Vincent's use of the word 'if' makes his diagram and Figure 7.3 here correct. In this particular example, the null hypothesis is rejected because the sample mean for the experimental group lies within the area of the estimated distribution of sample means for the control group above the critical value of 50.3. We would not actually know whether we have made a Type I Error or not because the population is unknown. We would, however, know that a Type II Error has not been made because such an error could only be made if the null hypothesis was accepted.

The above example is a step towards helping readers understand the process of analysing data with inferential tests that leads to the p value that we see in so many research papers that we read. The sample statistics are used to make an inference about population parameters. The sample statistic used in the previous example was the mean and the difference between two groups would typically be compared using the t distribution rather than two standard normal distributions. Chi square and F have distributions that are used when testing other population parameters using sample statistics. A similar process to that outlined by Diamantopoulos and Schlegelmilch (1997: 135–47) occurs with test scores being compared to critical values within distributions of the test statistics.

SIGNIFICANCE, POWER AND EFFECT

A statistical test may give us 95 per cent confidence that an intervention improves performance in a written physical education exam. However, the mean observed difference in scores of 0.4 between those exposed to the intervention and the control group might not be a meaningful difference in real educational terms. Similarly, in sport there may be statistically significant differences found by research studies where the actual difference is not a meaningful difference. Effect size is used to represent the meaningfulness of a difference. Different tests use different effect size measures as we will see in Chapters 10 to 13 of this book. Where we are testing for differences, an effect size could express the difference as the number of standard deviations of the dependent variable that the difference represents. Such effect size measures include Cohen's d. Other effect size measures such as omega squared (ω^2) and eta squared (η^2) represent the percentage of variance in the dependent variable that is explained by the group membership or conditions being compared. This book encourages readers to report effect sizes as well as p values when presenting the findings of statistical tests.

Statistical power is the probability of a test rejecting the null hypothesis when the null hypothesis is actually false according to the population of interest. This is the conditional probability $1 - \beta$ shown in Table 7.6. Power can be determined from the α level set for a study, the number of participants involved in the study and the true effect size in the relevant population. Once again we have a situation where the population parameters of interest are unknown. However, from theory we may have reason to anticipate a particular effect in the population and we wish to ensure that our test will include enough participants to give a good chance (for example, $1 - \beta > 0.8$) of finding such an effect to be significant at our chosen α level if it really exists in the population. This will be covered in greater depth in Chapter 20 along with other applications of statistical power analysis. If we consider Figure 7.3, one obvious way to increase power is to increase the α level from 0.05 to 0.1. This would reduce β concurrently increasing $1 - \beta$. Another way of increasing the power of a study is to include more participants. Again, if we consider Figure 7.3, the width of the middle hump of each distribution shown would be narrowed by increasing the number of participants in each group. Note that it would be incorrect to say the spread of values in a normal distribution decreases because a normal distribution covers values from $-\infty$ to $+\infty$. With 30 participants in each group, the SEM was $1.0 / \sqrt{30} = 0.183$. If the number of participants in each group was to increase to 100, the SEM would be $1.0 / \sqrt{100} = 0.1$ which would reduce the critical value of x where we would reject the null hypothesis from 50.30 to 50.16. This combined with the narrowing of the SEM for the experimental group would reduce β increasing the power of the study, $1 - \beta$.

SPSS allows an 'observed power' value to be calculated using the sample values being analysed. The author has shown how to obtain these measures of 'observed power' for ANOVA tests in Chapters 11 to 13 in case some readers may wish to use these in their own analyses. The author prefers to use some hypothesized population parameter when discussing the power of a statistical test. Figure 7.4 is an example that will be revisited in Chapter 10 when the one-sample t-test is introduced and in Chapter 20 when statistical power is discussed. Here we wish to test whether the mean 2,000m time for entrants in a national indoor rowing championship is under seven minutes (420s).

There are different ways of showing α and β on diagrams such as Figure 7.4. Hinton (2004: 96–7) compared a known mean and an unknown mean while Tabachnick and Fidell (2007: 35) compared idealized distributions based on assumed population means. The author found showing sample means and then trying to work backwards to unknown population means caused too much confusion when trying to explain α and β. The author might as well admit that he became so confused using this approach that he had to redo the entire section starting with hypothesized means! We assume that there are least 30 rowing performances and that the standard deviation, σ, is 25s. Therefore, the standard error of the mean is 25 / √30 = 4.564. The fact that there are 30 performances permits us to use the normal distribution. The distribution on the right of Figure 7.4 is for an idealized sampling distribution where the null hypothesis is true and the sample mean is exactly seven minutes (μ = 420s). We have chosen an α level of 0.05 and are performing a one-tailed test. Therefore, the area of 0.05 to the left of this distribution represents where we would reject the null hypothesis and hence make a Type I Error. Remember, that this is a sampling distribution derived from a population mean of 420s where the null hypothesis is true. The critical 2,000m time under which we would reject the null hypothesis is 412.5s.

In determining the effect of interest in the population, the author did not use an effect size value based on a number of standard deviations (such as Cohen's d) or on the percentage variance explained (such as ω^2). Instead, the author considered rowing paces for the 1,000m, 2,000m and 5,000m to identify a speed increase where we would expect a completely different mix of energy sources to be used. This lead to a decision that 6 minutes and 52s (412s) would be comfortably under seven minutes for the 2,000m. The distribution to the left of Figure 7.4 is a sampling distribution for an idealized population with a mean, μ, of 412s and a standard deviation, σ, of 25s. Therefore, the distribution of the sample mean has a mean±SD of 412±4.564s. The area of this distribution above the critical time of 412.5 is where we would falsely accept the null hypothesis making a Type II Error. The area under the curve where we would accept the null hypothesis, β, is 0.456. Therefore, the power

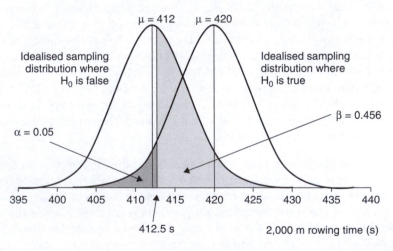

Figure 7.4 Comparing distributions of sample means based on a point estimate (known from a sample) and a hypothesized value given that the null hypothesis is true.

of the test assuming a population mean of 412s and a standard deviation of 25s is $1 - \beta =$ 0.544. Statistical power will be discussed further in Chapter 20.

Once we have determined the hypotheses for the study and set a level for α, we can choose the statistical test that we wish to apply to the data. The next section of this chapter is specifically devoted to this topic and is arguably the most important section of the book. The choice of test depends on the purpose of the study which may be to analyse a single sample, compare different groups, different conditions or to assess the relationship between variables. The distribution of each variable also influences the decision made as we may need to use a parametric procedure or a non-parametric procedure. Once the statistical test to be used has been decided, we apply the test using a computerized statistical analysis package such as SPSS and interpret the results. These last two stages are the subject of Chapters 8 to 18 of this book.

SELECTING A TEST

Single variable testing

Contrary to what many believe about published research methods and statistics books, there is plenty of coverage of what test should be used in different situations (Diamantopoulos and Schlegelmilch, 1997: 174; Salkind, 2004: 180–1; O'Donoghue, 2010: 180). Fallowfield *et al.* (2005: 153, 182, 217, 262) provided a series of decision trees to assist readers in selecting statistical tests to use. The inside front cover of Hinton's (2004) book, inside back cover of Hinton *et al.*'s (2004) book and the inside back cover of Vincent's (1999) book all provide similar decision trees.

This section expands on the decision structure used by the author in his first book on performance analysis of sport (O'Donoghue, 2010: 180) to incorporate tests used in other disciplines of sport and exercise science. This makes it necessary to break down the decision structure into different decision trees for individual types of analysis. First, we will deal with situations where single variables are being tested using some criteria. Figure 7.5 shows the tests that can be used with data measured on different scales. The purpose of the chi square goodness of fit test is to test an observed distribution of participants against some theoretically expected distribution of participants. For example, if approximately 50 per cent of the population are female and 50 per cent are male, then we may wish to test if the gender breakdown of spectators at a given sport reflects that of the wider population. The chi square goodness of fit test is covered in Chapter 15. A one sample t-test compares an observed mean with some hypothesized standard value of interest. For example, comparing indoor rowing performances with an expected seven minute standard has been described in the current chapter. The one sample t-test is covered in Chapter 10.

Testing for relationships between variables

Figure 7.6 shows the main tests that can be used to test the relationship between two variables. Associations between nominal variables are tested using the chi square test of independence. This test compares the distribution of cases with respect to one variable for

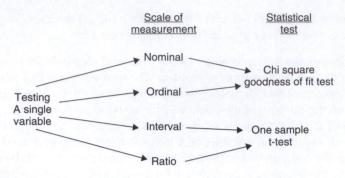

Figure 7.5 Tests of single variables.

each level of the other. If the sample proportions are similar, then the variables are concluded to be independent. The chi square test of independence is covered in Chapter 15. Ordinal variables as well as numerical variables that do not satisfy the assumptions of Pearson's r can be correlated using non-parametric, or ranked, correlations. There are two non-parametric correlation techniques and the choice of which to use depends on the number of pairs of values there are to be correlated. Pearson's r is the most commonly used correlation technique for interval and ratio scale variables in sport and exercise science. Pearson's r, Kendall's τ and Spearman's ρ are covered in Chapter 8. Not every relationship is tested using variables measured on the same scale. Therefore, the variable with the lowest scale of measurement dictates the test used. Here we consider nominal to be the lowest scale of measurement with ratio scale being the highest. There are occasions where we may need to control for the confounding influence of some third variable on the relationship between two other variables. This can be done using partial correlations which are also covered in Chapter 8.

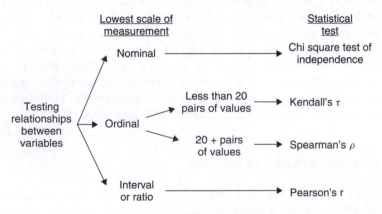

Figure 7.6 Tests of relationships between pairs of variables.

Testing for differences between independent groups

Figure 7.7 shows the four main tests used to test for differences between independent groups. These tests are used where we have some nominal grouping variable and a numerical scale dependent variable. The Mann–Whitney U test and the Kruskal–Wallis H test are non-parametric tests that can be used with any data that can be ranked. Therefore, these two tests can be used when the dependent variable is measured on an ordinal scale. The independent samples t-test and the one-way analysis of variances (ANOVA) test are parametric tests that can only be used when the dependent variables are measured on interval or ratio scales, are normally distributed and where there is approximately equal variances between the samples. Where these assumptions are violated, the non-parametric alternatives should be used. The independent t-tests is covered in Chapter 10, the one-way ANOVA test is covered in Chapter 11 and the non-parametric tests are covered in Chapter 14. The tests shown in Figure 7.7 can be used when we have a single independent variable. Sometimes we need to examine the effect of multiple independent variables and their interaction on some dependent variable of interest. This can be done using factorial ANOVA tests which are covered in Chapter 12.

Testing for differences between conditions

Figure 7.8 shows the main tests used to test differences between different conditions related to the same group of participants. The tests shown compare repeated measures of some dependent variable. For example, we could have two repeated measures representing tests before and after some experimental period. We could have three or more repeated measures; for example, the four quarters of a basketball match. In each case, there is a parametric procedure and a non-parametric alternative. When we have two

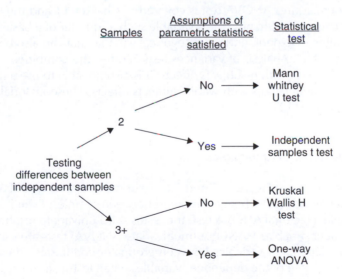

Figure 7.7 Tests of differences between independent groups.

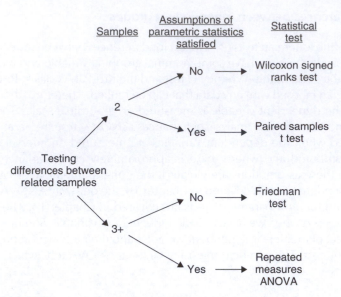

Figure 7.8 Tests of differences between related groups.

repeated measures, we can either use the parametric paired samples t-test or the non-par-ametric Wilcoxon signed ranks test. If the assumptions of parametric statistics are satisfied, the paired samples t-test can be used, otherwise we should use the Wilcoxon signed ranks test. Similarly, when we have three or more repeated measures to compare, we could use the repeated measures ANOVA if the data satisfy the assumptions of parametric statistics, otherwise we should use the Friedman test. Paired samples t-tests are covered in Chapter 10, the repeated measures ANOVA test is covered in Chapter 11 and the two non-para-metric tests are covered in Chapter 14. A variable such as quarter of a basketball match or time (before or after an experiment) is a categorical variable that can also be referred to as a within-subjects effect. Analysis of variances tests can test the combined and interactive effects of more than one within-subjects effects. It is also possible to use a mixed ANOVA that combines grouping factors with within-subjects effects. These factorial ANOVA tests are covered in Chapter 12.

Testing multiple dependent variables

The ANOVA tests shown in Figures 7.7 and 7.8 are univariate ANOVA tests because there is only one dependent variable being compared between samples. Even if there are mul-tiple factors within a factorial ANOVA test, the test is still a univariate test because there is only one dependent variable. MANOVA (multivariate ANOVA) tests are used to compare a set of numeric dependent variables between groups or conditions. These tests use an optimal linear composite of the dependent variables when testing if the factors of interest have an influence on the variables as a whole. If there is a significant influence, univariate

ANOVA tests are used to examine the effect of the factors on the individual dependent variables. MANOVA tests can also incorporate covariates, and when they do so, they are referred to as MANCOVA tests. MANOVA and MANCOVA tests are covered in Chapter 13.

Predictive modelling

Figure 7.9 shows different predictive modelling techniques covered in this book. Predictive modelling has two stages: model creation and prediction. A predictive model is created by analysing known cases to assess if there is a strong enough relationship between predictor variables and some dependent variable of interest. For example, if there is a strong correlation between a predictor variable, x, and some logically dependent variable, y, bivariate linear regression can be used to model y in terms of x. This model can then be used for further cases where the value of x is known and the value of y is to be predicted. Multiple linear regression can be used when we a modelling some dependent variable, y, using several independent variables. Both bivariate linear regression and multiple linear regression are covered in Chapter 9.

Binary logistic regression produces a model for some dichotomous variable and can be viewed as a t-test in reverse. In other words, the categorical variable is hypothesized to be influenced by some numerical independent variable(s) rather than the other way round. The predictor variables in binary logistic regression can include nominal and ordinal variables as well. Discriminant function analysis is similar to binary logistic regression except the dependent variable has three or more values. Binary logistic regression and discriminant function analysis are covered in Chapter 16.

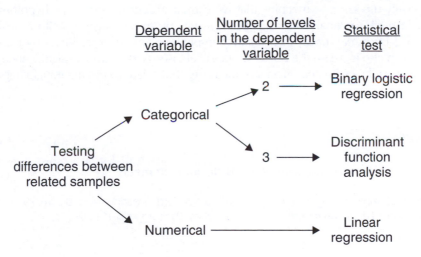

Figure 7.9 Predictive modelling techniques.

Other statistical procedures

There are other statistical procedures covered in this book that do not fall under the categories of testing relationships, testing differences or predictive modelling. Principal components analysis is covered in Chapter 18 and is a data reduction technique that explores correlations between variables allowing redundant variables to be identified. Ultimately, a large set of variables can be replaced by a smaller set of broader dimensions that allow for a more concise analysis of the data.

Cluster analysis is used to identify different groups of participants based on the data collected. These groupings will not be as obvious as gender, position on a field hockey team or level of participation in sport. In sports development and sports business areas, the groupings identified by cluster analysis can be cross-tabulated with demographic factors to provide very useful information about market sectors. Cluster analysis is covered in Chapter 17.

As has already been mentioned in the introduction to this book, if the variables used in a study are inaccurate, subjective and unreliable, then the whole study is compromised. Before applying the statistical analysis procedures covered in this book, the researcher should ensure that the variables in a study are valid and reliable. There are various tests for reliability that depend on the scale of measurement of the variable. Chapter 19 is one of the largest chapters in the book and is devoted to the statistical procedures used in the evaluation of reliability.

SUMMARY

This chapter has covered hypothesis testing and introduced the readers to the various statistical tests that are available and when they should be used. Hypotheses are anticipated outcomes that form a logical framework for research studies. The research studies themselves typically use samples meaning that we cannot accept or reject any hypothesis with certainty. Instead, we qualify any significant results of an investigation with a probability that a Type I Error may have been made. Sampling error can be reduced by using large samples which also increase the power of statistical tests used. Statistical significance should be accompanied by estimated effect size measures that relate to the meaningfulness of any results found.

EXERCISES

Exercise 7.1. Determine the sampling distribution of mean

A population has a mean height, μ, of 1.75m with a standard deviation, σ, of 0.03m. Determine the mean and standard deviation for the sample mean where the sample size is 64.

132
hypothesis testing

Exercise 7.2. Determine the confidence interval

Determine the 95 per cent confidence interval for an unknown population mean for height where a random sample of 120 people has a sample mean, $\bar{x}$, of 1.74m and a standard deviation, s, of 0.04m.

Exercise 7.3. What test to use when?

Consider the following research questions and in each case identify the independent and dependent variables, the scales of measurement of these variables, the test to be used if the data satisfy the assumptions of parametric statistics as well as the alternative nonparametric procedure.

a) To compare the percentage of service points that are aces between male club players and male professional players.
b) To compare player type (serve-volley or baseliner) between male club players and male professional players.
c) To compare the percentage of points won on serve between points when the first serve is in and points where a second serve is required.
d) To compare the percentage of points where a set of players attack the net between three score-line states (level on breaks, ahead on breaks and behind on breaks).
e) To compare the percentage of points where a set of elite players attack the net between three court surfaces (clay, grass, cement).
f) To compare the percentage of points where players attack the net between three different levels (club, county, national).
g) To compare the percentage of points where players attack the net between the winning and losing players within a set of matches.
h) To determine if there is a relationship between the percentage of points won when the first serve is in and when a second serve is required.
i) To determine if there is a relationship between the court surface of a tournament (clay, grass, cement, acrylic) and the type of player making the third round (baseliner, serve-volley).

PROJECT EXERCISE

Exercise 7.4. Coin tossing exercise

Perform the coin-tossing experiment in your class with the class representing the population. Each member of the class should toss a coin 10 times and count the number of 'heads' that result. Ask how many people in the class got a result of 0, 1, 2, 3, up to 10 in turn. For each result, count the number of people with their hands up. Enter the frequencies into the cells B2 to B12 of the 'Exercise' sheet of the 07-heads_or_tails.XLS spreadsheet; these are the grey cells. When the frequencies are entered, the population mean and standard deviation will be calculated in the cells D16 and G16 respectively. Now perform the following sampling exercises:

a) Select six groups of two people and ask them to determine their sample means. Enter these into the cells B19 to G19. The mean and standard deviation of the sample means are shown in the cells H19 and I19 respectively.
b) Select six groups of four people and ask them to determine their sample means. Enter these into the cells B20 to G20. The mean and standard deviation of the ample means are shown in the cells H20 and I20 respectively.
c) Select six groups of six people and ask them to determine their sample means. Enter these into the cells B21 to G21. The mean and standard deviation of the ample means are shown in the cells H21 and I21 respectively.

What differences do you see in the sample means as sample size grows from two to six? Columns P and Q show the mean and maximum sampling error respectively. What patterns do you observe as sampling error increases? You can try even larger sample sizes and/or look at greater numbers of samples to further explore sampling error.

hypothesis testing

CHAPTER 8

CORRELATION

INTRODUCTION

Correlation coefficients give a numerical value to the strength and direction of a relation between variables. This chapter covers correlation techniques used with interval, ratio scale and ordinal scale variables. The chi square test of independence covered in Chapter 15 can test the association between two nominal variables. The current chapter covers Pearson's r, Spearman's ρ and Kendall's τ. The correlation statistics that measure the strength and direction of a relationship are called correlation coefficients and their values range from −1.0 to +1.0. The sign of the correlation coefficient (+ or −) tells us whether any relation is positive or negative. Figure 8.1 uses scatter plots to show some examples of relationships between two arbitrary variables, x and y. A scatter plot contains points representing individuals by their values for the variables x and y. These points are called co-ordinates. A positive correlation, such as Figure 8.1(a), means that as one variable, x, increases we see an increase in the

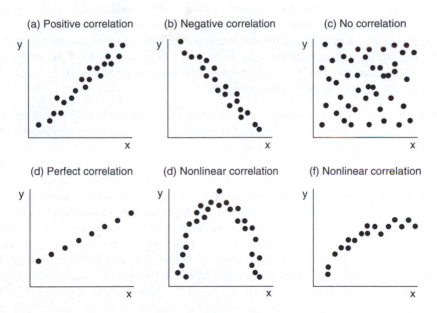

Figure 8.1 Different types of correlation

other variable, y. A negative correlation, such as Figure 8.1(b), means that as one variable, x, increases we see a decrease in the other variable, y. The absolute value of a correlation, which is when the sign is removed, is between 0.0 and 1.0. Figure 8.1(c) shows two variables that have no correlation; this would typically have a correlation coefficient of about 0.0. Figure 8.1(d) is an example of a perfect correlation where a straight line could be drawn through all the co-ordinates. It is not necessary for a straight line of best fit to pass through the origin of the graph for there to be a perfect correlation. Furthermore, perfect correlations can be positive, as is the case in Figure 8.1(d), or negative. Where a perfect correlation exists, the correlation coefficient will be −1.0 if it is a negative correlation or +1.0 if it is a positive correlation. Figures 8.1(a) and 8.1(b) are examples of correlations that are strong but not perfect. The coefficients of correlation here would have absolute values higher than 0.85.

So what is a good absolute value for a correlation coefficient? This depends on the purpose of the correlation being done and the variables involved. If correlation techniques were used to see if there was a relationship between the exact same variable when measured on different occasions or by different persons, we would require a much higher correlation than if we were assessing the relationship between two different variables. Relative reliability of a variable is where independent observations of the variable have a strong positive correlation. This use of correlation will be covered in Chapter 19. Returning to the question of how to interpret values of correlation coefficients, slightly different guidelines have been proposed by different authors (Vincent, 1999: 97; Fallowfield et al., 2005: 137). Vincent stated that absolute correlations greater than 0.9 were high, absolute values of less than 0.7 were low while values between 0.7 and 0.8 could be considered as moderate. Fallowfield et al. (2005: 137) classified absolute values of over 0.7 as representing strong correlations, with 0.45 to 0.7 representing moderate correlations, 0.2 to 0.45 representing weak correlations and 0.0 to 0.2 representing no correlation. The correlation coefficients are important because they give a numerical value to the strength and direction of the relationship between two variables. Imagine that you are inspecting two similar scatter plots between variables x and y that are based on different data. If you need to make a decision as to which scatter plot shows the stronger relationship, it may not be possible to give a definite answer. If, however, you had the correlation coefficient for each set of data, you could make a definite decision choosing the data set giving the higher absolute correlation.

Figures 8.1(e) and 8.1(f) are examples of non-linear relationships between pairs of variables. In each case, there is clearly a relationship between x and y, but the correlation techniques covered in this chapter would not be appropriate to evaluate these. In the case of Figure 8.1(f), there may be a linear relationship between y and log(x). The use of logarithmic transformations is viewed by many as over-complicating data analysis by taking a valid measured variable with known units and replacing it with a log value that cannot be compared with values reported in previous research. However, there are occasions where logarithmic transformations are very useful and do represent valid concepts. For example, in tennis the ability gaps between the following pairs of world rankings may be similar: 1 and 10, 10 and 100, 100 and 1,000. Therefore, a player who has improved from 1,000 in the world to 100 will need to make a similar improvement just to reach the world's top 10. The logs (to the base 10) of 1, 10, 100 and 1,000 are 0, 1, 2 and 3. Therefore, the log may represent the concept of ability in a way that reflects a growing number of players as we look at lower ability levels.

PEARSON'S r

Purpose of the test

The first coefficient of correlation covered in this chapter is Pearson's r, sometimes referred to as Pearson's product moment. Pearson's r is used to measure the direction and strength of relation between two interval or ratio scale variables. Anderson *et al.* (1994: 557) stated that correlations could be used where neither of the two variables of interest could be seen as being dependent on the other. In such cases, the two variables are merely co-related. However, correlation techniques are also useful to determine the strength of relationship between pairs of variables where one variable could logically be influenced by the other. Pearson's r is derived from the covariance of two variables. Readers interested in the full equation for covariance and how r is computed from this are directed to Anderson *et al.*'s (1994: 558–63) textbook. Essentially, each co-ordinate is considered in terms of the difference to the mean values for the variables x and y. This will compute a positive value if one variable increases as the other increases and a negative value if one variable decreases while the other increases. If there is no correlation between the variables, those co-ordinates with x values below the mean of x and y values above the mean for y will be cancelled out by those co-ordinates with x values above the mean of x and y values below the mean for y.

Coefficient of determination, r^2

The square of Pearson's r is called the coefficient of determination, r^2. The coefficient of determination is always positive and represents the common variance between the two variables of interest; that is, the variance of each variable that can be explained by their relationship. An r value of 0.8 would, therefore, give an r^2 value of 0.64 meaning that 64 per cent of the variance in the data is explained by the relationship between the two variables with the remaining 36 per cent of the variance being explained by other factors.

Example: Relationships between anthropometric measures and estimated $\dot{V}O_2$ max

The stature and body mass of a set of 236 students (96 females and 140 females) are measured during their induction week. During the same period, the students perform a multi-stage fitness test (Ramsbottom *et al.*, 1988) to obtain an estimate of their $\dot{V}O_2$ max. We wish to determine if there is a relationship between any pair of these three variables as well as the directions and strengths of any relationships.

SPSS

The data for this example are found in the file 08-fitnesstests.SAV. There are also variables for the gender, main sport and age of the students. Before determining Pearson's r, it is worth inspecting the scatter plots for the pairs of variables of interest (Diamantopoulos and Schlegelmilch, 1997: 203). These scatter plots will allow us to determine if the relationships are linear or not and if logarithmic or other transformations of the variables might be useful. We will use the relationship between body mass and estimated $\dot{V}O_2$ max as an example. To produce a scatter plot in SPSS we use **Graph → Chart Builder** which makes the pop-up window shown in Figure 8.2 appear.

Click on scatter plot in the 'Choose from' area and drag the simple scatter tile from the 'gallery' into the *Chart preview* area. Now we drag 'body mass' into X-axis and 'estimated $\dot{V}O_2$ max' into Y-axis in the *Chart preview* area.

We have some flexibility in the presentation of the chart. If we wished the male and female participants to be shown using different symbols, then we can click on the 'Groups/Point ID' tab and put a tick in the Grouping/stacking variable check box as shown in Figure 8.2. Then

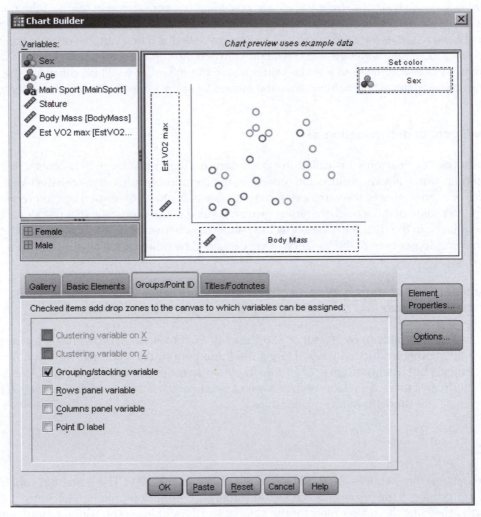

Figure 8.2 The Chart Builder pop-up window

we can drag 'sex' into set colour in the *Chart preview* area. Now when we click on **OK**, the scatter plot will appear in the output viewer.

If we don't like the colours or symbols used, the chart produced in the output viewer can be edited in Chart Editor. This is activated by double clicking on the chart in the output viewer which causes the Chart Editor pop-up window to appear as shown in Figure 8.3. If we click on one of the symbols in the Sex legend (at the top right-hand side of the chart), we get the pop-up window shown in Figure 8.4 allowing us to select different symbols, border colours, fill colours and apply these. When we have finished editing, we can close the Chart Editor window and any pop-up windows used for changing point properties and the reformatted chart will replace the previous version in the output viewer. It is worth noting that we could

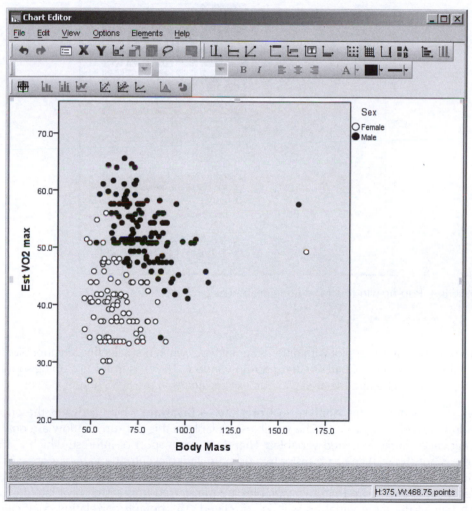

Figure 8.3 The SPSS Chart Editor

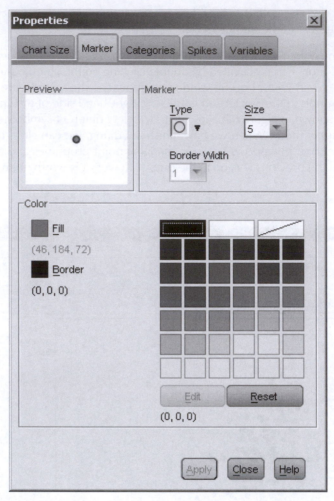

Figure 8.4 Pop-up window for editing properties of chart elements

have requested a scatter plot with different point types when first selecting a scatter plot type from the gallery in the Chart Builder pop-up window. The reason for not doing so in this example was to illustrate the ability to edit existing charts in the SPSS output viewer.

Pearson's r is done using **Analyse → Correlate → Bivariate** which activates the pop-up window shown in Figure 8.5. The variables available in this pop-up window are only the numerical or numeric-coded variables. Therefore, main sport of interest, which is a text string, is not shown. We can transfer two or more variables into the *Variables* area; in this example, 'stature', 'body mass' and 'estimated $\dot{V}O_2$ max' are transferred into the *Variables* area. Even if more than two variables are transferred, this is still 'bivariate' because each pair of the three variables will be correlated. The default correlation coefficient is Pearson's r.

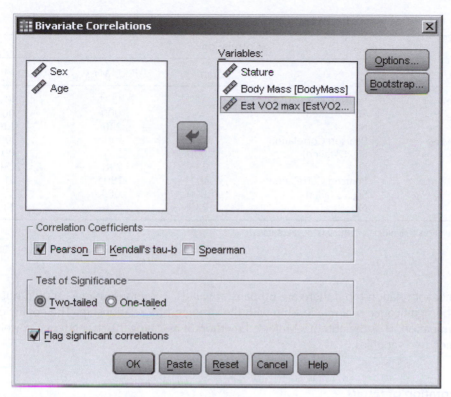

Figure 8.5 Bivariate correlations pop-up window

When we click on **OK**, we obtain the output shown in Table 8.1. The first thing to observe in the SPSS output is that each variable is correlated with each other variable including itself. The perfect correlations (r = +1.000) on the diagonal simply state the obvious that a variable has a perfect correlation with itself. Each correlation is effectively done twice; for example, body mass is correlated with stature giving the same result as when stature is correlated with body mass (r = +0.434). These correlations are identical because the scatter plots will simply be mirror images about the x = y diagonal line on the chart. Imagine the scatter plot of body mass plotted against stature printed on a sheet of transparent paper. When the transparent paper is turned over, the points will be just as close to an imaginary straight line of best fit as they were before. Where we have many variables to correlate, we may prefer to give some direction to the correlation output rather than simply correlating every single pair of variables. Ntoumanis (2001: 117) shows how a command can be created in SPSS to correlate the specific pairs of variables of interest. Table 8.1 also shows that p (Sig.) values are produced for each correlation. The p value indicates the chance that a relationship between a pair of variables could be due to chance; for example, a p value less than 0.05 might be considered significant because there is less than a one in 20 chance that the correlation is due to sampling error. The p value not only depends on the coefficient of correlation but also on the number of pairs of values used to determine the correlation. A p value of under 0.05 is found with an absolute correlation coefficient of 0.576 if there are 10 pairs of values. However, an absolute correlation coefficient of 0.250 is sufficient to indicate

Table 8.1 SPSS output for Pearson's r

		Stature	Body Mass	Est $\dot{V}O_2$ max
		Correlations		
Stature	Pearson Correlation	1	.434*	.401*
	Sig. (2-tailed)		.000	.000
	N	236	236	236
Body Mass	Pearson Correlation	.434*	1	.190*
	Sig. (2-tailed)	.000		.003
	N	236	236	236
Est VO2 max	Pearson Correlation	.401*	.190*	1
	Sig. (2-tailed)	.000	.003	
	N	236	236	236

* Correlation is significant at the 0.01 level (2-tailed).

a significant relationship if there are 60 pairs of values. Therefore, readers should not confuse the significance of a relationship with the strength of a relationship. The coefficient of determination, r^2, is the author's preferred method of assessing the strength of a correlation between two variables.

Presentation of results

The results of this example could be reported using a combination of scatter plots and supporting values of Pearson's r. We could show three scatter plots (stature vs body mass, stature vs estimated $\dot{V}O_2$ max and body mass vs estimated $\dot{V}O_2$ max). Each scatter plot would look like the one in Figure 8.3. We could choose to only show the scatter plots for those pairs of variables with strong correlations. The supporting text in the results section could state the r values found as follows:

> There were positive correlations between stature and body mass (r = .434), stature and estimated $\dot{V}O_2$ max (r = .401) and between body mass and estimated $\dot{V}O_2$ max (r = .190).

PARTIAL CORRELATIONS

Purpose of the test

A partial correlation correlates two variables, x and y, controlling for confounding influence of some third intervening variable, z, on the relationship (Ntoumanis, 2001: 119–20). Ntoumanis (2001: 119–20) used an example of the relationship between bench press performance (weight lifted) and body mass where there could be confounding influence of weight training experience.

Example: Confounding influence of distance covered by soccer players on the relationship between body mass and the number of path changes performed

This example is based on the PhD study of Gemma Robinson which analysed three different types of path change performed in English FA Premier League soccer (Robinson *et al.*, 2011). These path changes were:

- A path change of 45° to 135° to the side of the dominant leg (Sharp dom).
- A path change of 45° to 135° to the side of the non-dominant leg (Sharp non-dom).
- A path change of greater than 135° to either side (V-cut).

One of the analyses done on this data was to determine if there was a relationship between body mass and the number of path changes performed. The frequency per match of the three types of path change was determined for 25 players who were observed for at least six 90-minute performances without being substituted or dismissed. The only path changes included in the study were those where the player had moved at 4m.s^{-1} or faster during the 1s before the point of path change or during the 1s after. The frequency of each type of path change and the frequency of path changes as a whole were correlated with body mass. However, the researchers felt that those players who covered the greatest distances when playing soccer might perform more path changes than other players simply because there was more scope for path changes to be made within the distance covered. The partial correlation allows the relationship between body mass and frequency of path changes to be determined controlling for any confounding influence of distance covered.

SPSS

The data used in this example is found in the file 08-turning.SAV and body mass was negatively correlated with sharp path changes to the side of the dominant leg (r = –0.307), path changes made to the side of the non-dominant leg (r = –0.364), V-cut path changes (r = –0.456) and path changes as a whole (r = –0.379) as shown in Table 8.2. These correlations are bivariate correlations that do not control for the influence of overall distance travelled. The influence of distance travelled is significant for all four path change variables having strong positive correlations with distance travelled: sharp path changes to the side of the dominant leg (r = 0.951, p < 0.001), sharp path changes made to the side of the non-dominant leg (r = 0.974, p < 0.001), V-cut path changes (r = 0.879, p < 0.001) and path changes as a whole (r = 0.987, p < 0.001).

To perform a partial correlation in SPSS we use **Analyse → Correlate → Partial** which activates the pop-up window shown in Figure 8.6. 'Bodymass', 'VCut', 'Sharp Dom', 'Sharp Non Dom' and 'TotalTurns' are transferred into the *Variables* area while 'total_km' is transferred into the *Controlling for* area. When we click on the **OK** button, the output shown in Table 8.3 is produced.

Once total distance covered is controlled for, the magnitude of the partial correlations is reduced from the misleading values shown in Table 8.2 to the values shown in Table 8.3. Indeed, one of the partial correlations is positive rather than negative once distance covered by the player is controlled for.

Table 8.2 Bivariate correlations between body mass and path changes in soccer

		Bodymass	Sharp Dom	Sharp Non Dom	VCut	TotalTurns
				Correlations		
Bodymass	Pearson Correlation	1	−.307	−.364	−.456*	−.379
	Sig. (2-tailed)		.135	.074	.022	.062
	N	25	25	25	25	25
Sharp Dom	Pearson Correlation	−.307	1	.917**	.809**	.967**
	Sig. (2-tailed)	.135		.000	.000	.000
	N	25	25	25	25	25
Sharp Non Dom	Pearson Correlation	−.364	.917**	1	.843**	.976**
	Sig. (2-tailed)	.074	.000		.000	.000
	N	25	25	25	25	25
VCut	Pearson Correlation	−.456*	.809**	.843**	1	.904**
	Sig. (2-tailed)	.022	.000	.000		.000
	N	25	25	25	25	25
TotalTurns	Pearson Correlation	−.379	.967**	.976**	.904**	1
	Sig. (2-tailed)	.062	.000	.000	.000	
	N	25	25	25	25	25

* Correlation is significant at the 0.05 level (2-tailed).
** Correlation is significant at the 0.01 level (2-tailed).

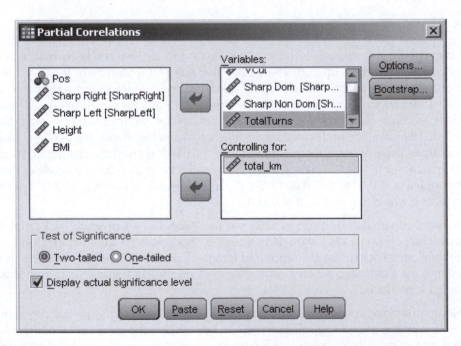

Figure 8.6 Pop-up window for partial correlations

Table 8.3 Partial correlations between body mass and path changes in soccer controlling for distance covered

				Bodymass	Sharp Dom	Sharp Non Dom	VCut	TotalTurns
				Correlations				
Control Variables								
total_km	Bodymass	Correlation		1.000	.164	−.005	−.290	−.075
		Significance (2-tailed)		.	.443	.981	.169	.729
		df		0	22	22	22	22
	Sharp Dom	Correlation		.164	1.000	−.124	−.182	.569
		Significance (2-tailed)		.443	.	.564	.394	.004
		df		22	0	22	22	22
	Sharp Non Dom	Correlation		−.005	−.124	1.000	−.120	.413
		Significance (2-tailed)		.981	.564	.	.575	.045
		df		22	22	0	22	22
	VCut	Correlation		−.290	−.182	−.120	1.000	.470
		Significance (2-tailed)		.169	.394	.575	.	.020
		df		22	22	22	0	22
	TotalTurns	Correlation		−.075	.569	.413	.470	1.000
		Significance (2-tailed)		.729	.004	.045	.020	.
		df		22	22	22	22	0

Presenting results

The results of partial correlations can be presented in a similar way to the results of bivariate correlations. This could include reporting relationships between the correlated variables and the variable being controlled for. The results might be expressed as follows:

> Table 8.4 shows there were negative correlations between body mass and the frequency of path changes performed by the 25 soccer players. However, when partial correlations were used to control for the confounding influence of total distance covered by the players, the magnitude of these correlations was reduced.

Table 8.4 Correlations (r) between body mass and frequency of path changes performed by soccer players.

Path change	Correlation with body mass	Partial correlation with body mass controlling for distance travelled
Sharp to the side of the dominant leg	−0.307	+0.164
Sharp to the side of the non-dominant leg	−0.364	−0.005
V-Cut path changes	−0.456	−0.290
Total path changes	−0.379	−0.075

145

correlation

NON-PARAMETRIC CORRELATIONS

Purpose of the tests

There are two non-parametric or rank correlation coefficients: Spearman's ρ and Kendall's τ. The purpose of these correlation coefficients is to give a numerical value to the correlation between a pair of interval or ratio scale variables. These correlation techniques use ranks rather than values meaning that they can be used to correlate any variables that can be ranked; this includes ordinal variables. The techniques can also be used to correlate an ordinal variable with an interval or ratio scale variable (Diamantopoulos and Schlegelmilch, 1997: 206–7). Spearman's ρ is used if there are at least 20 pairs of values being correlated, otherwise Kendall's τ is recommended. We will illustrate the use of Spearman's ρ and Kendall's τ using the same example as was used to illustrate Pearson's r.

SPSS

Spearman's ρ and Kendall's τ are done using **Analyse → Correlate → Bivariate** which activates the bivariate correlations pop-up window shown in Figure 8.5. We use the same file as before (ex8.1-fitnesstests.SAV) and enter the same three variables into the *Variables* area ('stature', 'body mass' and 'estimated $\dot{V}O_2$ max'). To obtain non-parametric correlations in the SPSS output, we tick the check boxes for the two nonparametric correlation coefficients and remove the tick from Pearson's r. Table 8.5 shows the output produced.

Table 8.5 SPSS output for nonparametric correlations

			Stature	Body Mass	Est VO2 max
Correlations					
Kendall's tau_b	Stature	Correlation Coefficient	1.000	.481*	.300*
		Sig. (2-tailed)	.	.000	.000
		N	236	236	236
	Body Mass	Correlation Coefficient	.481*	1.000	.130*
		Sig. (2-tailed)	.000	.	.004
		N	236	236	236
	Est VO2 max	Correlation Coefficient	.300*	.130*	1.000
		Sig. (2-tailed)	.000	.004	.
		N	236	236	236
Spearman's rho	Stature	Correlation Coefficient	1.000	.653*	.447*
		Sig. (2-tailed)	.	.000	.000
		N	236	236	236
	Body Mass	Correlation Coefficient	.653*	1.000	.210*
		Sig. (2-tailed)	.000	.	.001
		N	236	236	236
	Est VO2 max	Correlation Coefficient	.447*	.210*	1.000
		Sig. (2-tailed)	.000	.001	.
		N	236	236	236

* Correlation is significant at the 0.01 level (2-tailed).

Presentation of results

The results of doing a non-parametric correlation of numerical variables are presented in a similar way to when Pearson's r is used: a scatter plot (like that shown in Figure 8.3) is supported with the relevant non-parametric correlation coefficient. If ordinal variables with six or fewer possible values are correlated, it is better to use a cross-tabulation of frequencies for the two variables supported by the relevant non-parametric correlation coefficient. In the example used in this chapter, we have three pairs of variables being correlated using data from more than 20 participants. Therefore, Spearman's ρ should be used with the results expressed as follows:

> There were positive correlations between stature and body mass (ρ = .653), stature and estimated $\dot{V}O_2$ max (ρ = .447) and between body mass and estimated $\dot{V}O_2$ max (ρ = .210).

SUMMARY

Sometimes the purpose of a research study is to determine whether or not relationships exist between variables. Correlation coefficients provide numerical measures for the strength and direction of any relationship between a pair of variables. Pearson's r is used with interval and ratio scale variables. The coefficient of determination, r^2, is the proportion of the common variance of two variables that is explained by their relationship. Where we have a restricted set of values for one or more variables being correlated, or where one or both of the variables is measured on an ordinal scale, non-parametric correlation coefficients can be used. There are two such correlation coefficients: Spearman's ρ and Kendall's τ. These correlation coefficients are computed using ranks rather than values. All three correlation coefficients have values between -1.0 and $+1.0$ with absolute values of over 0.7 representing strong relationships between a pair of variables and absolute values of under 0.2 representing no correlation between a pair of variables.

EXERCISES

Exercise 8.1. Anthropometric variables and estimated $\dot{V}O_2$ max

Use **Data → Split Data** to logically partition the file ex8.1-fitnesstests.SAV by gender. Now produce Pearson's r values for the female and male students separately for each pair of variables from 'stature', 'body mass' and 'estimated $\dot{V}O_2$ max'.

Exercise 8.2. Variables related to margin of victory in soccer matches

The file ex8.2-pool-matches-2007–2010.SAV contains data for a set of 153 soccer matches from international tournaments played between 2007 and 2010. There are three difference variables of interest within the current exercise that represent the gap between the higher and lower ranked teams in a match:

■ 'GD' is how much the higher ranked team in FIFA world rankings won the match by (it this value is zero the match was a draw and if the value is negative the higher ranked team lost the match).

- 'RP_Diff' is how many more FIFA ranking points the higher ranked team has than the lower ranked team.
- 'Dist_Diff' is how much further it is from the capital city of the higher ranked team to the capital city of the host nation than it is from the capital city of the lower ranked team to the capital city of the host nation.

Due to restricted number of values for 'GD', use Spearman's ρ to determine which of 'RP_Diff' and 'Dist_Diff' is most correlated with 'GD'.

Exercise 8.3. Serving in tennis

The file ex8.3-tennis.SAV contains data about 252 singles tennis matches. There are two serving variables of interest: 'the percentage of points where the first serve is in' and 'the percentage of points that are won when the first serve is in'. Produce a scatter plot for these two variables using different symbols to represent men's and women's singles matches. Use **Data → Split File** to make sure any correlation is applied to men's and women's singles matches separately. Determine the value of Pearson's r and Spearman's ρ between the two variables for men's and women's singles. Explain the direction of the correlation found. Now analyse all cases together (that is, men's and women's singles matches together); explain any problem arising from not separating the genders.

PROJECT EXERCISE

Exercise 8.4. Correlation between stature, body mass and estimated V̇O₂ max

Measure stature and body mass for your class mates, and also use the multistage fitness test to determine an estimate of V̇O₂ max for each of your class mates. Produce separate scatter plots and Pearson's r values for each pair of variables for the males and females.

Exercise 8.5. Efficacy of World ranking in professional tennis

For each of the Grand Slam tennis tournaments, examine the world rankings of the players reaching each round of the tournament. Use 1 to represent losing in round 1, 2 to represent losing in round 2, . . ., 6 to represent losing a semi-final, 7 to represent losing the final and 8 to represent winning the tournament. Use Spearman's ρ to determine which tournament has the strongest relationship between round of elimination and World ranking of the player. Do this for both the men's and women's singles events. The data for players reaching the different rounds can be found in the draws sections of the following websites:

- www.ausopen.org
- www.frenchopen.org
- www.wimbledon.org
- www.usopen.org

CHAPTER 9

LINEAR REGRESSION

INTRODUCTION

In Chapter 8, we covered correlation techniques. Where there is a very strong linear correlation between two variables and one can be hypothesized to depend on the other, we can model the relation between the two variables as a straight line that fits the correlated data when presented on a scatter plot. This line is called a regression line or a line of best fit and allows us to predict values of the dependent variable where we know the value of the independent variable. The process of producing the model is referred to as bivariate regression. It is also possible to model a dependent variable in terms of more than one independent variable; this is called multiple linear regression. This chapter covers both bivariate regression and multiple linear regression, the assumptions of the techniques and how the models produced can be used for predictive purposes.

BIVARIATE LINEAR REGRESSION

Purpose of the test

The purpose of bivariate regression is to determine an equation for some dependent variable in terms of some independent variable given cases where the values of the dependent and independent variables are known. The equation is that of the straight line of best fit; this is the straight line such that there is no other straight line that is closer to the set of co-ordinates for the known cases. The equation is in the form shown in equation 9.1 where x is the independent variable, y is the dependent variable, b is the gradient of the line of best fit and a is the value of y where x is zero. The coefficient b tells us that increasing x by one will increase y by b.

$$y = a + b.x \tag{9.1}$$

Figure 9.1 shows a graph of such a line. In bivariate regression, the independent variable, x, and the dependent variable, y, are both interval or ratio scale variables. Linear regression is sometimes referred to as the 'least squares method' (Anderson *et al.*, 1994: 502) and the regression equation produces the straight line that minimizes the sum of the squared differences the coordinates are from the line. These differences are differences between

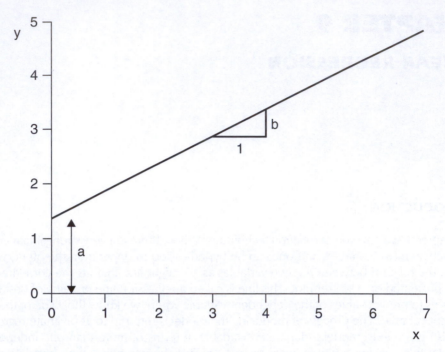

Figure 9.1 Regression line $y = a + b.x$

an observed y value and the predicted value, y', for the given value of x. Fallowfield *et al.* (2005: 158) described a three-phase process consisting of producing a scatter plot, determining the correlation and then performing linear regression.

Interpolation and extrapolation

Once an equation for some dependent variable y has been determined in terms of x, we can use the equation to predict values for y for known values of x. Interpolation is where a value of y is being determined for a known value of x that is within the range of the values of x that was used to produce the equation during the original regression analysis. Extrapolation, however, is where we attempt to predict a value for y given a known value of x that is outside the range of x values used to construct the regression equation. There is a danger that extrapolation will produce inaccurate predictions where the relationship between x and y does not follow the same straight line of best fit. The line may curve up or down for values outside the range used to produce the regression equation. A very good example of inappropriate extrapolation was highlighted by Graham (2006: 97–9) where it can be shown that female sprinters will be running faster than their male counterparts by AD 2252. Maybe they

linear regression

will! Maybe they won't. However, a flawed use of linear regression does not tell us whether they will or will not.

Another example of inappropriate extrapolation was a regression equation of international soccer matches made by O'Donoghue *et al.* (2004). The data used to create the regression equation included distances travelled by teams competing in international tournaments between 1994 and 2001. This regression equation was then used to predict the outcomes of matches of the 2002 FIFA World Cup. This led to a prediction that Japan would defeat South Korea in the World Cup final. This was because competing teams had not travelled the sorts of differences they had to travel to Japan and South Korea in previous tournaments. Extrapolating the effect of home advantage (represented by low values for distance travelled) to this extent overstated the chances of Japan, South Korea and China, although South Korea did actually reach the semi-finals of the tournament.

Uses of linear regression

Linear regression can be used to validate a measurement against a gold standard measurement. Cross-validation is where a regression equation is constructed using half of the known cases available and then tested using the remaining known cases. Linear regression is also used within ANCOVA tests (which will be covered in Chapter 11), path analysis, structured equation modelling and factor analysis (Allison, 1999: 176–83). Linear regression is also a specific version of the General Linear Model on which analysis of variances tests are based (Allison, 1999: 183–5).

Assumptions

Linear regression assumes that any relationship between the independent variable and the dependent variable is linear (Newell *et al.*, 2010: 140). Ntoumanis (2001: 120–1) identified two assumptions that must be satisfied by data used to produce a bivariate regression equation.

- There must be no outliers in the independent variable, x, or the dependent variable, y. Outliers and extreme values have a 'leverage' on the regression line which may distort the regression line (Tabachnick and Fidell, 2007: 124). This is because the squared distance from the line of best fit increases the further away the outlier is from the regression line. These outliers and extreme values either need to be excluded from the regression analysis or transformed. Furthermore, there must be no outliers in residual values as these can have an undue influence on the regression equation.
- Residuals should be independent, homoscedastic and normally distributed. If residual values are independent of x, there is little correlation between x and the residuals. Independence also means that the order of measurement is unrelated to residual values. Homoscedasticity means that the variance in residual values (and hence in y values) is similar for all values of x (Anderson *et al.*, 1994: 521; Vincent, 1999: 111). Residuals should be normally distributed (Vincent, 1999: 111); this can be tested using a Kolmogorov–Smirnov test or a Shapiro–Wilk test.

There were two other assumptions listed by Ntoumanis (2001: 120–1) which are relevant to multiple linear regression and will be covered later in this chapter. The assumptions can be tested by inspecting plots of residuals against x (Anderson *et al.*, 1994: 540–8, Tabachnick and Fidell, 2007: 125–7).

Significance

The regression line is based on a sample of previous cases and, therefore, the line is an estimate rather than a true model that would be based on the whole population (Anderson *et al.*, 1994: 522–5). Therefore, standard errors and significance values (p values) are determined for y′ and for the individual regression coefficients (Tabachnick and Fidell, 2007: 147–8; Hinton, 2004: 278–9). The standard errors and p values will be illustrated with the aid of an example.

Example: Middle distance running

Athletics coach Frank Horwill (1982) proposed equations to set targets for middle distance runners. The purpose of this example is to show how linear regression can be used to produce such equations. In this example, we have a set of middle distance runners with known personal best times for the 800m and 1,500m. We wish to determine an equation for 1,500m time in terms of 800m time, assuming that most athletes would be doing 800m competition earlier in their careers than 1,500m competition.

SPSS

The file 09-middledistance.SAV is used in this example and contains personal best times (PBs in s) for the 800m, 1,500m and 3,000m for a set of 20 male middle distance runners. The first step in the modelling process is to explore the relationship between the two variables of interest ('800m PB' and '1500m PB'). This is done using Chart Builder as described in Chapter 8; the scatter plot for 800m and 1,500m time is shown in Figure 9.2.

To apply a fit line to the chart, we double click on it to activate the Chart Editor window as shown in Figure 9.3. Use **Elements → Fit Line** to activate the Properties pop-up window shown in Figure 9.4. When we select linear and click on the Apply button, a straight line of best fit is drawn through the points as shown in Figure 9.5. This also shows the value of the coefficient of determination, r^2.

If we are satisfied that there is a strong linear relationship between two variables, we can use **Analyse → Regression → Linear** to determine the equation of the line of best fit. The pop-up window shown in Figure 9.6 is used to place '1500m PB' into the *Dependent* area and '800m PB' into the *Independent(s)* area.

The **Statistics** pop-up window (Figure 9.7) can be used to produce information needed to test the assumptions of linear regression (Ntoumanis, 2001: 122–3). There are alternative

linear regression

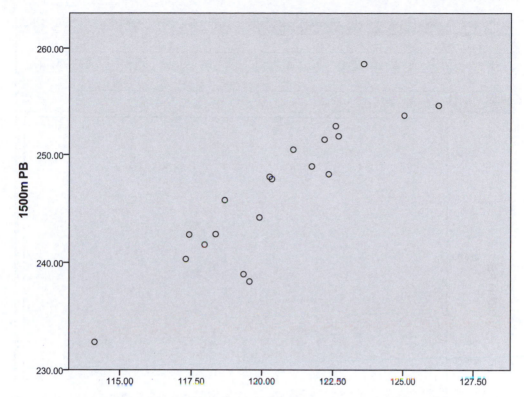

Figure 9.2 Relation between 800m PB and 1,500m PB

ways to test the assumptions of linear regression. This author prefers to **Save** unstandardized residual values and predicted values. These new variables can then be explored for normality and correlations just like any other variables. When we close the pop-up windows and click on **OK** for the main regression analysis pop-up window, the results of the regression analysis appear in the output viewer. Tables 9.1 to 9.3 show the main tables of the output produced by SPSS.

The r^2 value (R Square in SPSS) in Table 9.1 is 0.799 which is strong enough to permit regression analysis to be done. The p (Sig.) value in the Table 9.2 shows that the independent variable, '800m PB', is a significant predictor of the dependent variable, '1500m PB'. Table 9.3 shows the regression coefficients to be used in a predictive model (regression equation or line of best fit). The regression coefficients b_0 and b_1 are 5.075 and 2.004 respectively, meaning that our regression equation is $y = 5.075 + 2.004 x$ where y is 1500m PB and x is 800m PB. Therefore, an athlete with a personal best of 2 minutes and 4s for the 800m would be expected to have a personal best for the 1,500m of 4 minutes 13.6s (5.075 + 2.004 x 124 = 253.6s). Roughly, double your 800m time and add 5s. Table 9.3 also shows the significance of the two regressions coefficients; b_0 is not significant (p = 0.861) but b_1 is (p < 0.001). The standardized regression coefficient for '800m PB' is the amount the '1500m PB' increases for each standard deviation of the variable '800m PB'.

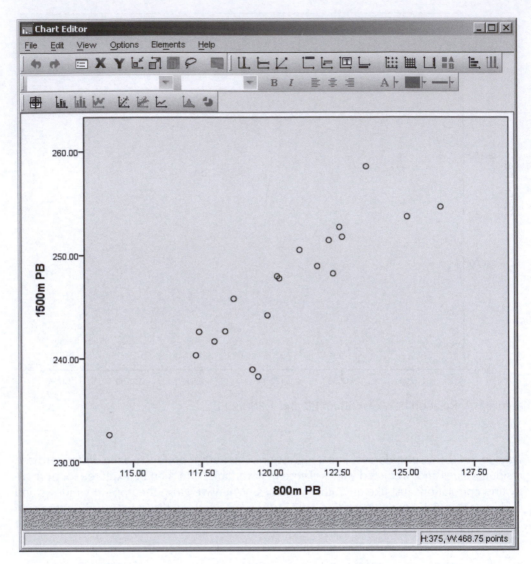

Figure 9.3 Chart Editor

Table 9.1 SPSS output for linear regression (model summary)

Model Summary				
Model	R	R Square	Adjusted R Square	Std. Error of the Estimate
1	.894[a]	.799	.788	2.97905

a. Predictors: (Constant), 800m PB

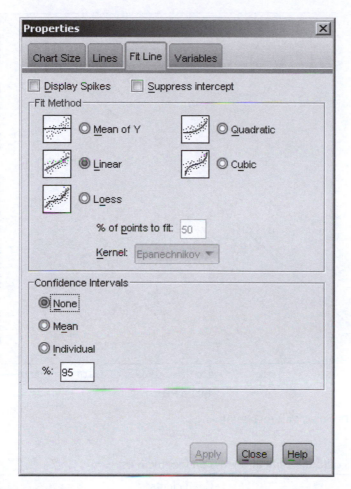

Figure 9.4 Chart Properties

Table 9.2 SPSS output for linear regression (ANOVA)

		Sum of Squares	Df	Mean Square	F	Sig.
Model						
1	Regression	636.214	1	636.214	71.688	.000[a]
	Residual	159.746	18	8.875		
	Total	795.960	19			

ANOVA[b]

a. Predictors: (Constant), 800m PB
b. Dependent Variable: 1500m PB

The coefficients for the intercept and slope of the line of best fit are estimates because the line is derived from a sample rather than a full population (Newell *et al.*, 2010: 136). The standard errors for the intercept and the gradient are 28.539 and 0.237 respectively. The standard error for the intercept seems very large and the 95 per cent confidence interval

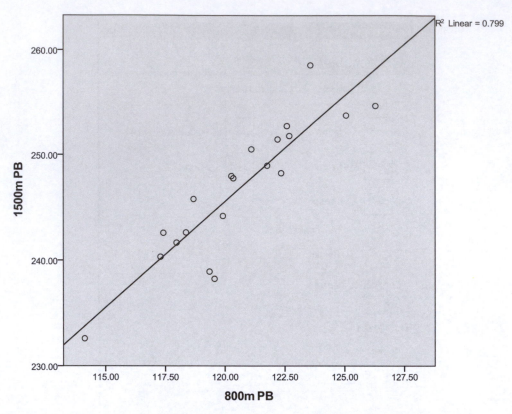

Figure 9.5 Scatter plot with regression line

Table 9.3 SPSS output for linear regression (coefficients)

	Coefficients[a]					
Model		*Unstandardized Coefficients*		*Standardized Coefficients*		
		B	*Std. Error*	*Beta*	*t*	*Sig.*
1	(Constant)	5.075	28.539		.178	.861
	800m PB	2.004	.237	.894	8.467	.000

a. Dependent Variable: 1500m PB

based on this standard error is 5.075 ± 1.96 x 28.539 = −50.861 to +61.011. The use of 1.96 ($z_{0.025}$) assumes the residuals are normally distributed. Remember that the scatter plot in Figure 9.5 does not include the origin (0,0) and to do so would result in the coordinates of the scatter plot being restricted to a small area of the top right-hand corner of the chart. A small change in the gradient of a line drawn through these points would be multiplied into a much larger change when the line is drawn right back to the y axis (the line x = 0). The standard error for the gradient of the line is 0.237, meaning that the 95 per cent confidence

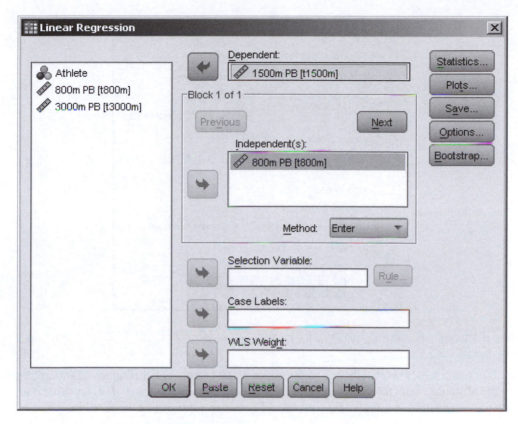

Figure 9.6 Linear Regression pop-up window

interval for the gradient is $2.004 \pm 1.96 \times 0.237 = 1.539$ to 2.469. This is quite a difference in real athletic terms. Taking 1.539s off one's 1,500m time for each 1s of improvement in 800m time could predict an athlete running the 1,500m at a faster pace (m.s^{-1}) than the 800m depending on the intercept of the line.

Having done the interesting bit of producing the regression equation for '1500m PB' in terms of '800m PB', we will now turn to the assumptions of linear regression. These would usually be checked earlier. When setting up the linear regression analysis, we use **Save** to create new variables for unstandardized residual values and unstandardized predicted values. This is done using the pop-up window shown in Figure 9.8.

The Explore facility in the descriptive statistics menu is used to perform a Shapiro–Wilk test on the residuals showing that they are sufficiently normal as shown in Table 9.4 ($p > 0.05$). If there had been 50 or more values, we would have used a Kolmogorov–Smirnov test. To check if there is any heteroscedasticity in the residuals with respect to the independent variable, it was necessary to produce a new variable for the magnitude of the residuals using **Transform → Compute**. If there is a positive correlation between x and the magnitude of the residuals, then there will be differing standard deviations of residuals for different subranges of the independent

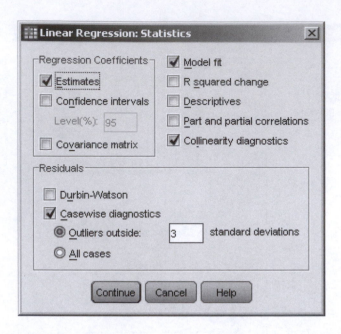

Figure 9.7 Statistics pop-up window for linear regression

variable. Table 9.5 shows that there is a correlation of 0.243 between the independent variable and the absolute values for the residuals which is a concern as it reflects a 'shotgun' effect very close to a correlation of +0.25 where we would consider transforming the variable.

Table 9.4 Results of testing the assumption of normality of residuals

Tests of Normality

	Kolmogorov-Smirnov[a]			Shapiro-Wilk		
	Statistic	df	Sig.	Statistic	Df	Sig.
Unstandardized Residual	.128	20	.200[*]	.962	20	.592

a Lilliefors Significance Correction
* This is a lower bound of the true significance.

Table 9.5 Results of testing homoscedasticity of the residuals

		Correlations	
		800m PB	abs_res
800m PB	Pearson Correlation	1	.243
	Sig. (2-tailed)	.302	
	N	20	20
abs_res	Pearson Correlation	.243	1
	Sig. (2-tailed)	.302	
	N	20	20

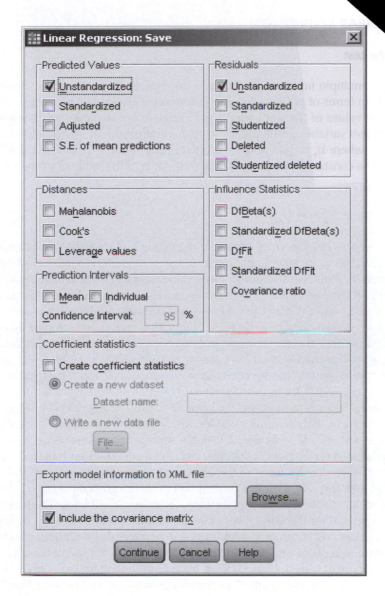

Figure 9.8 The Save pop-up window for linear regression

Presentation of results

Ntoumanis (2001: 132) showed how to report regression results in a table. A scatter plot (such as Figure 9.5) and the regression equation ($y = 5.075 + 2.004\ x$) would typically be sufficient to communicate the key results to the readers of research reports. Where the researcher also wishes to show the significance of the regression equation and the standard errors for the regression coefficients, a table would be recommended.

lear regression is to determine an equation for some depend-
N (two or more) independent variables, $x_1, x_2, x_3, \ldots x_N$, given
the dependent and independent variables are known. If there
es, multiple linear regression produces an equation in the form
to b_N are the multiplying coefficients for the independent vari-
it of the regression equation; b_0 is the value of y when all x values
are 0.

$$y = b_0 + b_1.x_1 + b_2.x_2 + b_3.x_3 + \ldots + b_N.x_N \qquad (9.2)$$

One of the differences between multiple linear regression and bivariate regression is that the independent variables in multiple linear regression are not limited to interval and ratio scale variables. Nominal variables with three or more values should not be included because they have no order and this makes it impossible to produce a meaningful multiplying coefficient to determine their contribution to the value of y. An ordinal variable such as level of agreement (with values strongly disagree, disagree, undecided, agree and strongly agree) should not be included because there is no fixed interval between the values. Dichotomous variables, however, can be used in multiple linear regression (Allison, 1999: 9–10). For example, a variable x_i representing gender could be represented as 0 and 1 for female and male. The regression coefficient B_i for this variable would be positive if being male raised the value of y and negative if being female raised the value of y. Other categorical variables could be replaced by a series of dichotomous variables which could be included in multiple linear regression if we wished. For example, a variable representing position on a soccer team (with values goalkeeper, defender, midfielder and forward) could be replaced by separate dichotomous variables for the three of the individual positions (goalkeeper, defender and midfielder) that take the value 1 if the player plays in the given position or 0 if he doesn't. We do not need a variable to represent forward because we will know the player is a forward if his values are 0 for the other three positional roles. It is this ability of multiple linear regression to represent dichotomous variables as well as interval and ratio scale variables that allows it to form the basis of the General Linear Model on which many other tests including analysis of variances tests are based. If we have N independent variables including any dichotomous variables, the form of the General Linear Model is given in equation 9.3 where E is an error value.

$$y = b_0 + b_1.x_1 + b_2.x_2 + b_3.x_3 + \ldots + b_N.x_N + E \qquad (9.3)$$

Equations 9.1 and 9.2 that were shown earlier in this chapter should really be equations for the predicted dependent variable, y', rather than the observed dependent variable, y. The error value is the difference between the observed and expected values as shown in equation 9.4. This error value is also referred to as a residual value.

$$E = y - y' \qquad (9.4).$$

Assumptions

Ntoumanis (2001: 120–1) listed four assumptions that must be satisfied by data used to produce a regression equation when there are two or more independent variables:

- There should be at least 20 cases for each independent variable.
- There must be no outliers in individual independent variables, the dependent variable or residuals. As well as considering outliers within individual variables, we also need to check multivariate outliers. Distance measures such as Mahalanobis distances can be used to identify outliers within the multivariate space (Ntoumanis, 2001: 124–5).
- Multicollinearity should be avoided in the independent variables. This means that no pair of independent variables should be highly correlated (the absolute values of r should be less than 0.9). If we do find two variables are highly correlated, one could be excluded from the analysis. Another solution is to use a hierarchy of regression equations (Allison, 1999: 137–8). One level is used to produce an equation involving our correlated independent variables. For example, if we wished to determine a regression equation $y' = b_0 + b_1.x_1 + b_2.x_2 + b_3.x_3$ but x_2 and x_3 were highly correlated ($|r| \geq 0.9$), we might determine an equation for x_2 in terms of x_3. This would mean logically justifying x_2 depends on x_3. The predicted values from the equation $x_2' = a + b.x_3$ could then be incorporated into a higher level regression equation for y' using x_1 and x_2 only.
- As with bivariate regression, residuals should be independent, homoscedastic and normally distributed. Rather than testing the distribution of the residuals for different subranges of each independent variable, the predicted value for y can be used (y'). Therefore we test that there is little correlation between y' and the absolute residual values to show homoscedasticity. Independence can be checked by checking the correlation between the residuals and a variable representing the order of measurement of the cases. Normality of the residuals can be tested using a Kolmogorov–Smirnov test or a Shapiro–Wilk test.

Significance

As with bivariate linear regression, the multiple liner regression equation is based on a sample of previous cases and so significance values (p values) are used to determine the significance of the model as a whole and each regression coefficient. Confidence intervals for each regression coefficient can also be produced using the standard error for each providing the residuals are normally distributed. The coefficient of determination indicates how good a predictive model the regression equation is (Newell et al., 2010: 139–40). In multiple linear regression, we use R^2 rather than r^2 to represent the coefficient of determination. With more than one independent variable, there will be values of r representing the correlation of each independent variable with the dependent variable. Therefore, R^2 is not determined by merely squaring a correlation coefficient. Allison (1999: 13–14) discussed the calculation of R^2 which is beyond the scope of this book. R^2 still represents the variation in the dependent variable that is explained by the independent variables. Other statistics produced by multiple linear regression will be discussed with the aid of an example done in SPSS.

Example: Predicting the outcomes of international soccer matches

The outcomes of international soccer matches are influenced by the quality of the teams who take part and there is plenty of evidence that home advantage also has an influence on match outcome (Courneya and Carron, 1992; Carron et al., 2005). Therefore, this example attempts to produce a predictive model for match outcome. Each match is between two teams, one of which will be ranked higher than the other in the FIFA world rankings. The dependent variable, 'GD', is the number of goals the higher ranked team wins by. If this value is 0, then the match is a draw, and if the value is negative, then the lower ranked of the two teams won the match. The two independent variables are 'RP_Diff' and 'Dist_Diff'. 'RP_Diff' is the difference between the two teams' FIFA ranking points. This is always positive as the higher ranked team will have more ranking points than the lower ranked team. 'Dist_Diff' is a crude measure of home advantage and is the difference in distance travelled to the tournament between the two teams. Each team's distance travelled is assumed to be the giant circle distance between its capital city and the capital city of the host nation. 'Dist_Diff' is the difference between the two teams' distance travelled estimates (the higher ranked team's value – the lower ranked team's value). We wish to produce an equation in the form of Equation 9.5 where b_0 to b_2 are the regression coefficients.

$$GD' = b_0 + b_1.RP\_Diff + b_2.Dist\_Diff \qquad (9.5)$$

SPSS

The example of international soccer match outcomes is illustrated using data for 153 matches from international soccer tournaments that are in the file 09-pool-matches-2007–2010.SAV. Figure 9.9 shows how the regression pop-up window is used to set up a multiple linear regression; basically, we transfer more than one variable ('RP_Diff' and 'Dist_Diff') into the *Independent(s)* area. Figure 9.10 shows the **Statistics** pop-up window which can be used to request the Durbin-Watson test and tests of other assumptions such as analysis of collinearity. In this example, the Durbin–Watson test result of 1.835 does not deviate greatly from 2 and so the residuals sufficiently independent (Ntoumanis, 2001: 123).

As with bivariate regression, we can save unstandardized residual and predicted values and use these new variables when testing the assumptions of multiple linear regression. Tables 9.6 to 9.8 show the main output produced by SPSS for multiple linear regression. Table 9.7 shows that using the two independent variables together makes a significant prediction of GD (p < 0.001). Table 9.8 shows that 'RP_Diff' is a significant predictor of GD but that 'Dist_Diff' is not. However, the R^2 value in the Table 9.6 should convince readers not to place any bets on soccer matches using this particular regression equation: only 10.1 per cent of the variance in GD results is explained by ranking points and distance travelled! Table 9.8 shows the regression coefficients in the B column: b_0 is 0.114; b_1 is shown to three decimal places and so if we wish to see further significant digits we can double click on this table in the SPSS output and take a closer look at the value as shown in Figure 9.11. This gives a value for b_1 of 0.00246; b_2 is given as '–4.012E-5', where 'E-5' means the value of 4.012 is multiplied by 10^{-5}. Therefore, we need to shift the decimal point five places to the left giving –0.0000401. So our regression equation for GD' is as shown in equation 9.6.

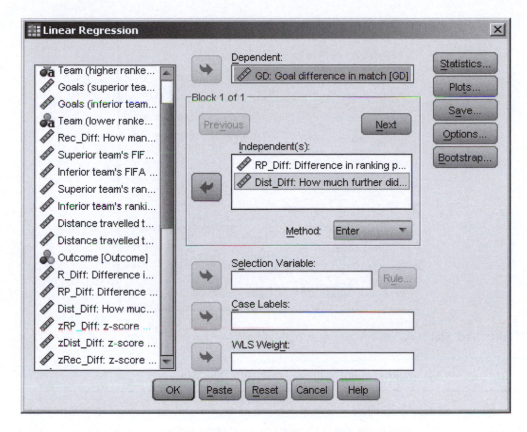

Figure 9.9 Linear Regression pop-up window using more than one independent variable

Table 9.6 SPSS output for multiple linear regression (Model Summary)

Model Summary[b]

Model	R	R Square	Adjusted R Square	Std. Error of the Estimate	Durbin-Watson
1	.318[a]	.101	.089	1.791	1.835

a. Predictors: (Constant), Dist_Diff: How much further did superior team travel to the tournament than the inferior team, RP_Diff: Difference in ranking points
b. Dependent Variable: GD: Goal difference in match

$$GD' = -0.114 + 0.00246 \times RP\_Diff - 0.0000401 \times Dist\_Diff \qquad (9.6)$$

The **Save** button allows residuals and predicted values to be saved so that the assumptions of linear regression can be tested. These variables can be explored in SPSS using **Analyse → Descriptive Statistics → Explore**. Table 9.9 shows that the residuals only just satisfy the assumption of normality ($p > 0.05$ from the Kolmogorov–Smirnov test) and Table 9.10

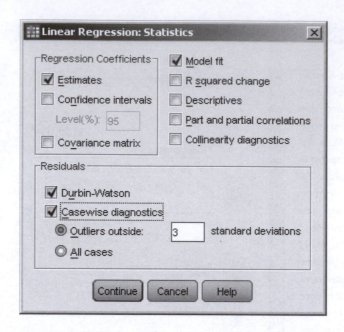

Figure 9.10 Statistics pop-up for linear regression

Table 9.7 SPSS output for multiple linear regression (ANOVA)

		ANOVA[b]			
Model	Sum of Squares	df	Mean Square	F	Sig.
1 Regression	54.173	2	27.086	8.441	.000[a]
Residual	481.330	150	3.209		
Total	535.503	152			

a. Predictors: (Constant), Dist_Diff: How much further did superior team travel to the tournament than the inferior team, RP_Diff: Difference in ranking points
b. Dependent Variable: GD: Goal difference in match

Table 9.8 SPSS output for multiple linear regression (Coefficients)

	Coefficients[a]				
Model	Unstandardized Coefficients		Standardized Coefficients		
	B	Std. Error	Beta	t	Sig.
1 (Constant)	−.114	.235		−.484	.629
RP_Diff	.002	.001	.318	4.092	.000
Dist_Diff	−4.012E-5	.000	−.061	−.785	.433

a. Dependent Variable: GD: Goal difference in match

Table 9.9 results of testing the assumption of normality of residuals for multiple linear regression

| | Tests of Normality | | | | | |
| | Kolmogorov–Smirnov[a] | | | Shapiro–Wilk | | |
	Statistic	df	Sig.	Statistic	df	Sig.
Unstandardized Residual	.070	153	.066	.995	153	.881

a. Lilliefors Significance Correction

Table 9.10 Results of testing the assumption of uncorrelated independent variables

| | Correlations | | |
		RP_Diff	Dist_Diff
RP_Diff	Pearson Correlation	1	.101
	Sig. (2-tailed)		.213
	N	153	153
Dist_Diff	Pearson Correlation	.101	1
	Sig. (2-tailed)	.213	
	N	153	153

shows that the absolute correlation between the two independent variables is not high (r = 0.101) while Table 9.11 shows that the residuals are homoscedastistic (r = 0.011). As with the bivariate regression, it was necessary to use **Transform → Compute** to determine the absolute values of the residuals to test this assumption.

The regression equation can be used to make a prediction, but the R^2 value suggests that such a prediction will not be accurate. Consider the FIFA 2010 World Cup match between France, ranked 7th in the World, and Mexico, ranked 15th, played in South Africa. France had 1,122 FIFA world ranking points at the time of the match compared with Mexico's 931 (RP_Diff =191). The distance from Paris to Pretoria is 8,659km and the distance from Mexico City to Pretoria is 14,588km (Dist_Diff = –5,929). Therefore, our model produces

Table 9.11 Results of testing the assumption of homoscedasticity of the residuals

| | Correlations | | |
		Unstandardized Predicted Value	abs_res
Unstandardized Predicted Value	Pearson Correlation	1	.011
	Sig. (2-tailed)		.895
	N	153	153
abs_res	Pearson Correlation	.011	1
	Sig. (2-tailed)	.895	
	N	153	153

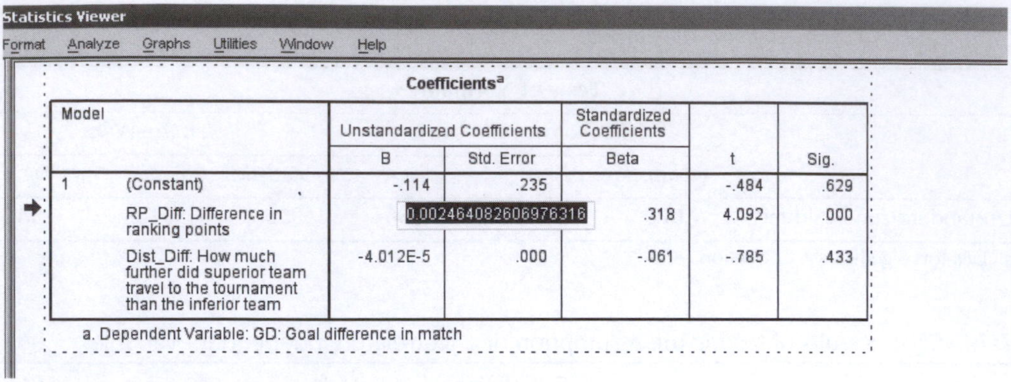

Statistics Viewer

Format Analyze Graphs Utilities Window Help

Coefficients[a]

Model		Unstandardized Coefficients B	Unstandardized Coefficients Std. Error	Standardized Coefficients Beta	t	Sig.
1	(Constant)	-.114	.235		-.484	.629
	RP_Diff: Difference in ranking points	0.002464082606976316		.318	4.092	.000
	Dist_Diff: How much further did superior team travel to the tournament than the inferior team	-4.012E-5	.000	-.061	-.785	.433

a. Dependent Variable: GD: Goal difference in match

Figure 9.11 More precise inspection of values in SPSS output

a predicted goal difference in favour of France, GD = –.114 + .00246 (191) –.0000401 (–5,929) = .822 which rounds to an expected one goal win for France (sorry Thierry, but Mexico won that match two–nil!).

Presentation of results

The presentation of results for multiple linear regression differs slightly to that of bivariate regression. The equation of the regression line can be presented as in (equation 9.6), but no single scatter plot will show the relationship between the dependent variable and each independent variable. We could opt to show the relationship with the main independent variables or use a three-dimensional representation to show the surface formed by the regression equation. Some researchers reporting linear regression results also choose to include the standard error values from the SPSS coefficients table. When a coefficient is divided by its standard error, it gives the t value shown in Table 9.12.

Stepwise and sequential methods

Hinton *et al.* (2004: 311–37) described enter and stepwise methods of including independent variables within a regression model. In the main Linear Regression pop-up window

Table 9.12 Summary of analysis of multiple linear regression

Independent Variable	*b*	*t*	*p*
Constant	–.114	–.48	0.629
Ranking Points difference	.00246	4.09	<0.001
Distance difference	–.000401	–.79	0.433

shown in Figure 9.9, there is a dropdown menu for the method of including independent variables in the regression model. The choices are Enter (which is the default), Stepwise, Remove, Backward and Forward. The Enter method is where all independent variables are entered at once; this is the method used in the previous examples in this chapter. The standard method of linear regression can make some independent variables look unimportant even when they are highly correlated with the dependent variable (Tabachnick and Fidell, 2007: 136–8). This is because credit is only given to the variance in the dependent variable explained exclusively by each independent variable. Some variance in the dependent variable may be explained by more than one independent variable. Including independent variables in the regression equation one at a time allows the full contribution of the first variable entered to be recognized. The next variable is entered by performing a regression analysis on the residuals left over after the first variable was included. The Stepwise and Sequential methods are two different ways of including variables one at a time.

The Stepwise method adds variables in descending order of how well they correlate with the dependent variable. The **Options** pop-up menu within linear regression allows us to set p values that can be used for entry and removal of variables as shown in Figure 9.12. Applying Stepwise entry to the data for the 153 international soccer matches with p values of 0.05 and 0.10 for entry and removal respectively gives the output shown in Figure 9.12. The SPSS output is shown in Table 9.13 to 9.16. In this example, 'RP_Diff' is included but 'Dist_Diff' is excluded as it is not a significant predictor of GD (p = 0.433). The exclusion of non-significant predictors can lead to a more parsimonious solution (Atkinson and Nevill, 2001).

Tabachnick and Fidell (2007: 138–44) compared the Stepwise method with the Sequential method where the system operator chooses the order in which the independent variables

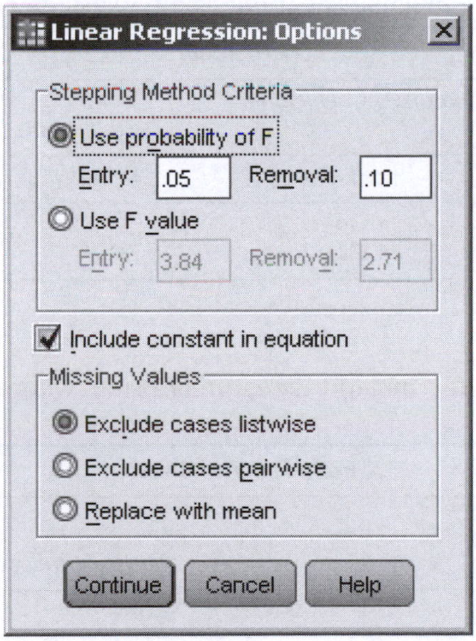

Figure 9.12 Options pop-up for linear regression

are included into the model. The advantage of this approach is that the operator can apply logic and knowledge of the area of investigation when including variables rather than allowing SPSS to use an entry order based purely on statistics.

Table 9.13 SPSS output for multiple linear regression with stepwise entry (Model Summary)

Model Summary[b]

Model	R	R Square	Adjusted R Square	Std. Error of the Estimate	Durbin-Watson
1	.312[a]	.097	.091	1.789	1.844

a. Predictors: (Constant), RP_Diff: Difference in ranking points
b. Dependent Variable: GD: Goal difference in match

Table 9.14 SPSS output for multiple linear regression with stepwise entry (ANOVA)

ANOVA[b]

Model		Sum of Squares	df	Mean Square	F	Sig.
1	Regression	52.193	1	52.193	16.307	.000[a]
	Residual	483.310	151	3.201		
	Total	535.503	152			

a. Predictors: (Constant), RP_Diff: Difference in ranking points
b. Dependent Variable: GD: Goal difference in match

Table 9.15 SPSS output for multiple linear regression with stepwise entry (Coefficients)

Coefficients[a]

Model		Unstandardized Coefficients		Standardized Coefficients		
		B	Std. Error	Beta	t	Sig.
1	(Constant)	−.110	.234		−.468	.640
	RP_Diff	.002	.001	.312	4.038	.000

a. Dependent Variable: GD: Goal difference in match

Table 9.16 SPSS output for multiple linear regression with stepwise entry (Excluded Variables)

Excluded Variables[b]

Model					Collinearity Statistics
	Beta In	t	Sig.	Partial Correlation	Tolerance
1 Dist_Diff	−.061[a]	−.785	.433	−.064	.990

a. Predictors in the Model: (Constant), RP_Diff: Difference in ranking points
b. Dependent Variable: GD: Goal difference in match

SUMMARY

Regression analysis goes beyond assessing the relation between a dependent variable and hypothesized independent variable(s) and produces an equation that can be used to predict future cases where the value of the dependent variable is unknown but the values of independent variables are known. This regression equation is based on a sample of previous cases and so there will be sampling error meaning that the regression equation is provided with significance results.

EXERCISES

Exercise 9.1. Predictive model of 3000m time in terms of 1500m time

Using the 1,500m and 3,000m PBs in the file ex9.1-middledistance.SAV, determine an equation for 3,000m PB in terms of 1,500m PB. Using your equation, what time would a 4 minute 30s 1,500m runner be expected to run for the 3,000m?

Exercise 9.2. Multiple linear regression prediction of international soccer matches (knock out stages)

The file ex9.2-ko-matches-2007–2010.SAV contains details of 53 matches played during the knockout stages of international soccer tournaments. Each match is between two teams with the higher ranked team according to FIFA being named first and the lower ranked team being named second. The dependent variable of interest is how many goals the higher ranked team won the match by, GD. If this is zero, then the match was level after 90 minutes and would have required extra time and possibly a penalty shoot-out to decide the result. If GD is negative, then the lower ranked team won. There are two independent variables that we wish to use to model GD: 'RP_Diff' and 'Dist_Diff' which are defined the same way as they were for the previous example of international soccer matches described in this chapter. Produce regression models using the Enter and Stepwise methods. With the Stepwise method, use p values of 0.05 and 0.10 for inclusion and exclusion of variables respectively. Do the data used to create the models satisfy all the assumptions of linear regression? In each case, what is the predicted outcome for a round two match in the 2010 FIFA World Cup between Spain ranked 1st and Portugal ranked 5th. Spain had 1,622 ranking points at the time of the match while Portugal had 1,181. The distance between Madrid and Pretoria is 8,038km while the distance between Lisbon and Pretoria is 8,123km. Does the result agree with the actual GD value in the match of 1 (a one goal win for Spain)?

PROJECT EXERCISE

Exercise 9.3. Relation between different event performances

Using publically available official sources of data, take two events within a timed sport and determine a regression equation to predict a competitor's best time for one event in terms of the other. This could be done for sprint events in athletics such as the 100m and 200m or swimming events using the same stroke, such as the 100m freestyle and 200m freestyle.

Exercise 9.4. Performance prediction in rugby union

Produce a predictive model for the Rugby Union World Cup using IRB world ranking points and distances between capital cities as we have done in this chapter for soccer. Is the R^2 value obtained in your rugby union model higher than the 0.101 achieved in the soccer example in this chapter?

CHAPTER 10

T-TESTS

INTRODUCTION

There are three different types of t-test: the one-sample t-test, the independent samples t-test and the paired samples t-test. The one-sample t-test compares a sample mean with some hypothesized mean. The other two t-tests are used to compare the means for an interval or ratio scale variable between two samples. The variable distinguishing the samples is the independent variable, while the interval or ratio scale variable hypothesized to be influenced by sample is the dependent variable. The main difference between independent and paired samples t-tests is that the independent samples t-test compares independent samples whereas the paired samples t-test compares related samples. Recall from Chapter 7 that independent samples are drawn from different groups of individuals who we wish to compare. For example, if we were comparing the mean of some variable between males and females, gender would be the independent variable with the male and female groups forming independent samples. Related samples are drawn from the same group of individuals under different conditions. For example, if we were comparing the mean of some performance variable between teams' home records and their away records, venue would be the independent variable with home and away being two conditions related to the set of teams included in the study.

No matter which t-test is being used, the t-statistic is determined using a t-distribution which is almost identical to the normal distribution when there are 30 or more degrees of freedom (Hinton, 2004: 66). The way in which the t-statistic is computed differs between the three types of t-tests. This chapter describes the three different t-tests, their assumptions, how they can be done in SPSS and how their results should be reported. The chapter also describes how effect sizes can be determined to support t-test results.

THE ONE-SAMPLE T-TEST

Purpose of the test

The purpose of the one sample t-test is to compare a sample mean against some hypothesized mean in order to make an inference about the population mean in comparison to that hypothesized mean. There are one-tailed and two-tailed versions of the one-sample t-test.

The one-tailed version assumes that any difference between the population mean and the hypothesized mean will be in a particular direction. The two-tailed version of the test will reject the null hypothesis (that the population mean is the same as the hypothesized mean) if the sample mean is outside some confidence limit of the hypothesized mean on either side of it. An effect size for the one-sample t-test is Cohen's d which is given by equation 10.1 where x_{diff} is the mean difference between the sample mean and the hypothesized mean and SD_{diff} is the standard deviation of that difference. A value of d of 0.2 indicates a small effect, 0.5 indicates a medium effect and 0.8 indicates a large effect (Cohen, 1988).

$$d = \overline{x}_{diff}/SD_{diff} \tag{10.1}$$

Assumptions

The assumptions of the one-sample t-tests are that the variable of interest is measured on an interval or ratio scale and is normally distributed. The scale of measurement for a variable can be determined using the definition of the variable and inspection of the values that have been recorded. The assumption of normality is tested using a Kolmogorov–Smirnov test if there are 50 or more values in the combined samples or a Shapiro–Wilk test if there are fewer than 50 values in the combined sample. These tests of normality compare an observed distribution with a theoretically expected distribution. When used to test normality, the observed distribution is compared with a perfect normal distribution of the same mean and standard deviation. If a p value of 0.05 or greater is produced, then the sample is considered to be sufficiently normal.

Example: Indoor rowing performances at national championships

Indoor rowing has become a popular sport in recent years with regional, national, continental and world championships taking place annually. These events include lightweight and heavy-weight men's and women's races for a large range of age groups with the 2010 British Indoor Rowing Championship seeing a new world record for men over 100 years old. Taking part in a national indoor rowing championship involves travelling to the event venue, possibly staying in a hotel the night before if the participant's race is early and, of course, months of hard training. Therefore, we can anticipate that those participating are not doing so just to row a time that they could do in their local gym. So we shall hypothesize that the average entrant in the open lightweight men's 2,000m race will complete the race in under seven minutes (420s). This example will use data from 2010 British Indoor Rowing Championship which are published on the internet (www.concept2.co.uk accessed 14 July 2011). The author is a relative beginner who has clocked 7 minutes 9.4s in his second year of rowing in the 45 to 49-year-old age lightweight group. Therefore, the author wants to be 99 per cent confident that the open men are averaging under seven minutes. We can test this using a one-tailed one-sample t-test.

Table 10.1 shows the finishing times of the 22 athletes in the men's open lightweight 2,000m race. Readers should already be able to see an outlier in the data with the 22nd finisher finishing 33.2s behind the penultimate finisher. This immediately raises the following questions about what we are doing:

Table 10.1 Finishing times for the open lightweight men's 2,000m at the 2010 British Indoor Rowing Championships (mins:s)

Pos	Time	Pos	Time	Pos	Time	Pos	Time
1st	6:26.9	7th	6:34.3	13th	6:46.8	19th	7:09.8
2nd	6:28.9	8th	6:34.9	14th	6:49.3	20th	7:22.8
3rd	6:31.8	9th	6:39.1	15th	6:52.7	21st	7:27.6
4th	6:32.0	10th	6:41.7	16th	7:00.5	22nd	8:00.8
5th	6:32.7	11th	6:42.5	17th	7:01.8		
6th	6:33.8	12th	6:44.9	18th	7:05.1		

1 If we remove the 22nd finisher, we may satisfy the assumptions of the one-sample t-test, but we are also making it easier to achieve a mean significantly below seven minutes. Will the removal of this value increase the real chance of making a Type I Error?

2 Are these 22 values a sample or the whole population? If the scope of the study is restricted to the 2010 British Indoor Rowing Championships open lightweight men's 2,000m, we do not need to use p values at all. If, however, we are using this as a cluster sample of a national championship, can we generalize from the British Indoor Rowing Championship to any national championship?

We will apply the one-sample t-test with and without the 22nd value as an example of how to perform the one-sample t-test. In doing so, we will consider the first of the questions asked above.

SPSS

The performances (total time in s) for the open lightweight men's 2,000m at the 2010 British Indoor Rowing Championships are found in the file 10-indoor_rowing.SAV. Exploring the variable as described in Chapter 7 confirms that the 22nd finishing time is an outlier with a Shapiro–Wilk test showing that the data are not normally distributed ($p = 0.003$). Even when the 22nd value is removed, the Shapiro–Wilk test still shows that the data are not normally distributed ($p = 0.031$). To perform the one sample t-test, we use **Analyse →
Compare Means → One sample t-test** which activates the pop-up window shown in Figure 10.1. 'Time' is transferred into the *Test Variable(s)* area and the *Test Value* is set to 420s (seven minutes exactly).

To set the confidence level, we click on **Options** which activates the pop-up window shown in Figure 10.2. The default one-sample t-test is the two-tailed version. In order to perform a one-tailed test, we need to ensure that we have a confidence limit that we can be 99 percent certain the true mean is less than. Therefore, we actually enter 98 into the *Confidence Interval Percentage* instead of 99. This will set two 0.01 tails at either end of the distribution, but we ignore the lower one to achieve a one-tailed test with α being 0.01.

The output when all 22 times are included is shown in Tables 10.2 and 10.3. Table 10.2 shows that the mean time is over 9s under the seven minute hypothesized mean (410.9s).

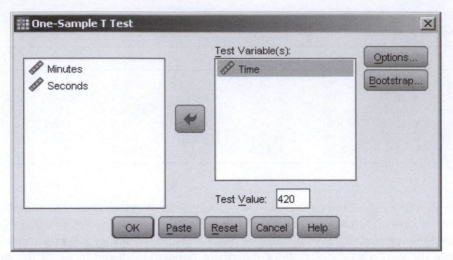

Figure 10.1 One-sample t-test pop-up window in SPSS.

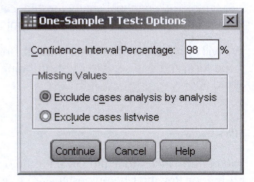

Figure 10.2 Options pop-up window for the one-sample t-test.

However, the value that 99 per cent of sampling distribution of the mean is below 3.4s above the hypothesized mean of 420s. Therefore, the null hypothesis that the true mean is greater than or equal to 420s cannot be rejected. The p-value in Table 10.3 is 0.081 but this is the p-value if we had been using a two-tailed test. With our test being a one-tailed test, the p value is actually 0.041, which is still greater than our decision criteria ($\alpha = 0.01$). Cohen's d is 0.39 (= 9.06 / 23.21), which suggests that the difference of 9.06s between the observed and hypothesized means represents a small to medium effect.

Table 10.2 Descriptive statistics for 22 indoor rowing performances

		One-Sample Statistics		
	N	Mean	Std. Deviation	Std. Error Mean
Time	22	410.941	23.2077	4.9479

174

t-tests

Table 10.3 One-sample t-test output for 22 indoor rowing performances

One-Sample Test

Test Value = 420

	t	df	Sig. (2-tailed)	Mean Difference	98% Confidence Interval of the Difference	
					Lower	Upper
Time	−1.831	21	.081	−9.0591	−21.516	3.398

The SPSS output for the one-sample t-test when the 22nd finisher's time is excluded is shown in Tables 10.4 and 10.5. This reduces the sample mean to 407.6s and 99 per cent of the sampling distribution of the mean is less than 417.3s (2.7s less than the hypothesized mean). This would allow us to reject the null hypothesis and conclude that the average performance was significantly below seven minutes ($p = 0.002$) if we were prepared to overlook the fact that the data are not quite normally distributed. A word should also be said about the p value of 0.004 reported by SPSS. This is technically the p value for a two-tailed test with α set at 0.02. Because we are performing a one-tailed test, we can divide this p value by two giving $p = 0.002$. Cohen's d is 0.70 (= 12.39 / 17.60) which is approaching a large effect.

What many, including the author, would be uncomfortable about is that the excluded value was not an outlier due to measurement error but a genuine finishing time in a 2,000m race that should be included in the study. When the removal of an outlier increases a p value but the result is still significant, we have increased confidence in that significant difference. However, in a case like this where removing a real performance turns a non-significant difference into a significant difference, we should exercise some caution in drawing conclusions.

Table 10.4 Descriptive statistics for 21 indoor rowing performances

One-Sample Statistics

	N	Mean	Std. Deviation	Std. Error Mean
Time	21	407.614	17.6038	3.8415

Table 10.5 One-sample t-test output for 21 indoor rowing performances

One-Sample Test

Test Value = 420

	T	df	Sig. (2-tailed)	Mean Difference	98% Confidence Interval of the Difference	
					Lower	Upper
Time	−3.224	20	.004	−12.3857	−22.097	−2.675

Reporting results

Given the difficulties described above caused by the exclusion of the 22nd finishing time, the results would be reported with some caution using text such as the following:

> The 22 finishing times were not normally distributed (Shapiro–Wilk test: $p = 0.003$) due to the 22nd finishing time being a statistical outlier. Once this value was removed, the data were close to being normally distributed (Shapiro–Wilk test: $p = 0.031$) and so a one-tailed one-sample t-test compared the sample mean of 6 minutes 47.6s with the 7 minute standard that the mean was hypothesized to be under. The time that 99 per cent of the sampling distribution of the mean was less than was 6 minutes 57.3s, meaning that the mean was significantly below the hypothesized value of 7 minutes ($t_{20} = 3.2$, $p = 0.002$, $d = 0.70$). However, it should be noted that if the 22nd finishing time was included in the analysis, 99 per cent of the sampling distribution of the mean would be less than was 7 minutes 3.4s, and the mean of 6 minutes 50.9s would not be significantly below 7 minutes ($t_{21} = 1.8$, $p = 0.041$, $d = 0.39$).

Note in reporting the results, the key figures from the SPSS output are the mean, the upper confidence limit, the t statistic, the degrees of freedom and the p (Sig.) value. In a one-sample t-test, the number of degrees of freedom is one less than our sample size. If we were performing a two-tailed test, we would also use the lower confidence limit. In this example, the p values were divided by two because we were performing a one-tailed test and we required p values of less than 0.01 to conclude a significant difference having set a requirement for 99 per cent confidence.

THE INDEPENDENT SAMPLES T-TEST

Purpose of the test

The purpose of the independent samples t-test is to compare two independent samples in terms of some numerical dependent variable. Independent samples come from different samples of participants that can be of differing sizes. Two sample means could be significantly different or not depending on the difference between the means, the standard deviations of the samples and number of values in the samples. This is illustrated by equation 10.2 where $\bar{x}_1$ and $\bar{x}_2$ represent the two sample means, s_1 and s_2 represent the two standard deviations and n_1 and n_2 represent the two sample sizes. This book is most concerned with the researcher's task in selecting the correct test, performing the test in SPSS, understanding the key results from the SPSS output and presenting the results scientifically. However, the equation of the independent samples t-test has been shown here to allow some discussion of the nature of inferential statistics. Where a t-statistic exceeds some critical value associated with our chosen α level, a significant result will be found. With the numerator of the equation 10.2 being the difference between the means and the denominator involving the pooled standard deviations and sample sizes, it is apparent that significant differences depend on having:

176

1 A large enough difference between the sample means.
2 Small enough standard deviations for the two samples.
3 Large enough sample sizes.

$$t = \frac{\overline{x}_1 - \overline{x}_2}{\sqrt{s^2 \left(\frac{1}{n_1} + \frac{1}{n_2} \right)}} \qquad (10.2)$$

$$s_2 = \frac{(n_1 - 1)s_1^2 + (n_2 - 1)s_2^2}{n_1 + n_2 - 2} \qquad (10.3)$$

The null hypothesis of the independent samples t-test is that there is no difference between the two population means. The t-value is used to determine p which is the probability of a Type I Error. That is p is the probability that we would be making a mistake if we concluded that any difference we see in the sample represented a difference in the wider population. Vincent (2005: 126) described the p value as the probability that two random samples drawn from the same population, distinguished by some hypothesized factor, differing by no more than would be expected by chance. As we can see from equation 10.2, the t-score and hence the p value is determined purely from the sample statistics (sample means, sample standard deviations and numbers of values in the samples). The size of the relevant population of interest could be thousands of people, millions of people or even billions of people. This will have some impact on the probability of making a Type I Error but is not accounted for in the equation for the independent samples t-test. This is also the case with the ANOVA tests to be covered in Chapters 11 and 12. Some readers may not like the sight of equations and feel that this textbook should only cover how to perform the tests in SPSS and how to report the results. However, for this one test, the author believes it is beneficial to consider how the t-statistic is calculated and the limitations of using such tests to make inferences about populations.

The number of degrees of freedom for an independent samples t-test is the combined size of the two samples less 2. Thus, there must be at least three values within the combined samples otherwise there will be no variance in either sample meaning that the t-score cannot be computed using 10.3. Cohen's d can be used as a measure of effect size using equation 10.4 and interpreting values of d the same way as they are interpreted for the one-sample t-test. An alternative to Cohen's d is ω^2 which is the proportion of variance in the dependent variable that is explained by our independent grouping variable (equation 10.5).

$$d = \frac{\overline{x}_1 - \overline{x}_2}{SD} \qquad (10.4)$$

$$\omega^2 = \frac{t^2 - 1}{t^2 + n_1 + n_2 - 1} \qquad (10.5)$$

Assumptions

The author has already expressed some views about the limitations of t-tests based on an inspection of equation 10.2. However, this should not prevent anyone from applying the test and reporting the results according to the test. There are three main assumptions of the independent samples t-tests: the dependent variable should be measured on an interval or ratio scale, the dependent variable should be normally distributed and there should be homogeneity of variances for the dependent variable between samples being compared. The scale of measurement for a variable can be determined using the definition of the variable and inspection of the values that have been recorded. The assumption of normality is tested using a Kolmogorov–Smirnov test if there are 50 or more values in the combined samples or a Shapiro–Wilk test if there are fewer than 50 values. The assumption of homogeneity of variances is tested using Levene's test, which is satisfied if a p value of 0.05 or greater is produced.

Example: A quasi-experimental study on the effectiveness of specific intermittent high intensity training

To illustrate the use of the independent samples t-test, we use a training study for under-14 Gaelic footballers (King and O'Donoghue, 2003). Two groups of 15 participants were tested before and after a training period of 13 weeks. One group of participants performed a once-weekly specific training session which was designed using research findings about the intermittent nature of high intensity activity in Gaelic football. The other group was a control group who performed an alternative skill-based session at a moderate intensity. All participants were tested at the beginning and at the end of the experimental period using a vertical jump test, a 20m sprint test, a multiple sprint test of eight 40m sprints with a 30s recovery (Baker et al., 1993) and a multistage fitness test (Ramsbottom et al., 1988). Precise details of the testing conditions are described in King and O'Donoghue's (2003) paper.

The difference between the pre- and post-test results for each fitness test was determined for each participant. This meant that instead of comparing four means for each fitness test (the pre- and post-tests for the control and training groups), two means would be compared (the pre-post difference for the control and experimental groups). The comparison of the mean pre/post differences between these two groups can be done with an independent samples t-test providing the data satisfy the assumptions of the test.

SPSS

The data for King and O'Donoghue's (2003) investigation is found in the file 10-specific_training_experiment.SAV. We wish to perform an independent samples t-test to compare change in estimated $\dot{V}O_2$ max test performance between the control and training groups. There are three assumptions that need to be tested. First, the dependent variable needs to be measured on an interval or ratio scale. This is confirmed by simply inspecting the definition of the variable and the values recorded. Second, we need to test that the dependent variable is normally distributed. This is done using a Shapiro–Wilk test because we have less than 50 values. The tests of normality are obtained using **Analyse → Descriptive Statistics → Explore** and transferring our dependent variable ('change in estimated $\dot{V}O_2$ max') into the *Dependent*

List. The display option that we choose is plots. It is necessary to click on the **Plots** button to then ask for normality plots and tests to be provided. When we finalize the test (clicking on **Continue** to come out of the **Plots** pop-up window and **OK** to come out of the Explore pop-up window), the output shown in Table 10.6 appears in the output viewer. The p value (Sig.) for the Shapiro–Wilk test is greater than 0.05 and, therefore, the data satisfy the assumption of normality.

The third assumption is equality of variances which is tested using Levene's test, which is an output of the independent samples t-test in SPSS. This means that we have to perform the independent samples t-test in SPSS but might not be able to use the results of the test if the assumption of equal variances is violated. The independent samples t-test is done using **Analyse → Compare Means → Independent Samples t Test**. The Independent t-Test pop-up window appears as shown in Figure 10.3. 'Change in estimated $\dot{V}O_2$ max' is entered into the *Tests Variable(s)* area while 'group' is entered into the *Grouping Variable* area. It is necessary to **Define** our groups as '1' and '2' (Group 1 = '1' and Group 2 = '2'). The values '1' and '2' are already labelled as training group and control group respectively. The independent t-test requires the control variable to be a numerical variable which can then be defined as nominal with the appropriate labels created as we have done here.

Table 10.6 Results of checking the assumption of normality of a pre-/post-change variable

| | Tests of Normality | | | | | |
| | Kolmogorov–Smirnova | | | Shapiro–Wilk | | |
	Statistic	Df	Sig.	Statistic	df	Sig.
Change in estimated $\dot{V}O_2$ max based on multistage fitness test	.126	30	.200*	.940	30	.088

a Lilliefors Significance Correction
* This is a lower bound of the true significance.

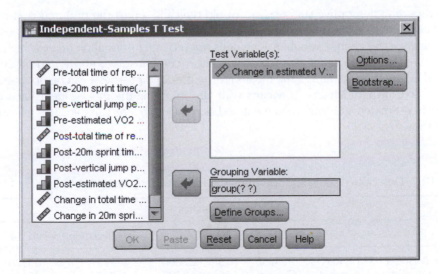

Figure 10.3 Pop-up window for the independent samples t-test.

The four change variables represent the pre- to post- change in the four fitness characteristics being tested. When we click on **OK**, the SPSS output for the Independent samples t-test appears as shown in Tables 10.7 and 10.8. The group statistics table shows that estimated $\dot{V}O_2$ max increased by 2.8 ± 3.4 mL.kg^{-1}.min^{-1} for the training group while it increased by 0.3 ± 0.8 mL.kg^{-1}.min^{-1} for the control group. The independent samples t-test table provided in the output tells us whether there is a significant difference in pre/post-change between the two groups of players. There are three different p (Sig.) values shown. The first p value is actually for Levene's test of homogeneity of variances rather than for the independent t-test. This particular p value ($p = 0.065$) shows that the variances for the dependent variable are not significantly different between the two groups and, therefore, equal variances can be assumed. This means that we use the first row of t-test results rather than the second row. If the assumption of equal variances had been violated, we could still perform an independent t-test. As we see in the second row of results, the number degrees of freedom can be reduced to cope with the violation of the assumption of homogeneity of variances (Wright, 2002: 71–3). This reduces the chance of a Type I error by making it look as though there are fewer participants in the study. This is not an issue in the current example, because the dependent variable is normally distributed and the variances of the two samples are similar. The p value reported for the independent t-test in this example is 0.022 which is less than the value of 0.05 used by King and O'Donoghue (2003) to indicate significance.

Cohen's d can be computed for the independent samples t-test using equation 10.4 but with the standard deviation being the pooled standard deviation of the two groups. In an experiment, or quasi-experimental study such as the current example, the standard deviation of the control group can be used in equation 10.4 (Vincent, 1999: 134). This gives an effect size, d, of 1.35 (= 2.42 / 1.79), which is a large effect. The pooled standard deviation is 2.73 according to equation 10.3, which still gives a large effect (d = 0.89) if used in equation 10.4.

When reporting descriptive statistics, we would typically report the pre- and post-test values for each group rather than just the amount of change between the pre- and post-tests. This is because the change variable alone does not provide information on how high or low the tests were. This information is essential to allow comparison to norms and other values reported in scientific literature. The means for the pre- and post-tests can be obtained using **Analyse → Compare Means → Means** with 'group' being the *Independent Variable* and 'change in estimated $\dot{V}O_2$ max' being entered as the *Dependent Variable*. This provides the output shown in Table 10.9.

Table 10.7 Descriptive statistics supporting the independent samples t-test

	Group Statistics				
	group	*N*	*Mean*	*Std. Deviation*	*Std. Error Mean*
Change in estimated $\dot{V}O_2$ max based on multistage fitness test	Training	15	2.7600	3.42278	.88376
	Control	15	.3400	1.78838	.46176

Table 10.8 Independent samples t-test results

Independent Samples Test

		Levene's Test for Equality of Variances		t-test for Equality of Means						95% Confidence Interval of the Difference	
		F	Sig.	t	df	Sig. (2-tailed)	Mean Difference	Std. Error Difference		Lower	Upper
Change in estimated $\dot{V}O_2$ max based on multistage fitness test	Equal variances assumed	3.692	.065	2.427	28	.022	2.42000	.99712		.37749	4.46251
	Equal variances not assumed			2.427	21.114	.024	2.42000	.99712		.34706	4.49294

Table 10.9 Individual group descriptive statistics for the pre- and post-tests

		Report	
Group		Pre-estimated $\dot{V}O_2$ max based on multistage fitness test	Post-estimated $\dot{V}O_2$ max based on multistage fitness test
Training	Mean	41.2133	43.9733
	N	15	15
	Std. Deviation	4.56819	5.26531
Control	Mean	39.5400	39.8800
	N	15	15
	Std. Deviation	4.62289	5.49379
Total	Mean	40.3767	41.9267
	N	30	30
	Std. Deviation	4.59518	5.68221

Reporting results

The results of this test could be reported as follows with a clustered bar graph being used to illustrate the means and error bars being used to show the standard deviations:

> Figure 10.4 shows that there was a greater pre/post-increase in estimated $\dot{V}O_2$ max for the training group than for the control group. An independent samples t-test found the difference in pre/post-increase in estimated $\dot{V}O_2$ max to be significantly different between the two groups (t_{28} = 2.4, p = 0.022, d = 1.35).

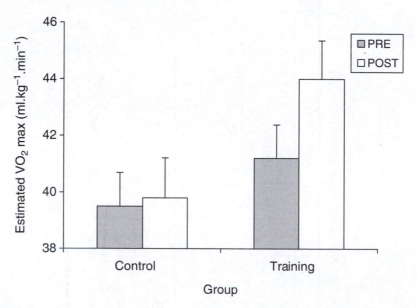

Figure 10.4 Presentation of descriptive results relating to the independent t-test.

Some choose to use error bars to represent standard error of the mean (SEM) rather standard deviations. The variation in practice means that researchers reporting their results should be transparent about what error bars within bar graphs represent.

THE PAIRED SAMPLES T-TEST

Purpose of the test

The paired samples t-test is used to compare the means of two samples related to the same group of participants. For example, we might wish to compare some variable for a set of subjects between two different times (night and day, or pre-study and post-study, or during childhood and when adults) or between two different conditions (for example, when under experimental treatment conditions and when not, or when playing at home and when playing away, or between two different measurement methods of the variable). In these examples, the independent variable is the time or condition variable of interest while the dependent variable is the numerical variable whose mean is being compared between the two samples. Because the two samples are related to the same group of participants, the two samples have the same number of values. Paired t-tests can also be used to compare different groups of participants where their values are related. For example, if we wished to know if some numerical variable was significantly different between the winning and losing teams within football matches, we could use a paired t-test if its the assumptions were satisfied. This is because the performances of the two teams in a football match are clearly related through opposition quality effect; they are not independent samples.

The independent samples t-test is significant if the difference between the two sample means is large enough in relation to the standard deviations of the samples and if there is a large enough number of values for generalization. Paired samples t-tests are different because significant differences can be revealed even with large standard deviations within the samples. Consider a set of people participating in an exercise programme with the objective of losing weight. The participants could have a wide range of body masses before and after the exercise programme but with all successfully losing 0.5 to 1.0 kg. Where an exercise programme results in very consistent weight loss values for all participants, it would be ridiculous if the statistical test failed to conclude this due to the high standard deviations of the pre- and post-measurements of body mass. Therefore, the paired samples t-test relates the mean of the pre/post-change to the standard deviation of the pre/post-change. When computing Cohen's d, the standard deviation in equation 10.4 is the standard deviation of the pre/post differences for the dependent variable.

Assumptions

The paired samples t-test has the same assumptions as the independent samples t-test except the samples are related rather than independent.

Example: Effect of instructional and motivational self-talk on sit-up performance

Theodorakis *et al.* (2000) found that instructional self-talk helped improve performance of motor tasks. However, the participants used the same self-talk words provided by the researchers rather than using their own self-talk words. It is possible that the participants own words might be more meaningful to them resulting in better task performance. The purpose of this example investigation is to compare a gross movement (sit-up performance) between two conditions: when instructional self-talk is used and when motivational self-talk is used. In each case, participants use their own self-talk words. Sit-up performance is measured by counting the number of completed sit-ups that are performed within 60s. There are 47 participants in the study. To combat any temporal effects on sit-up performance, a crossover design is used with 24 participants performing the sit-up test following motivational self-talk first and then performing the sit-up test following instructional self-talk. The other 23 participants do the sit-up task twice but with the self-talk conditions being the other way round. Each group of participants has a 15-minute recovery between the two sit-up tests. Given that this experiment produces two samples related to the same set of participants, a paired samples t-test is used to compare sit-up performance between the two self-talk conditions.

SPSS

When performing a paired samples t-test, we conceptually have an independent variable (for example, self-talk condition) and a dependent variable (for example, the number of sit-ups that can be completed in 60s). However, when performing a paired samples t-test in SPSS, it is necessary to have two versions of the dependent variable, one under each condition of interest. Consider the file 10-self_talk.SAV; this datasheet does not contain a variable for self-talk condition but contains two versions of the dependent variable (sit-up task with motivational self-talk and sit-up task with instructional self-talk). This is because the SPSS datasheet is a variable by case matrix where there is one row per participant with all data for that participant including repeated measures of any variable placed in that row of the datasheet.

Placing repeated measurements of a variable in different columns makes them look like different variables and the SPSS package cannot be expected to understand our language and variable names when the datasheet is created. At the point at which statistical tests are performed, it is possible to have options to test the assumptions of those tests. In version 18.0 of SPSS, the paired samples t-test does not contain an option to perform Levene's test of homogeneity of variances or a test of normality. There are different ways in which the assumptions of the paired samples t-test can be tested. A colleague of the author's, Laurence Llewelyn, used a separate copy of the datasheet for the purpose of testing the assumptions. An example of this is the file 10-self_talk_assumptions.SAV where we essentially trick SPSS into thinking that there are 94 participants rather than 47. The 47 values for sit-up task with instructional self-talk have been copied and pasted below the 47 values for sit-up task with instructional self-talk. A Kolmogorov–Smirnov test can be applied to this set of 94 values giving the results shown in Table 10.10. This indicates that the data are normally distributed (p > 0.05).

Table 10.10 Results of normality test prior to the paired samples t-test

	Kolmogorov–Smirnova			Shapiro–Wilk		
Tests of Normality	Statistic	Df	Sig.	Statistic	df	Sig.
Sit Up Task with Motivational Self Talk	.075	94	.200*	.981	94	.189

a Lilliefors Significance Correction
* This is a lower bound of the true significance.

Some may prefer not to merge the two versions of the sit-up variable when testing for normality due to non-independence in the data. An alternative decision would be to test the variable separately for the two conditions of interest. However, when testing for homogeneity of variance, it will still be necessary to produce a Llewelyn-style copy of the file if versions of SPSS up to and including version 18 are being used. Levene's test can be done by creating a dummy grouping variable showing two groups of 47 participants and then performing an independent samples t-test on this variable just so that we can obtain the results of Levene's test as shown in Table 10.11. In this example, we have approximately equal variances between the pair of related samples (p = 0.466). Note that if this assumption had been violated, SPSS would not adjust the degrees of freedom within the paired samples t-test to cope with this and so we would not be able to use the test validly.

Once the assumptions have been tested, we can perform the paired samples t-test using **Analyse → Compare Means → Paired Samples t test**. The related variables being tested are transferred into the *Paired Variables* area of the pop-up window as shown in Figure 10.5.

Table 10.11 An independent t-test can be used to 'trick' SPSS into doing a homogeneity of variance test for paired samples

Independent Samples Test		Levene's Test for Equality of Variances		t-test for Equality of Means					95% Confidence Interval of the Difference	
		F	Sig.	T	df	Sig. (2-tailed)	Mean Difference	Std. Error Difference	Lower	Upper
Sit Up Task with Motivational Self Talk	Equal variances assumed	.536	.466	−.595	89	.554	−1.315	2.212	−5.710	3.079
	Equal variances not assumed			−.597	88.61	.552	−1.315	2.202	−5.691	3.060

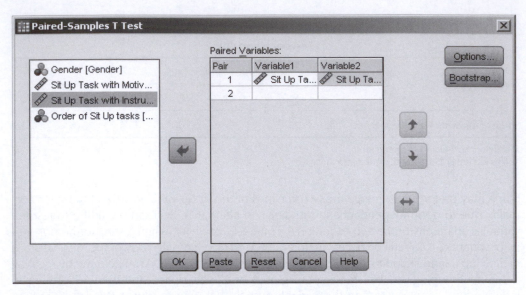

Figure 10.5 Pop-up window for the paired samples t-test.

The output produced for the paired samples t-test contains three tables as shown in Tables 10.12 to 10.14. Table 10.12 provides the all-important descriptive statistics. Table 10.13 shows us that there is a strong positive correlation for sit-up performance between the two self-talk conditions (r = 0.945). Table 10.14 shows the p value (Sig.) for the test in the right most column. In this case, there is no significant difference in sit-up performance between motivational or instructional self-talk conditions (p = 0.226). Cohen's d is 0.18 (=0.66/3.69)

Table 10.12 Descriptive statistics supporting the paired samples t-test

		Mean	N	Std. Deviation	Std. Error Mean
	Paired Samples Statistics				
Pair 1	Sit Up Task with Motivational Self Talk	44.38	47	10.301	1.503
	Sit Up Task with Instructional Self Talk	45.04	47	11.196	1.633

Table 10.13 Correlation between two related samples

		N	Correlation	Sig.
	Paired Samples Correlations			
Pair 1	Sit Up Task with Motivational Self Talk & Sit Up Task with Instructional Self Talk	47	.945	.000

Table 10.14 Paired samples t-test results

| | | Paired Differences | | | | | | | |
| | | Mean | Std. Deviation | Std. Error Mean | 95% Confidence Interval of the Difference | | t | df | Sig. (2-tailed) |
					Lower	Upper			
Pair 1	Sit Up Task with Motivational Self Talk – Sit Up Task with Instructional Self Talk	−.660	3.685	.537	−1.741	.422	−1.227	46	.226

meaning that the difference of 0.66 sit-ups between the two self-talk conditions is just short of a small effect.

It is worth mentioning that when the paired t-test is used in a crossover design like this, the paired variables are for the conditions under which the dependent variable was measured rather than the order in which they were measured. If we wished to examine the difference between the two conditions for those who did motivational self-talk first and instructional self-talk second, we could use the fourth variable in the 10-self_talk.SAV file which tells us which condition was done first by each participant. We can use **Analyse → Compare Means → Means** to obtain separate descriptive statistics for those doing different self-talk conditions first. The *Independent Variable* is 'Order of Sit Up tasks' and the *Dependent variable(s)* are 'Sit Up Task with Motivational Self-Talk' and 'Sit Up Task with Instructional Self-Talk'. The results in Table 10.15 show that both groups performed better when using instructional

Table 10.15 Descriptive statistics when using a cross-over design

	Report		
Order of Sit Up tasks		*Sit Up Task with Motivational Self Talk*	*Sit Up Task with Instructional Self Talk*
motivational first	Mean	43.33	44.25
	N	24	24
	Std. Deviation	10.507	11.946
instructional first	Mean	45.48	45.87
	N	23	23
	Std. Deviation	10.197	10.559
Total	Mean	44.38	45.04
	N	47	47
	Std. Deviation	10.301	11.196

self-talk irrespective of the order in which the self-talk preparations were done. In Chapter 12, we will discuss the mixed ANOVA which can include order of testing as a factor during inferential testing.

Reporting results

The results could be presented in graphical form (Figure 10.6), tabular form or in paragraph text as follows:

The 45.0 ± 11.2 (mean ± SD) sits performed in 60s when instructional self-talk was used was not significantly greater than the 44.4 ± 10.3 done when motivational self-talk was used (t_{46} = 1.2, p > 0.05, d = 0.18).

SUMMARY

This chapter has covered the one sample t-test, independent samples t-test and paired samples t-test. These are all used to compare the means of some interval or ratio scale dependent variable. The one sample t-test is used to compare a sample mean with a hypothesized mean for a population. The independent samples t-test is used to compare the mean of two independent groups while the paired samples t-test is used to compare the means for the same group under different conditions.

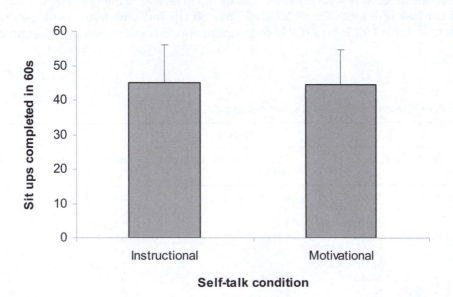

Figure 10.6 Sit ups performed in 60s under using different methods of self-talk.

EXERCISES

Exercise 10.1. Specific training experiment

Using the other three change variables (20m sprint, total time for the multiple agility run test and vertical jump performance) performed by the Gaelic footballers in the file ex10.1-specific_training_experiment. SAV do the following:

a) Determine if they satisfy the assumption of normality.
b) For those change variables that are normally distributed, compare the control and training groups using independent samples t-tests reporting the appropriate descriptive and inferential statistics.

Exercise 10.2. Comparing the percentage of points won in tennis when the first serve is in and when a second serve is required

Using the file ex10.2-tennis.SAV, use a paired samples t-test to determine if the percentage of points won is similar or different between points where the first serve is in and points where a second serve is required. Use Llewelyn's approach to test the assumptions of normality and homogeneity of variance. In this exercise use **Data → Split File** in order to perform the paired t-test for each combination of gender and surface ($2 \times 4 = 8$ groups).

Exercise 10.3. Dominant vs non-dominant Y balance test

Using the file ex10.3-Ybalance-test.SAV, use a paired samples t-test to determine if there is a difference in test results for the Y-balance test when standing on the dominant and non-dominant foot. Use Llewelyn's approach to test the assumptions of normality and homogeneity of variance.

PROJECT EXERCISE

Exercise 10.4. Gender effect on indoor rowing strategy

Concept2 produce row ergometers that are used in indoor rowing competitions. The results of some competitions together with split times are published on the Concept2 website (www.con-cept2.co.uk). The purpose of this project exercise is to compare the rowing strategy of male and female competitors in these events in 2,000m. The indicator of strategy is the percentage of 2,000m time that is taken to cover the final 500m. This indicator of strategy is relative to finishing time so meaningful comparisons can be made between male and female rowers even though their finishing times may be different. A value of 25 per cent indicates that the final 500m is done at the same average speed as the first 1,500m. A value of less than 25 per cent indicates that the rower does the last 500m at a faster speed than the first 1,500m while a value of greater than 25 per cent indicates that the rower does the last 500m at a slower speed than the first 1,500m. Using a 2,000m event with male and female competitors in the same age and weight category, where 500m split times are provided, determine the percentage of the total time taken for the last 500m and compare this between the top 20 male and top 20 female competitors. This project could also be done to compare the 2,000m rowing strategy of rowers from two different age groups or two different weight divisions.

Exercise 10.5. Home advantage in soccer

Examine a football league table at the end of the season and based on three points for a win and one point for a draw, work out the number of points each team earned at home and away from home. Enter the home and away points into SPSS and perform a paired samples t-test to determine whether the number of points earned at home and away from home are significantly different.

CHAPTER 11

ANALYSIS OF VARIANCES

INTRODUCTION

Analysis of variances (ANOVA) tests are used to compare three or more samples in terms of some numerical dependent variable. ANOVA tests will determine whether or not there are any significant differences between the samples. An immediate question that comes to mind is why can a series of t-tests not be used to compare the samples? After all, if there are differences between the samples, then we will wish to compare each pair of samples anyway. Indeed, in this age of computing power, a series of t-tests can be conducted relatively quickly even if there are multiple samples being compared. There are two main reasons for using a single ANOVA test to compare three or more samples. The first reason is that if there are no significant differences between the samples, then this one single ANOVA test will determine this and the result can be expressed concisely. The second reason for using an ANOVA test is to avoid inflating the probability of making a Type I Error when comparing multiple pairs of samples using t-tests. Imagine if we were comparing the six different socio-economic groups with respect to some numerical variable. There are 15 different pairs of groups here. If the probability of making a Type I Error is 0.05 (one chance in 20 of making a mistake when concluding there is a difference) each time we compare a pair of socio-economic groups, then our chance of avoiding making a Type I Error somewhere decreases to 0.95^{15} which is 0.463. Therefore, the chance of making a Type I Error somewhere in our series of 15 t-tests has inflated to 0.537. Another way of thinking about multiple pairwise comparisons is a person trying to cross a river using 20 stepping stones between the two banks. Each time the person steps on a stepping stone, there is one chance in 20 that the person will fall in the river. So with 20 stepping stones to be used, we can expect that the person will probably fall in the river once when they try to cross. Therefore, an ANOVA test provides a means of restricting the chances of making a Type I Error. This is because it is a single test which determines whether or not there are differences between the samples.

There are three types of ANOVA test covered in this chapter: the one-way ANOVA, the repeated measures ANOVA and the analysis of covariance test (ANCOVA). The one-way ANOVA is used to compare three or more independent samples while the repeated measures ANOVA is used to compare three or more related samples. It is worth reminding ourselves that independent samples are different groups of participants; for example, soccer players of different positional roles. Related samples are samples taken from the same group of participants but under different conditions or at different times. The ANCOVA test is used to compare samples in terms of some numerical dependent variable adjusting it for the known influence of some covariate that cannot be experimentally controlled.

If any of these ANOVA tests reveals significant differences between the samples, then 'post hoc' tests are used to compare the different pairs of samples of interest. This chapter commences with the one-way ANOVA but then talks about the general linear model (GLM) that was mentioned in Chapter 9 before covering the repeated measures ANOVA and ANCOVA tests. The GLM is covered after the one-way ANOVA so that readers will have an understanding of the concept of analysis of variances testing before underlying GLM is discussed.

THE ONE-WAY ANOVA TEST

Purpose of the test

The one-way ANOVA test is used to compare three or more independent samples in terms of some numerical variable. The samples are distinguished by some categorical variable which is a factor measured at three or more levels. For example, we may have a variable 'positional group' which is a factor classifying soccer players into four different positions: goalkeepers, defenders, midfielders and forwards. This factor has four levels because it has four possible values. We may wish to investigate the influence of positional group on height. This can be done using a one-way ANOVA test. ANOVA tests use an F distribution and determine a single significance level for the differences between the three or more groups being compared. The F ratio is the ratio of the between-samples variance (MSB) to the within-samples variance (MSW). This division of variances is where the name 'analysis of variances' comes from. An F-ratio of one or less can never be significant because it indicates that the differences between subjects within different samples are smaller than the differences we see between subjects within the same samples. The F-ratio has to be sufficiently above one for the number of samples being compared and the number of participants in the study for the ANOVA test to deem the differences between samples to be significant. This is why there are two degree of freedom values used with ANOVA tests. In the one-way ANOVA, the first degree of freedom value is for our factor of interest and would be one less than the number of levels of the factor. For example, if we had K positional roles, there would be $K - 1$ degrees of freedom for this factor. The other degree of freedom value is the error degrees of freedom and is the difference between the number of participants, n, and the number of samples, K. Therefore, our two degrees of freedom values in a one-way ANOVA test are $K - 1$ and $n - K$.

If there are differences between the independent samples compared by a one-way ANOVA test, then it is desirable to investigate differences between individual pairs of groups. This can be done by employing post hoc tests; there are many different types of post hoc tests that can be used with different disciplines of sport and exercise science preferring different post hoc tests. The least significant difference test (LSD) is a very liberal post hoc procedure which uses an α level of 0.05 for each pairwise comparison. Bonferroni-adjusted, post hoc tests use an α level of 0.05 divided by the number of pairwise comparisons to be made. This avoids inflating the probability of making a Type I Error during the post hoc tests. Therefore, we achieve more significant differences between pairs of samples when using LSD post hoc tests than when using Bonferroni-adjusted, post hoc tests. A lot of ANOVA tests performed in sports psychology research use Tukey's post hoc procedure. Readers are encouraged to examine journal papers in their own discipline of sport and exercise science to see which post hoc tests are preferred by the leading researchers in those disciplines.

Assumptions

The one-way ANOVA test is a parametric procedure and should only be used where data satisfy the assumptions of the test. There are four main assumptions of the one-way ANOVA test. The first assumption is that the dependent values are independent and taken from a random sample (Newell *et al.*, 2010: 183). Second, the dependent variable should be measured on an interval or ratio scale. This can be tested by simply inspecting the values of the variable as well as the definition of the variable and how it is measured. The third assumption is that the variable must be normally distributed (Vincent, 1999: 152). This is tested using either the Shapiro–Wilk test, if there are fewer than 50 participants, or the Kolmogorov–Smirnov test if there are 50 or more participants. The fourth assumption is that the variance of the dependent variable should be similar between the different samples being compared. This can be tested using Levene's test of homogeneity of variances. If the data do not satisfy these assumptions, then values should be transformed so as the assumptions are satisfied (Nevill, 2000; Tabachnick and Fidell, 2007: 86–8) or the non-parametric Kruskal–Wallis H test should be used instead.

Example: Dietary intake of prepubescent female aesthetic athletes

The example used to illustrate the one-way ANOVA test is based on the research of Soric *et al.* (2008) into energy intake of four different groups of prepubescent female aesthetic athletes. Readers are encouraged to examine the paper to see how the data analysis methods and statistical results are reported. The data used in this example are fictitious and use a subset of the dependent variables used by Soric *et al.* (2008). In the current example, there are four groups of prepubescent female aesthetic athletes: controls, artistic gymnasts, rhythmic gymnasts and ballet dancers. These groups are compared in terms of their daily energy intake which is estimated using a quantitative food frequency questionnaire.

SPSS

For this example, we will use the file '11-food_freq_quest.SAV'. The dependent variable of interest (energy intake) is measured in kcals and inspection of the values show that they are on a ratio scale. When we explore the variable in SPSS, asking for normality plots and tests, we find that the variable is normally distributed (p > 0.05) according to the Shapiro–Wilk test which we would use because there are fewer than 50 subjects (n = 42). To test whether the variance of this variable are similar between the four groups, we actually have to perform the one-way ANOVA test asking for Levene's test to be done as an option.

193

To do a one-way ANOVA in SPSS we use **Analyse** → **Compare Means** → **One-Way ANOVA** which causes the one-way ANOVA pop-up, shown in Figure 11.1, to appear. The variable 'Group' is our independent *Factor* that we have hypothesized as influencing 'Energy intake', which is our dependent variable that we transfer into the *Dependent List*. We use the **Options** button to provide a further pop-up window shown in Figure 11.2.

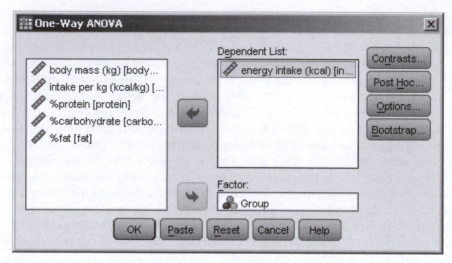

Figure 11.1 Pop-up window for the one-way ANOVA

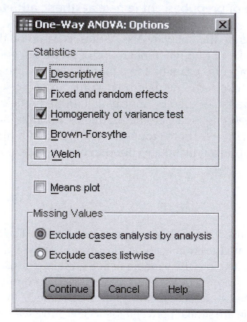

Figure 11.2 Options for the one-way ANOVA test

194

analysis of variances

Place a tick in Descriptive statistics and in Homogeneity of variance test before clicking on **Continue** to close the Options pop-up window. Now click on the **Post Hoc** tests button to activate the Post Hoc pop-up window shown in Figure 11.3. As you can see, there are 14 different post hoc tests that can be used. We will use Bonferroni-adjusted, post hoc tests but we will also ask for LSD post hoc tests so that we can make a comparison between the two.

Tables 11.1 to 11.4 show the output of the one-way ANOVA test. Note how the SPSS output in Tables 11.1 to 11.4 is verbose in comparison to how the results are expressed above or how they might be expressed with the aid of summary tables or charts. The key to using SPSS successfully is knowing what to look for and extracting the most important statistics from the output provided. As we can see in Table 11.2, Levene's test has been passed (the sig (p) value of .872 is well above .05). The important details of the output for the one-way ANOVA are the F-ratio, the between groups and within groups degrees of freedom as well as the significance level. A t-test only requires a single value for degrees of freedom because we know we are comparing two samples. However, with an ANOVA test we need to represent the number of samples that are being compared as well as the total number of subjects. In a one-way ANOVA, the first degree of freedom value is one less than the number of groups (in this example $4 - 1 = 3$). The second degree of freedom is the number of participants minus the number of groups (in this example $42 - 4 = 38$). The one-way ANOVA test expresses the ratio of the variance between different groups and the variance within groups. The F-ratio shown in Table 11.3 is 3.453 which we can round to 3.5. The p value (Sig.) associated with

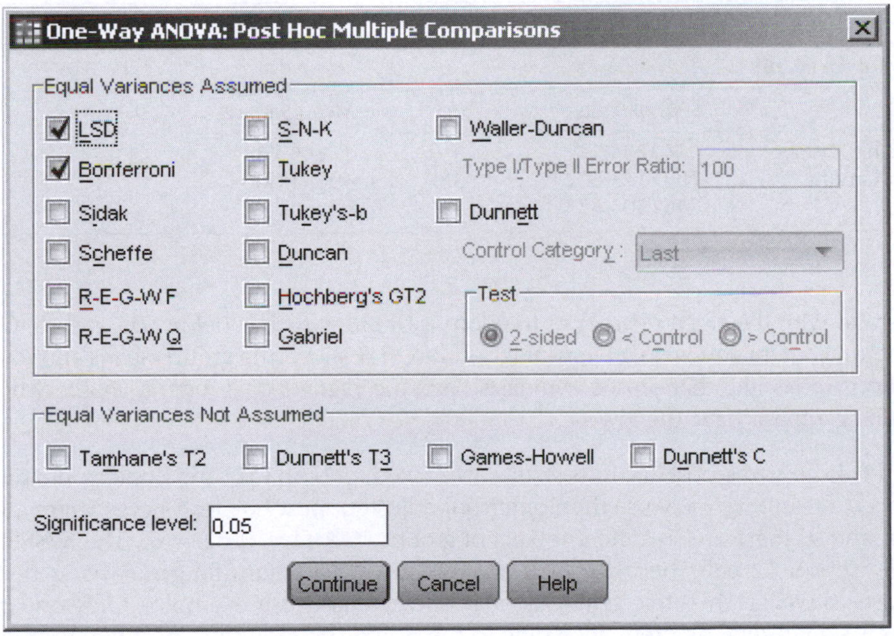

Figure 11.3 Post hoc test selection

Table 11.1 SPSS output for the one-way ANOVA test (Descriptive Statistics)

Descriptives

energy intake (kcal)

	N	Mean	Std. Deviation	Std. Error	95% Confidence Interval for Mean		Minimum	Maximum
					Lower Bound	Upper Bound		
Control	11	1854.82	198.540	59.862	1721.44	1988.20	1514	2114
Artistic gymnast	7	1947.86	232.330	87.812	1732.99	2162.73	1687	2398
Rhythmic gymnast	13	1647.85	254.238	70.513	1494.21	1801.48	1201	2127
Ballet dancer	11	1727.18	198.353	59.806	1593.93	1860.44	1538	2162
Total	42	1772.83	242.019	37.344	1697.41	1848.25	1201	2398

Table 11.2 SPSS output for the one-way ANOVA test (Levene's test)

Test of Homogeneity of Variances

energy intake (kcal)

Levene Statistic	df1	df2	Sig.
.235	3	38	.872

Table 11.3 SPSS output for the one-way ANOVA test (ANOVA results)

ANOVA

energy intake (kcal)

	Sum of Squares	df	Mean Square	F	Sig.
Between Groups	514378.011	3	171459.337	3.453	.026
Within Groups	1887125.822	38	49661.206		
Total	2401503.833	41			

this F-ratio with the given degrees of freedom is 0.026, which is below .05 and so 'Group' has a significant influence on energy intake. Table 11.1 shows the group means and standard deviations indicating that artistic gymnasts have the highest daily energy intake while the rhythmic gymnasts have the lowest.

Table 11.4 shows the post hoc tests results. We asked for both LSD and Bonferroni-adjusted, post hoc tests but will only use the Bonferroni-adjusted, post hoc tests here. There are four groups and so there are six different pairs of groups (4 × (4 − 1) / 2 = 6). The SPSS output actually shows 12 pairs because each pair is shown twice (artistic gymnasts vs rhythmic gymnasts as well as rhythmic gymnasts vs artistic gymnasts, for example). Only one pair of groups is significantly different according to the Bonferroni-adjusted, post hoc tests (artistic gymnasts vs rhythmic gymnasts, p = .040).

analysis of variances

Table 11.4 SPSS output for the one-way ANOVA test (Post hoc test results)

Dependent Variable: energy intake (kcal)

Multiple Comparisons

	(I) Group	(J) Group	Mean Difference (I-J)	Std. Error	Sig.	95% Confidence Interval	
						Lower Bound	Upper Bound
LSD	Control	Artistic gymnast	−93.039	107.746	.393	−311.16	125.08
		Rhythmic gymnast	206.972*	91.295	.029	22.16	391.79
		Ballet dancer	127.636	95.023	.187	−64.73	320.00
	Artistic gymnast	Control	93.039	107.746	.393	−125.08	311.16
		Rhythmic gymnast	300.011*	104.473	.007	88.52	511.50
		Ballet dancer	220.675*	107.746	.048	2.56	438.79
	Rhythmic gymnast	Control	−206.972*	91.295	.029	−391.79	−22.16
		Artistic gymnast	−300.011*	104.473	.007	−511.50	−88.52
		Ballet dancer	−79.336	91.295	.390	−264.15	105.48
	Ballet dancer	Control	−127.636	95.023	.187	−320.00	64.73
		Artistic gymnast	−220.675*	107.746	.048	−438.79	−2.56
		Rhythmic gymnast	79.336	91.295	.390	−105.48	264.15
Bonferroni	Control	Artistic gymnast	−93.039	107.746	1.000	−392.95	206.87
		Rhythmic gymnast	206.972	91.295	.175	−47.14	461.09
		Ballet dancer	127.636	95.023	1.000	−136.86	392.13
	Artistic gymnast	Control	93.039	107.746	1.000	−206.87	392.95
		Rhythmic gymnast	300.011*	104.473	.040	9.21	590.81
		Ballet dancer	220.675	107.746	.285	−79.23	520.58
	Rhythmic gymnast	Control	−206.972	91.295	.175	−461.09	47.14
		Artistic gymnast	−300.011*	104.473	.040	−590.81	−9.21
		Ballet dancer	−79.336	91.295	1.000	−333.45	174.78
	Ballet dancer	Control	−127.636	95.023	1.000	−392.13	136.86
		Artistic gymnast	−220.675	107.746	.285	−520.58	79.23
		Rhythmic gymnast	79.336	91.295	1.000	−174.78	333.45

* The mean difference is significant at the 0.05 level.

Presenting results

We might report the results of the ANOVA tests as follows or use a bar graph or a summary table of results. The effect size reported in these results (partial η^2) will be described in the General Linear Model section of this chapter.

> A one-way ANOVA revealed significant differences between the different types of prepubescent female aesthetic athletes ($F_{3,38} = 3.4$, $p = .026$, partial $\eta^2 = 0.214$) with Bonferroni-adjusted, post hoc tests revealing that the (mean$\pm$SD) 1948$\pm$232 kcal for artistic gymnasts was significantly greater than the 1648$\pm$254 kcal for rhythmic gymnasts ($p = .040$). There was no significant difference between any other pair of athlete groups. The energy intake for the ballet dancers and control subjects was 1727$\pm$198 kcal and 1855$\pm$199 respectively.

The impact of using Bonferroni-adjusted, post hoc tests is that only one pair of groups was found to be significantly different compared to three pairs of groups if LSD tests had been used: artistic vs rhythmic gymnasts ($p = .007$), artistic gymnasts vs ballet dancers ($p = .048$) and rhythmic gymnasts vs controls ($p = .029$).

Bonferroni-adjusted, post hoc tests

Closer inspection of the p (sig) values produced for the Bonferroni-adjusted, post hoc tests reveals that these are simply the p values from the least significant difference (LSD) tests multiplied by the number of pairwise comparisons being made (six). This can actually lead to p values greater than 1.0, which is not possible for a probability. Therefore, SPSS imposes an upper limit of 1.000 on the p values for Bonferroni-adjusted, post hoc tests. An alternative method of determining the α value is to determine what p value for each post hoc test would lead to an overall probability of making a Type I Error somewhere in the post hoc tests on .05. If we have K samples, then the number of pairs of samples L is given by equation 11.1.

$$L = K \times (K\text{-}1) / 2 \tag{11.1}$$

We wish the experiment-wise α level to be 0.05 but if we use $\alpha = 0.05$ for each of the L pairwise tests, then the maximum allowable probability of making a Type I Error somewhere during the post hoc tests will inflate beyond 0.05. Where $\alpha = 0.05$ for each post hoc test, then our minimum allowable probability for avoiding a Type I Error is 0.95. This means that the minimum allowable probability for avoiding a Type I Error in the post hoc tests together decreases to 0.95^L and the probability of making a Type I Error somewhere in the post hoc tests inflates to $1 - 0.95^L$. For example, if we had four samples (K=4) then we would have six pairs of samples (L=6) and the maximum allowable probability of making a Type I Error somewhere in the post hoc tests would inflate to $1 - 0.95^6 = 1 - 0.74 = 0.26$. We wish to use a maximum allowable probability of making a Type I Error of 0.05 and, therefore, need to know what α value to use within the post hoc tests. Equation 11.2 can be arranged as equation 11.3 to help determine the α value to use.

$$1 - (1\text{-}\alpha)^L = 0.05 \tag{11.2}$$

$$0.95 = (1-\alpha)^L \tag{11.3}$$

In equation 11.4, we take the natural logarithm of both sides of equation 11.3. The equation is then rearranged as shown in equation 11.5 before the exponential of each side is taken in 11.6. Equation 11.7 is then produced as an expression of α in terms of the number of pairwise comparisons, L, to be made.

$$\ln(0.95) = L \ln(1-\alpha) \tag{11.4}$$

$$\ln(1-\alpha) = \ln(0.95)/L \tag{11.5}$$

$$1-\alpha = e^{\ln(0.95)/L} \tag{11.6}$$

$$\alpha = 1 - e^{\ln(0.95)/L} \tag{11.7}$$

In our example, we have K = 4 groups meaning that there are L = K × (K-1) / 2 = 6 pairs of groups. Entering this value of L into equation 11.7 shows that an α level of .0085 for each pairwise comparison will restrict the probability of making a Type I Error within the full set of six post hoc tests to .05. When this critical p value of .0085 is used to classify the values from the LSD post hoc tests, we find that only one pair of groups are significantly different: the rhythmic gymnasts and the artistic gymnasts (p = .007 < .0085).

The General Linear Model

Before proceeding to the repeated measures ANOVA test, it is worth discussing the General Linear Model on which linear regression, analysis of variance and analysis of covariance are based (Rutherford, 2001: 3–4). It was mentioned in Chapter 9 that multiple linear regression could use independent variables that were numerical or dichotomous variables. The file 11-food_freq_quest.SAV contains three dichotomous variables: 'Control', 'Artistic' and 'Rhythmic'. For each of these variables, a participant will have a value of one if they are that type of athlete or zero if they are not. We do not need a dichotomous variable for ballet dancers because any athletes who are not members of the other three groups will be ballet dancers. When we perform a multiple linear regression (**Analyse → Regression → Linear**) including 'Energy intake' as *Dependent* variable and 'Control', 'Artistic' and 'Rhythmic' as the *Independent(s)*, we obtain the output shown in Tables 11.5 to 11.7. The coefficient for the constant in Table 11.7 is exactly equal to the mean for the ballet dancers shown in the Table 11.1. The coefficients for the three dichotomous variables ('Control', 'Artistic' and 'Rhythmic') are the differences between the means shown for these groups in Table 11.1 and the mean shown for ballet dancers. The F-ratio and associated p value shown in Table 11.6 are exactly the same as those shown in Table 11.3. Basically, the one-way ANOVA test uses the General Linear Model (GLM) and the predicted daily energy intake, DEI, for a participant can be determined using equation 11.8. The predicted energy intake is the mean of the group that the participant is in given that no other variables have been included in the analysis. The constant and multiplying coefficients in equation 11.8 come from the coefficients in Table 11.7.

$$\text{DEI} = 1727.2 + 127.6 \text{ Control} + 220.7 \text{ Artistic} - 79.3 \text{ Rhythmic} \tag{11.8}$$

Table 11.5 Linear regression using dichotomous independent variables (Model Summary)

Model Summary

Model	R	R Square	Adjusted R Square	Std. Error of the Estimate
1	.463a	.214	.152	222.848

a Predictors: (Constant), Rhythmic, Artisitic, Control

Table 11.6 Linear regression using dichotomous independent variables (ANOVA)

ANOVA[b]

Model		Sum of Squares	df	Mean Square	F	Sig.
1	Regression	514378.011	3	171459.337	3.453	.026a
	Residual	1887125.822	38	49661.206		
	Total	2401503.833	41			

a Predictors: (Constant), Rhythmic, Artistic, Control
b Dependent Variable: energy intake (kcal)

Table 11.7 Linear regression using dichotomous independent variables (Coefficients)

Coefficients[a]

Model		Unstandardized Coefficients		Standardized Coefficients		
		B	Std. Error	Beta	t	Sig.
1	(Constant)	1727.182	67.191		25.705	.000
	Control	127.636	95.023	.235	1.343	.187
	Artistic	220.675	107.746	.344	2.048	.048
	Rhythmic	−79.336	91.295	−.153	−.869	.390

a Dependent Variable: energy intake (kcal)

A limitation of the one-way ANOVA facility in the **Compare Means** menu is that we cannot obtain an estimate of effect size. This can be done if we use **Analyse → General Linear Model → Univariate** transferring 'Energy intake' into the *Dependent Variable* area and 'Group' into the *Fixed factor(s)* area. We can use **Options** to request an estimate of effect size which is included in the output shown in Table 11.8. Within **Options**, we will also request parameter estimates in order to make another point about the General Linear Model. The partial eta squared value (partial η^2) of 0.214 is identical to the R^2 value Table 11.6. This is because both partial η^2 and R^2 represent the proportion of the variance in the dependent variable that is explained by the independent variable(s). The values in the B column of the parameter estimates column in Table 11.9 are the same as those in the B column of Table 11.7. The GLM may not be suitable for all situations where we are analysing the effect of some categorical factor on numerical dependent variables. Where the dependent variable is a binomial response variable such as the proportion of events leading to a given outcome, logit models have been suggested (Nevill *et al.*, 2002). Log-linear models have also

Table 11.8 SPSS output for a univariate ANOVA test

Tests of Between-Subjects Effects

Dependent Variable: energy intake (kcal)

Source	Type III Sum of Squares	Df	Mean Square	F	Sig.	Partial Eta Squared	Noncent. Parameter	Observed Power[b]
Corrected Model	514378.011[a]	3	171459.337	3.453	.026	.214	10.358	.729
Intercept	1.283E8	1	1.283E8	2583.222	.000	.986	2583.222	1.000
Group	514378.011	3	171459.337	3.453	.026	.214	10.358	.729
Error	1887125.822	38	49661.206					
Total	1.344E8	42						
Corrected Total	2401503.833	41						

a R Squared = .214 (Adjusted R Squared = .152)
b Computed using alpha = .05

Table 11.9 GLM parameter estimates produced for a univariate ANOVA test

Parameter Estimates

Dependent Variable:energy intake (kcal)

Parameter	B	Std. Error	t	Sig.	95% Confidence Interval		Partial Eta Squared
					Lower Bound	Upper Bound	
Intercept	1727.182	67.191	25.705	.000	1591.160	1863.203	.946
[Group=1]	127.636	95.023	1.343	.187	−64.727	320.000	.045
[Group=2]	220.675	107.746	2.048	.048	2.556	438.795	.099
[Group=3]	−79.336	91.295	−.869	.390	−264.152	105.481	.019
[Group=4]	0[a]	.	.	.	.	.	.

a This parameter is set to zero because it is redundant.

been proposed for dependent variables derived from frequencies of discrete events in sport (Nevill et al., 2002; Taylor et al., 2008).

REPEATED MEASURES ANOVA

Purpose of the test

The repeated measures ANOVA test is used to compare the mean of some interval or ratio scale variable between three or more samples related to the same group of individuals. These related samples could be different times at which some measurement was made or different conditions under which some measurement was made. The dependent variable of the test is the interval or ratio scale variable that is hypothesized to be influenced by time or condition of measurement. The independent variable of the test is the categorical variable that distinguishes the different samples being compared. For example, the independent variable could be time of day (08:00, 12:00, 16:00, 20:00, 00:00) in a study of diurnal variation in some human performance characteristic. Where the independent variable is found to have a significant effect, each pair of repeated measures are compared by employing post hoc tests.

Assumptions

The assumptions of the repeated measures ANOVA test are that the dependent variable is measured on an interval or ratio scale, the dependent variable is normally distributed and that there is homogeneity of variance and covariance between samples. Homogeneity of variance and covariance is referred to as sphericity and is tested using Mauchly's test of sphericity. Homogeneity of variance means that the variances of the repeated measures are approximately equal. Homogeneity of covariance means that there must be similar covariance matrices between each pair of repeated measures. Chapter 16 discusses covariance in more detail. Students interested in this information are directed to equation 16.1 and Table 16.1(b) in Chapter 16.

Example: Work-rate during different quarters of a netball match

The example used to illustrate the repeated measures ANOVA test is time-motion analysis of netball match play. Specifically, we wish to compare the percentage of time a set of players spent performing high-intensity activity between the four quarters of a match. Altogether there were 28 players who have been observed during all four quarters of the match without changing positional role.

SPSS

This example uses the file 11-netball.SAV and on inspecting this datasheet we will observe a similar problem to that experienced when performing paired t-tests. Conceptually there is a single independent variable (quarter) and a single dependent variable (percentage of time spent performing high-intensity activity). However, we do not actually have a variable 'quarter' in the SPSS datasheet and instead of having a single variable for the 'percentage of time spent performing high intensity', we have four repeated measurements of this variable ('hi_q1', 'hi_q2', 'hi_q3' and 'hi_q4'). This means that the approach devised by Llewelyn to test the assumption of normality in the paired t-test can be extended for use with the repeated measures ANOVA.

The four repeated measures of high-intensity activity have been placed in a single column 'hi_q1' in the file 11-netball_assumptions.SAV. This allows a single Kolmogorov–Smirnov test to be applied to the data. Table 11.10 shows that the data satisfy the assumption of normality (p > 0.05). Even if we did separate Shapiro–Wilk tests for the four repeated measures in the 11-netball.SAV file (remember with fewer than 50 values we cannot use the Kolmogorov–Smirnov test), the data are still sufficiently normal (p > 0.05).

Once we have tested the assumption of normality, we can work with the file 11-netball.SAV to perform the repeated measures ANOVA. This is done using **Analyse → General Linear Model → Repeated Measures** which presents us with a pop-up window to define the repeated measures test as shown in Figure 11.4. Unlike the one-way ANOVA, we will not have some grouping variable and unlike the paired samples t-test, SPSS will not know how many repeated measures there are. Therefore, we need to set up a *Within-subjects factor name* called 'Qtr' and advise SPSS that it is measured at four *Levels*. When we click on **Add**, this factor is added to the model we are asking SPSS to perform.

Table 11.10 Results of a Kolmogorov–Smirnov test applied using Llewelyn's approach

	Tests of Normality					
	Kolmogorov–Smirnov[a]			Shapiro–Wilk		
	Statistic	df	Sig.	Statistic	df	Sig.
Qtr 1: %time spent performing high intensity activity	.078	112	.094	.979	112	.081

a Lilliefors Significance Correction

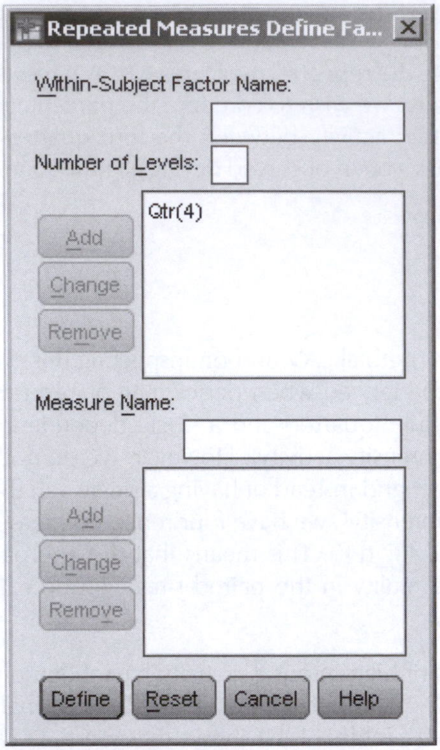

Figure 11.4 Defining a within-subjects factor when performing a repeated measures ANOVA test

Once we have added our independent factor, 'Qtr', we click on the **Define** button which activates the pop-up window shown in Figure 11.5. This is used to advise SPSS of the four 'variables' in the datasheet that represent the four repeated measurements of our conceptual dependent variable. The four variables ('hi_q1' to 'hi_q4') are transferred into the *Within Subjects Variables* area.

We use **Options** to make sure descriptive statistics, an estimate of effect size and observed power are included in the output as shown in Figure 11.6. While we are using the **Options** pop-up window, we can also select post hoc tests. In versions of SPSS up to and including version 18.0, the **Post Hoc** pop-up window does not allow post hoc tests to be selected for repeated measures. However, in the **Options** pop-up window we can transfer 'Qtr' into the *Display Means For* area and click on Compare main effects, choosing Bonferroni adjustment for Confidence interval adjustment. This is also illustrated in Figure 11.6.

Once we have completed setting up the options for the test, we can click on **Continue** to close down the Options pop-up window and click on **OK** to execute the test and observe the output. There are several output tables produced in SPSS for the repeated measures ANOVA test: Tables 11.11 to 11.14 shows the ones providing the results needed. Table 11.11 shows the all-important descriptive statistics. Table 11.12 contains the results of Mauchly's test of

analysis of variances

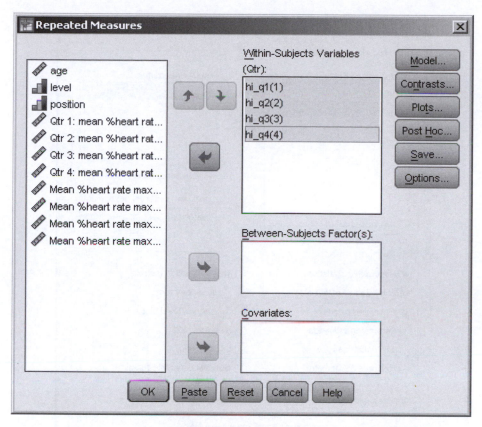

Figure 11.5 Identifying the repeated measurements of a dependent variable

Table 11.11 SPSS output for the repeated measures ANOVA test (Descriptive Statistics)

	Mean	Std. Deviation	N
	Descriptive Statistics		
Qtr 1	23.0475	5.28167	28
Qtr 2	22.0082	5.37944	28
Qtr 3	20.6757	5.77453	28
Qtr 4	19.2214	5.40273	28

spheririty. The p (Sig.) value for Mauchly's test is greater than 0.05 in this example meaning the sphericity is assumed. You will notice in Table 11.13 that there are four different sets of results for the repeated measures ANOVA test: sphericity assumed, Greenhouse Geisser, Huyhh-Feldt and Lower bound. If sphericity is assumed, we simply use the sphericity assumed row of results from the table of Tests of within subjects effects table. This gives us $F_{3,81} = 7.2$, $p < 0.001$. The three degrees of freedom for our main effect is given by the number of levels at which quarter is measured minus one. The 81 error degrees of freedom

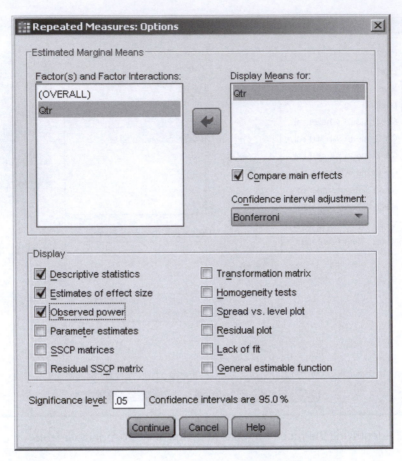

Figure 11.6 Options pop-up for the repeated measures ANOVA

Table 11.12 SPSS output for the repeated measures ANOVA test (Mauchly's test)

Mauchly's Test of Sphericity[b]

Measure: MEASURE_1

| Within Subjects Effect | Mauchly's W | Approx. Chi-Square | df | Sig. | Epsilon[a] | | |
					Greenhouse-Geisser	Huynh-Feldt	Lower-bound
Qtr	.714	8.682	5	.123	.816	.904	.333

Tests the null hypothesis that the error covariance matrix of the orthonormalized transformed dependent variables is proportional to an identity matrix.

a   May be used to adjust the degrees of freedom for the averaged tests of significance. Corrected tests are displayed in the Tests of Within-Subjects Effects table.
b   Design: Intercept
Within Subjects Design: Qtr

is given by $(4 - 1) \times (28 - 1)$ recalling that there were 28 players for whom four repeated measures of the dependent variable were measured. The F-ratio is 7.2 when expressed to one decimal place and the p (Sig.) value is not only less than 0.05 but also less than 0.001 which should be presented in our results.

If the assumption of sphericity is violated by the data, there is an increased chance of making a Type I error and we may need to use an alternative row of results from Table 11.13. These alternative results cope with violations of sphericity by reducing the number of degrees of freedom to make it more difficult to obtain a significant result meaning that we can be more confident in any significant result obtained. This is done by multiplying the degrees of freedom for the main effect and the error effect by the appropriate epsilon (ε) value shown in the Mauchly's test results. The process of selecting which row of results to use from the SPSS output is as follows:

- If sphericity is satisfied (Mauchly's test having a p values of greater than 0.05), then the Sphericity Assumed results are used.
- If sphericity is violated, but the result of the Sphericity Assumed ANOVA test is not significant ($p > 0.05$), then this row of results should be used. This is because it is impossible to make a Type I Error when we are concluding no significant difference and using a strict condition would actually increase our chance of making a Type II Error.
- If sphericity is violated, but the results of the Greenhouse Geisser adjusted ANOVA are significant ($p < 0.05$) then the Greenhouse Geisser results should be reported. The Greenhouse Geisser is a strict condition where we can be more confident in any significant result found.
- Where we have violated sphericity and the strictest condition (Greenhouse Geisser) reveals no significant difference, but the most liberal condition (Sphericity Assumed) does reveal a significant difference, then the Huynh Feldt ANOVA results should be used to arbitrate.

Table 11.13 SPSS output for the repeated measures ANOVA test (ANOVA results)

Tests of Within-Subjects Effects

Measure:MEASURE_1

Source		Type III Sum of Squares	df	Mean Square	F	Sig.	Partial Eta Squared	Noncent. Parameter	Observed Power[a]
Qtr	Sphericity Assumed	231.007	3	77.002	7.239	.000	.211	21.716	.979
	Greenhouse-Geisser	231.007	2.449	94.344	7.239	.001	.211	17.724	.956
	Huynh-Feldt	231.007	2.711	85.210	7.239	.000	.211	19.624	.969
	Lower-bound	231.007	1.000	231.007	7.239	.012	.211	7.239	.737
Error (Qtr)	Sphericity Assumed	861.659	81	10.638					
	Greenhouse-Geisser	861.659	66.111	13.034					
	Huynh-Feldt	861.659	73.198	11.772					
	Lower-bound	861.659	27.000	31.913					

a Computed using alpha = .05

Table 11.14 is the results of post hoc comparisons between pairs of repeated measures. These reveal that the percentage of time spent performing high-intensity activity was significantly different between quarters one and four and between quarters two and four.

Presenting results

The results might be presented as follows:

> Figure 11.7 shows the percentage of time spent performing high-intensity activity in the four quarters. Quarter had a significant influence on the percentage of time spent performing high intensity activity ($F_{3,81}$ = 7.2, p < 0.001, partial η^2 =0.211). The 19.2$\pm$5.4% of Q4 spent performing high-intensity activity was significantly less than the 23.0$\pm$5.3 of Q1 (p < 0.05) and the 22.0$\pm$5.4 of Q2 (p < 0.01).

ANALYSIS OF COVARIANCE (ANCOVA)

Purpose of the test

Analysis of covariance (ANCOVA) combines regression analysis with an ANOVA test (Wildt and Ahtola, 1978: 8). The ANCOVA test is used where a variable related to the dependent variable interferes with the effect of some independent sampling variable. For example, we may wish to compare the 60m sprint time of basketball players, soccer players and field hockey

Table 11.14 SPSS output for the repeated measures ANOVA test (Post hoc test results)

Pairwise Comparisons

Measure: MEASURE_1

(I) Qtr	(J) Qtr	Mean Difference (I-J)	Std. Error	Sig.a	95% Confidence Interval for Difference^a	
					Lower Bound	Upper Bound
1	2	1.039	.849	1.000	−1.377	3.456
	3	2.372	.918	.093	−.243	4.986
	4	3.826*	1.105	.011	.680	6.972
2	1	−1.039	.849	1.000	−3.456	1.377
	3	1.332	.789	.618	−.915	3.580
	4	2.787*	.740	.005	.681	4.893
3	1	−2.372	.918	.093	−4.986	.243
	2	−1.332	.789	.618	−3.580	.915
	4	1.454	.777	.432	−.757	3.666
4	1	−3.826*	1.105	.011	−6.972	−.680
	2	−2.787*	.740	.005	−4.893	−.681
	3	−1.454	.777	.432	−3.666	.757

Based on estimated marginal means
a Adjustment for multiple comparisons: Bonferroni.
* The mean difference is significant at the .05 level.

208

analysis of variances

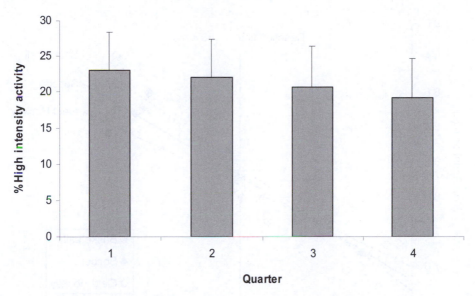

Figure 11.7 Percentage of netball match spent performing high-intensity activity

players. However, we are aware that there is a strong positive relationship between reaction time and 60m sprint time for the players. The different groups of athletes being compared in terms of 60m sprint time may have different means for reaction time and this will have an impact on the results of any ANOVA test that does not address the effect of reaction time. In this example, reaction time is a covariate and what we are really interested in is whether there are any differences between the three groups of games players for 60m sprint time once we have controlled for the relationship between reaction time and 60m sprint time. An ANCOVA test consists of two steps. First, a regression equation will be determined for the dependent variable in terms of the covariate for each of our samples. In our example, a regression equation for 60m sprint time in terms of reaction time is produced for basketball players, soccer players and field hockey players separately. The second step compares the samples using an ANOVA test on an adjusted dependent variable. The idea of the adjustment is to remove any differences in the covariate between the samples. This is shown in Figure 11.8 where the mean reaction time of 186.4ms is used for all three samples. Provided that the data are linear and homoscedastistic, it is valid to add the distribution of residuals to the expected 60m sprint time to the predicted 60m sprint time when reaction time is 186.4ms. Marginal means are the means for the dependent variable or each group given a reaction time of 186.4ms and applying the GLM which is a single model rather than three separate models.

Assumptions

The ANCOVA test assumes that the relationship between the covariate and the dependent variable is linear and that the regression lines for the different samples being compared have

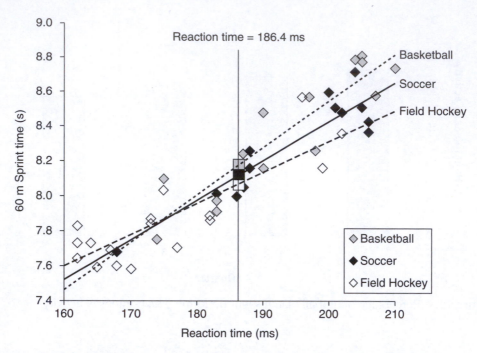

Figure 11.8 Relationship between reaction time and 60m sprint time (the grey, black and white squares are the predicted values for basketball players, soccer players and field hockey players respectively when the mean reaction time is applied

similar gradients (Newell *et al.*, 2010: 197). The ANCOVA can be implemented using a parallel lines model or a separate slopes model (Newell *et al.*, 2010: 198–201). The separate slopes shown in Figure 11.8 have gradients varying from 0.018 to 0.027. The lines are all positive, but the non-parallelism leads to the lines crossing over within the range of reaction times of the players in the study. This may be a concern as the specific General Linear Model determining the estimated marginal means applies a uniform gradient adding a coefficient value a group under consideration.

Example: 60m sprint time of team game athletes

The example that we have already touched on will be used to illustrate the ANCOVA test. Specifically, 46 fictitious male athletes participate in the study (14 basketball players, 15 soccer players and 17 field hockey players). Each performs a series of tests to estimate their reaction time before performing a 60m sprint that is electronically timed. It is impossible to control the participants' reaction times experimentally and if they are related to 60m sprint time, then we will need to adjust 60m sprint time for the effect of reaction time statistically. The data used in this exercise are shown in Figure 11.8 and we can see a strong positive relationship between reaction time and 60m sprint time for each group. The groups' 60m sprint times can be compared using an ANCOVA test that includes reaction time as a covariate.

SPSS

The data used in this example are found in the file 11-sprint60m.SAV. The Chart Builder tool  can be used to produce a scatter plot similar to Figure 11.8 and we can use bivariate correlations to determine the strength of the relationship between 'Reaction time' and 'Time 60m'. We will perform a univariate ANOVA test without using the covariate, 'Reaction time', and an ANCOVA test where the covariate is used. This will illustrate the impact of using a covariate on the results produced. Tables 11.15 to 11.17 show the output of the univariate ANOVA tests which reveals a significant difference between the three groups of players ($F_{2,43}$ = 7.3, p = 0.002, partial η^2 = 0.254) with Bonferroni-adjusted, post hoc tests showing that field hockey players performed the 60m sprint significantly faster than basketball players (p = 0.004) and soccer players (p = 0.011).

The ANCOVA test is performed using **Analyse → General Linear Model → Univariate** transferring 'Time 60m' into the *Dependent Variable* area, 'Group' into the *Fixed factor(s)* area and 'Reaction time' into the *Covariate(s)* area as shown in Figure 11.9. We can use the **Options** button to activate the Options pop-up window to request descriptive statistics, estimated effect size and observed power. The Options pop-up window is also used to transfer 'Time 60m' into the *Display Means for* area which will output estimated marginal

Table 11.15 Univariate ANOVA results without a covariate (Descriptive Statistics)

Descriptive Statistics

Dependent Variable: Time 60m

Group	Mean	Std. Deviation	N
Basketball	8.3069	.35943	14
Soccer	8.2555	.34135	15
Field Hockey	7.9029	.27754	17
Total	8.1408	.36783	46

Table 11.16 Univariate ANOVA results without a covariate (ANOVA results)

Tests of Between-Subjects Effects

Dependent Variable: Time 60m

Source	Type III Sum of Squares	df	Mean Square	F	Sig.	Partial Eta Squared	Noncent. Parameter	Observed Power[b]
Corrected Model	1.545[a]	2	.773	7.313	.002	.254	14.625	.921
Intercept	3039.576	1	3039.576	28768.464	.000	.999	28768.464	1.000
Group	1.545	2	.773	7.313	.002	.254	14.625	.921
Error	4.543	43	.106					
Total	3054.650	46						
Corrected Total	6.088	45						

a R Squared = .254 (Adjusted R Squared = .219)
b Computed using alpha = .05

Table 11.17 Univariate ANOVA results without a covariate (post hoc test results)

Multiple Comparisons

Time 60m
Bonferroni

(I) Group	(J) Group	Mean Difference (I-J)	Std. Error	Sig.	95% Confidence Interval Lower Bound	Upper Bound
Basketball	Soccer	.0514	.12079	1.000	−.2495	.3523
	Field Hockey	.4039*	.11731	.004	.1117	.6962
Soccer	Basketball	−.0514	.12079	1.000	−.3523	.2495
	Field Hockey	.3525*	.11515	.011	.0657	.6394
Field Hockey	Basketball	−.4039*	.11731	.004	−.6962	−.1117
	Soccer	−.3525*	.11515	.011	−.6394	−.0657

Based on observed means.
The error term is Mean Square(Error) = .106.
* The mean difference is significant at the .05 level.

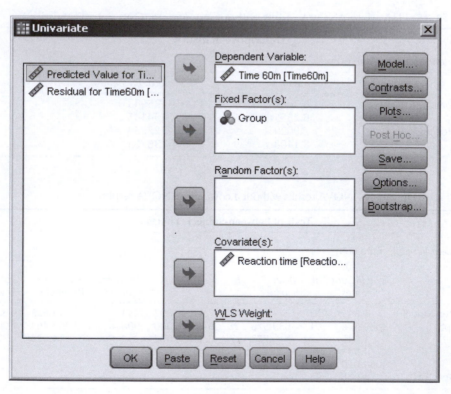

Figure 11.9 Setting up an ANCOVA test in SPSS

analysis of variances

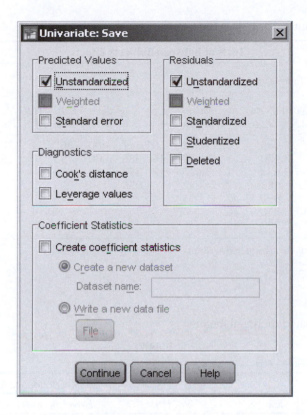

Figure 11.10 Saving predicted values and residuals when performing an ANCOVA test

means. The **Save** button activates the pop-up window shown in Figure 11.10 which we use to request that the predicted values and residual values for 'Time 60m' are saved as new variables which we can explore.

The main output provided by SPSS for the ANCOVA test is shown in Table 11.18. This shows that when 'Reaction time' is included as a covariate, 'Group' no longer has a significant effect on 'Time 60m' ($F_{2,42} = 2.0$, $p = 0.154$, partial $\eta^2 = 0.085$) with 'Reaction time' having a far greater effect (partial $\eta^2 = 0.802$). Note that the number of error degrees of freedom has reduced from 43 to 42 as a result of the covariate being included.

Table 11.19 shows the estimated marginal means which are computed using the parameter estimates in Table 11.20 applying a value of 186.4ms for reaction time. For example, the estimated marginal mean for the soccer players is $4.001 + 0.022 \times 186.4 + 0.103 \times 0 + 0.010 \times 1 = 8.12$. The estimated marginal means are not observed values but predicted values based on the specific linear model when the sample mean is used assumed for the covariate.

So why is there such a difference between the results of the univariate ANOVA test performed without a covariate and the ANCOVA test performed with a covariate? The mean$\pm$SD for

Table 11.18 Output from an ANCOVA test in SPSS

Tests of Between-Subjects Effects

Dependent Variable: Time 60m

Source	Type III Sum of Squares	df	Mean Square	F	Sig.	Partial Eta Squared	Noncent. Parameter	Observed Power[b]
Corrected Model	5.856[a]	3	1.952	91.521	.000	.867	274.563	1.000
Intercept	3.470	1	3.470	162.686	.000	.795	162.686	1.000
Reactiontime	3.633	1	3.633	170.358	.000	.802	170.358	1.000
Group	.083	2	.042	1.955	.154	.085	3.909	.382
Error	.896	42	.021					
Total	3058.276	46						
Corrected Total	6.752	45						

a R Squared = .867 (Adjusted R Squared = .858)
b Computed using alpha = .05

Table 11.19 Estimated marginal means

Estimates

Dependent Variable: Time 60m

Group	Mean	Std. Error	95% Confidence Interval	
			Lower Bound	Upper Bound
Basketball	8.213[a]	.041	8.131	8.295
Soccer	8.120[a]	.039	8.041	8.199
Field Hockey	8.110[a]	.040	8.029	8.191

a Covariates appearing in the model are evaluated at the following values: Reaction time = 186.43.

Table 11.20 Parameter estimates for the General Linear Model produced by an ANCOVA test

Parameter Estimates

Dependent Variable: Time 60m

Parameter	B	Std. Error	t	Sig.	95% Confidence Interval		Partial Eta Squared	Noncent. Parameter	Observed Power[a]
					Lower Bound	Upper Bound			
Intercept	4.001	.298	13.427	.000	3.400	4.603	.811	13.427	1.000
Reactiontime	.022	.002	13.052	.000	.019	.025	.802	13.052	1.000
[Group=1]	.103	.061	1.682	.100	-.020	.226	.063	1.682	.376
[Group=2]	.010	.059	.164	.870	-.110	.130	.001	.164	.053
[Group=3]	0[b]	.	.	.	.	.	.	.	.

a Computed using alpha = .05
b This parameter is set to zero because it is redundant.

reaction time was 193.4 ± 12.0ms for the basketball players, 192.6 ± 14.4ms for the soccer players and 175.2 ± 13.0ms for the field hockey players. When data are adjusted to apply the mean reaction time of 186.4ms, the adjusted 60m sprint times increase for the field hockey players and decrease for the other two groups. Table 11.21 shows the raw unadjusted values for 60m sprint time and the estimated marginal means assuming each player had a reaction time of 186.4ms. This example has turned a misleading significant result into a non-significant result. It is also possible that an ANCOVA test can turn a misleading non-significant result into a significant result (Newell et al., 2010: 198).

Presenting results

The results could use a scatter plot such as Figure 11.8, a table such as 11.21 and these results could be supported with the following text.

> Table 11.21 shows that field hockey players had quicker 60m sprint times than soccer players and basketball players. Reaction time explains 80.2 percent of the variance in 60m sprint time. Therefore, reaction time was included as a covariate within an ANCOVA test to compare 60m sprint time between basketball players, soccer players and field hockey players. Type of player had no significant influence on 60m sprint time once the effect of this variable was adjusted for its strong relationship with reaction time ($F_{2,42} = 2.0$, $p = 0.154$, partial $\eta^2 = 0.085$).

RANDOM FACTORS

Readers will have noticed in Figure 11.9 that there are areas for *Fixed factor(s)* and *Random factor(s)* that can be used when performing univariate ANOVA tests. A fixed factor is where every value of the independent variable is represented, while a random factor is a subset of the possible levels of the independent variable. There is a case for saying that random factors should have been used in two of the examples in this chapter. First, the four types of female prepubescent aesthetic athlete may not have represented all possible types of female prepubescent aesthetic athlete. Second, there are other team game players besides basketball players, soccer players and field hockey players. If we clearly restrict the scope of a study and how the results can be generalized to the sports and activities represented, then we can include the independent variables as fixed factors. If, however, we wish to make more general conclusions about type of sport in general, then we would have to include the independent variable as a random factor.

Table 11.21 Raw and adjusted 60m sprint times.

Group	60m Sprint time (s)	
	Observed mean±SD	Estimated marginal mean
Basketball players (n=14)	8.30±0.36	8.21
Soccer players (n=15)	8.26±0.34	8.12
Field hockey players (n=17)	7.90±0.27	8.11

SUMMARY

Analysis of variances tests are used to compare three or more samples. Where independent samples are compared, the one-way ANOVA test can be used. The repeated measures ANOVA test is used to compare different samples related to the same group of participants. Where an ANOVA test reveals a significant effect, post hoc tests can be employed to compare individual pairs of samples. The ANCOVA test allows the effect of some independent factors on a dependent variable of interest to be analysed adjusting the dependent variable for some covariate it is related to.

EXERCISES

Exercise 11.1. Body mass adjusted energy intake and protein, carbohydrate and fat within the diet of female prepubescent aesthetic athletes

The file 'ex11.1-food_freq_quest.SAV' contains five additional variables not used in the example in this chapter. Three of these are of interest in the current example: %protein in diet, %carbohydrate in diet and %fat in diet. Check if these variables satisfy the assumptions of the one-way ANOVA and perform a series of one-way ANOVA tests, employing Bonferroni-adjusted, post hoc tests where necessary, to compare the four groups of female prepubescent aesthetic athletes. Produce a summary results table that includes the key descriptive and inferential statistics for this analysis.

Exercise 11.2. 400m hurdle performance

This example is based on the work of Greene *et al.* (2008) on touchdown times in the 400m hurdles. The file ex11.2-hurdles.SAV is a datasheet containing split times for 56 athletes who competed in 400m hurdle races. There were 19 females and 37 males in the sample. The timings were taken at the point the lead leg touched down after each hurdle was cleared and the time at which the athlete crossed the finish line. This gives a split time from the start to the first hurdle (split_H1), the split times up to the remaining hurdles (split_H2 to split_H10) and the split time from the final hurdle to the finish of the race (split_FN). The distance between each pair of hurdles is 35.0m and the purpose of this exercise is to compare the times taken between each pair of hurdles; that is, the split times split_H2 to split_H10. Use split file to make sure female and male athletes are analysed separately and use repeated measures ANOVA tests and Bonferroni-adjusted, post hoc tests to determine if there are significant differences between any pair of hurdles.

Exercise 11.3. Daily energy intake adjusted for body mass

Use an ANCOVA to analyse the daily energy intake of the fictitious prepubescent female aesthetic athletes whose data are stored in the file ex11.3-food_freq_quest.SAV. The daily energy intake is to be compared between the four different groups using body mass as a covariate. Compare the result with the result using a one-way ANOVA with the data described earlier in the chapter. Also apply a one-way ANOVA to energy intake per kg of body mass and compare the results with the ANCOVA that you have done.

216

analysis of variances

PROJECT EXERCISE

Exercise 11.4. Positional effect on the height of soccer players

Go to the official internet site of a football squad of your choice and access the player information. Enter the positional group (defender, midfielder or forward) as well as the reported height (m) for each outfield player into an SPSS datasheet. Use a one-way ANOVA and Bonferroni-adjusted, post hoc tests to determine if there are significant differences between the positional groups for player height.

CHAPTER 12

FACTORIAL ANOVA

INTRODUCTION

The one-way and repeated measures analysis of variances (ANOVA) tests covered in Chapter 11 were single factor tests: the one-way ANOVA analysed the influence of a between-subjects factor while the repeated measures ANOVA analysed the influence of a within-subjects factor. The current chapter covers the purpose and use of ANOVA tests that include more than one factor. These are sometimes referred to as factorial ANOVA tests or multifactor ANOVA tests. These tests allow conclusions to be drawn about the effect of more than one factor on some numerical dependent variable. There are benefits to using a factorial ANOVA test instead of using a single factor ANOVA test for each factor of interest. The first benefit is that the factorial ANOVA test will not only analyse the effect of the individual factors (main effects), but will also analyse the combined effect of the factors (interaction effects). An interaction effect exists where the effect of one factor differs for the different levels of some other factor(s).

Where there is a significant interaction between two factors, there will be a non-parallel change in the dependent variable for one factor between the levels of the other factor. For an example of self-reported exercise by boys and girls of different age groups, a significant interaction would be where the change in self-reported exercise between age groups within boys was not the same as the change between age groups within girls. Where there is a significant interaction between factors, post hoc tests can be performed for all pairwise level combinations or for conceptually interesting pairs of factor levels (Newell *et al.*, 2010: 186). Alternatively, single factor ANOVA tests with any necessary post hoc tests can be done for one factor for each of the levels of the other factor.

The second benefit of using a factorial ANOVA test is that the effect of each factor is tested controlling any variance known to be due to other factors. Consider the example of analysing the effect of age and gender on self-reported physical activity. There may be a large difference between males and females which could result in a single factor ANOVA failing to show a significant effect of age group. This could simply be due to the high variability within each age group as both male and female participants are included. A factorial ANOVA test would statistically control for known variance due to gender making it more likely that a significant age group effect could be detected.

One area that the single factor tests covered in Chapter 10 and the factorial tests covered in the current chapter have in common is that they test the effect of factor(s) on a single

dependent variable. Such tests are referred to as univariate tests irrespective of how many independent variables are involved. Other terms used with analysis of variances tests are 'fixed effects' and 'random effects'. Fixed effects are where all values of a factor are used within the study, whereas random effects are where a subset of those values is used. For example, there may be five broad positions of outfield soccer player (centre backs, wide backs, centre midfielders, wide midfielders and forwards). Where players from all five positions are included in a study, the factor position is a fixed effect. If, however, we only used three of these positions to determine if position in general had an influence on some dependent variable, then the factor position would be a random effect.

Effect sizes can be determined for factorial ANOVA tests where population effects are known. This is done in a using eta squared (η^2) which is the percentage of variance in the dependent variable that is explained by the particular effect being investigated. Technically, the effect size is the effect within the population being studied and, therefore, studies based on samples produce an estimated effect size (partial η^2). Partial effect sizes can be determined for main effects as well as interaction effects.

There is a choice of models that can be used when applying factorial ANOVA tests. The default in a package such as SPSS is to analyse all main effects and all interaction effects. This is called a full factorial model. However, there are cases where we may wish to focus on a subset of possible interactions. In this textbook, the default scenario is covered. As factors are added, it can lead to a more dramatic increase in interaction effects to be explored. A two-way ANOVA will only involve one interaction effect: the interaction between the two factors. If we were analysing the effect of four factors – A, B, C and D – the multifactor ANOVA would have 11 interaction effects (A × B, A × C, A × D, B × C, B × D, C × D, A × B × C, A × B × D, A × C × D, B × C × D, A × B × C × D). The term 'factorial ANOVA' is used where all main and interaction effects are tested. There may be little conceptual rationale for analysing some of the interactions. Therefore, packages such as SPSS allow multifactor ANOVA tests to be customized to focus on a subset of main effects and interaction effects of interest.

THE BETWEEN–BETWEEN DESIGN

Purpose of the test

The between–between design uses a two-way factorial ANOVA where both factors are between-subjects effects. The purpose of the test is to determine if each of the main effects has an influence on some numerical dependent variable when controlling for variance due to the other factor as well as determining if the combined effect of the two factors has an influence on the dependent variable of interest.

Assumptions

The assumptions of a between–between ANOVA test are similar to those of a one-way ANOVA test. The dependent variable should be an interval or ratio scale variable, observations should be independent and there should be homogeneity of variances between groups. Different authors give different advice about the assumptions of between–between ANOVA tests. For example, Newell *et al.* (2010: 190) stated that the dependent variable should be normally distributed for each combination of the two factors. Ntoumanis (2001: 82–4) stated that the residuals should be normally distributed and have similar variances between factor combinations. A residual value is the difference between the dependent variable's value for the given participant and what would have been predicted for the groups of which the participant is a member. The advice of Newell *et al.* (2010: 190) and Ntoumanis (2001: 82–4) is reasonably consistent as they agree that normality should not be checked on the dependent variable for the entire sample. It is important that residuals are inspected rather than raw scores because genuine factor effects on a dependent variable could render the raw dependent variable non-normal. Even though data may be drawn from a normally distributed population, there may be legitimate reasons why values are higher for some sub-groups than others to the extent that normality may be compromised. This can certainly occur with random effects where a subset of possible groups is analysed leaving large sections of the population unrepresented by sample used in a study.

Multiple normality tests will give more chance of one failing to support the assumption than would a single test. However, the between–between ANOVA test is robust to violation of normality as long as the largest standard deviation of any factor combination does not exceed double the value of the smallest standard deviation (Newell *et al.*, 2010: 190).

Example: Children's activity in the playground during morning break

This example is based on the observational study of McLaughlin and O'Donoghue (2002), but using fictitious data. The activity of 64 children is observed during morning break in the school playground using a computerized time-motion analysis system. The children are classified according to two between subjects factors: class (or school year) and gender. The children analysed are either in year 1 (4–5 years old) or year 4 (7–8 years old). We wish to determine if class, gender and the interaction of class and gender have a significant influence on the percentage of time children spend performing high-intensity activity in the playground. If we used two separate independent samples t-tests (one for class and one for gender), there is a possibility of a Type II error occurring due to the variance within each group being compared which results from the factor not being analysed by the particular independent samples t-test. The other issue with using independent samples t-tests is that the combined effect of class and gender cannot be tested. Therefore, a two-way ANOVA should be used including class and gender as between-subjects effects.

SPSS

This example uses the file 12-Playground.SAV. The variable 'break_time' is the percentage of time a child spends performing high-intensity activity during morning break. This is the dependent variable for the current example. In order to test the assumptions of a two-way ANOVA, it is necessary to perform the ANOVA test in an exploratory way just to save the residual values. This is done using **Analyse → General Linear Model → Univariate**, which causes the pop-up window in Figure 12.1 to appear. We transfer our dependent

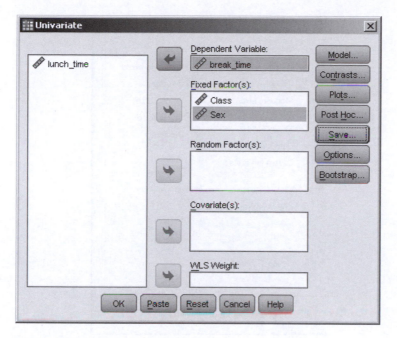

Figure 12.1 SPSS window for univariate ANOVA tests.

variable of interest ('break_time') into the *Dependent Variable* area and our factors of interest ('Class' and 'Sex') into the *Fixed Factor(s)* area. There is a discussion point here as to whether 'Class' should be a fixed effect or a random effect. There is a case for making it a random effect because there are only two out of seven primary school classes that are represented. However, if the purpose of our investigation is to compare these two classes rather than test the effect of class in general only using two classes, then we have a case for making class a fixed effect.

When we click on the **Save** button, the Save pop-up window appears as shown in Figure 12.2. This is used to request that predicted and residual values are saved as new variables within the datasheet.

The residual values can now be explored in the same way as any other variables can be explored. In this particular example, a Kolmogorov–Smirnov test reveals that the residual values are normally distributed (p > 0.05) as shown in Table 12.1.

Table 12.1 Results of normality tests for residuals

	Tests of Normality					
	Kolmogorov–Smirnov[a]			*Shapiro–Wilk*		
	Statistic	*df*	*Sig.*	*Statistic*	*df*	*Sig.*
Residual for break_time	.073	64	.200[*]	.979	64	.348

a Lilliefors Significance Correction
* This is a lower bound of the true significance.

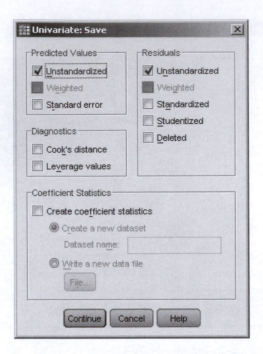

Figure 12.2 Saving predicted and residual values when performing a univariate ANOVA test.

A second exploratory two-way ANOVA test is done except this time the residual value is tested instead of the raw value as shown in Figure 12.3. The reason for doing this is to test homogeneity of variances of the residuals.

We use the **Options** pop-up to request a homogeneity of variances test as shown in Figure 12.4. The results of this are shown in Table 12.2, which reveal that the variances of the different factor level combinations are sufficiently normal ($p > 0.05$).

Tables 12.3 to 12.6 show the results of independent samples t-tests performed to analyse the effects of 'Sex' and 'Class' separately. As we can see in Tables 12.4 and 12.6, sex had a significant influence on the percentage of morning break spent performing high-intensity activity ($p = 0.025$) but class did not ($p = 0.328$).

We have already used the two-way ANOVA in an exploratory way to produce the residual values. It is now necessary to apply the test for its main purpose which is to analyse the effects of the two factors and their interaction on the percentage of morning break time spent performing high-intensity activity. We use **Analyse → General Linear Model → Univariate** as before, but this time when we select **Options** we request descriptive statistics, observed power and estimate of effect size as shown in Figure 12.5.

If we needed to perform post hoc tests for any factor, we would use the **Post Hoc** pop-up window to select the factors where post hoc tests are to be employed. In this particular example, post hoc tests are not necessary because each factor is only measured at two levels.

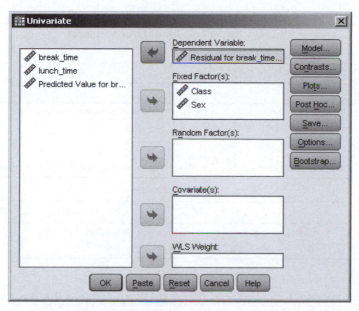

Figure 12.3 Applying an ANOVA test to the residuals to test homogeneity of variances.

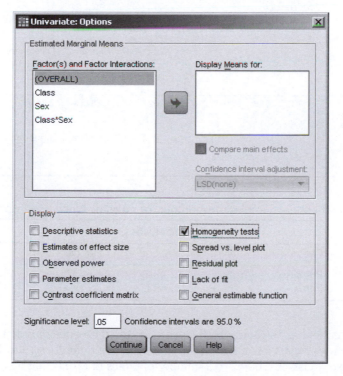

Figure 12.4 Options pop-up window for requesting homogeneity tests when performing univariate ANOVA tests.

Table 12.2 Result of the homogeneity test of the residual values

Levene's Test of Equality of Error Variances[a]

Dependent Variable: Residual for break_time

F	df1	df2	Sig.
1.127	3	60	.345

Tests the null hypothesis that the error variance of the dependent variable is equal across groups.
a Design: Intercept + Class + Sex + Class * Sex

Table 12.3 Independent Samples t-test output for morning break (Descriptive statistics)

Group Statistics

	Sex	N	Mean	Std. Deviation	Std. Error Mean
break_time	Female	33	11.2170	8.87887	1.54561
	Male	31	16.0587	7.84912	1.40974

Table 12.4 Independent Samples t-test output for morning break (t-test results)

Independent Samples Test

		Levene's Test for Equality of Variances		t-test for Equality of Means						
		F	Sig.	t	df	Sig. (2-tailed)	Mean Difference	Std. Error Difference	95% Confidence Interval of the Difference	
									Lower	Upper
break_time	Equal variances assumed	1.150	.288	−2.305	62	.025	−4.8417	2.10012	−9.0398	−.6437
	Equal variances not assumed			−2.314	61.78	.024	−4.8417	2.09196	−9.0238	−.6597

Table 12.5 Independent Samples t-test output for morning lunch (Descriptive statistics)

Group Statistics

	Class	N	Mean	Std. Deviation	Std. Error Mean
break_time	Primary 1	32	12.4919	8.62271	1.52429
	Primary 4	32	14.6325	8.73900	1.54485

Therefore, if gender has a significant influence, then the difference must between males and females. Similarly, if class has a significant influence, the difference must be between the Year 1 and Year 4 classes. There are a number of tables of results that are produced in the SPSS output viewer when we eventually click on **OK** to execute the two-way ANOVA test. Tables 12.7 and 12.8 show the main tables of results. In this example, sex has a significant

Table 12.6 Independent Samples t-test output for lunch break (t-test results)

		Independent Samples Test								
		Levene's Test for Equality of Variances		t-test for Equality of Means						
		F	Sig.	t	df	Sig. (2-tailed)	Mean Difference	Std. Error Difference	95% Confidence Interval of the Difference	
									Lower	Upper
break_ time	Equal variances assumed	.021	.886	−.986	62	.328	−2.14063	2.1703	−6.4789	2.1977
	Equal variances not assumed			−.986	61.99	.328	−2.14063	2.1703	−6.4789	2.1977

Figure 12.5 Options pop-up window for univariate ANOVA tests.

influence on the percentage of morning break spent performing high-intensity activity (p = 0.017) while class (p = 0.195) and the interaction of sex and class (p = 0.130) do not. The p values for both gender and class are lower when we use a two-factor ANOVA rather than separate independent t-tests. This is due to each factor being considered controlling for the variability due to the other factor. Even though the effect of gender is significant, the effect size is small (partial η^2 = .092).

Table 12.7 Two-way ANOVA output (Descriptive Statistics)

Descriptive Statistics

Dependent Variable: break_time

Class	Sex	Mean	Std. Deviation	N
Primary 1	Female	11.4753	10.15870	15
	Male	13.3888	7.20641	17
	Total	12.4919	8.62271	32
Primary 4	Female	11.0017	7.95603	18
	Male	19.3007	7.59055	14
	Total	14.6325	8.73900	32
Total	Female	11.2170	8.87887	33
	Male	16.0587	7.84912	31
	Total	13.5622	8.67918	64

Table 12.8 Two-way ANOVA output (ANOVA results)

Tests of Between-Subjects Effects

Dependent Variable: break_time

Source	Type III Sum of Squares	df	Mean Square	F	Sig.	Partial Eta Squared	Noncent. Parameter	Observed Power[b]
Corrected Model	644.878[a]	3	214.959	3.145	.032	.136	9.435	.703
Intercept	12054.085	1	12054.085	176.367	.000	.746	176.367	1.000
Class	117.138	1	117.138	1.714	.195	.028	1.714	.251
Sex	413.095	1	413.095	6.044	.017	.092	6.044	.677
Class * Sex	161.503	1	161.503	2.363	.130	.038	2.363	.328
Error	4100.794	60	68.347					
Total	16517.379	64						
Corrected Total	4745.672	63						

a R Squared = .136 (Adjusted R Squared = .093)
b Computed using alpha = .05

REPORTING RESULTS

The results of the two-way ANOVA are reported together with a table or chart showing the descriptive results. The accompanying text might be as follows:

> Figure 12.6 shows the percentage of time spent performing high-intensity activity during morning break by Year 1 and Year 4 girls and boys. Boys performed a significantly greater percentage of high-intensity activity than girls ($F_{1,60}$ = 6.0, p = 0.017, partial η^2 = .092). However, neither class ($F_{1,60}$ = 1.7, p = 0.195, partial η^2 = .028) nor the interaction of class and gender ($F_{1,60}$ = 2.4, p = 0.130, partial η^2 = .038) had a significant effect.

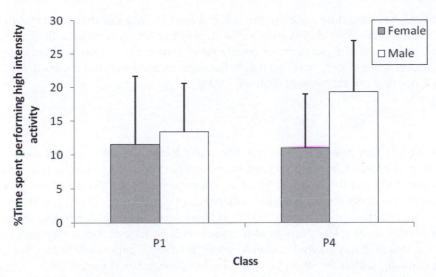

Figure 12.6 Percentage of morning break spent performing high-intensity activity.

THE WITHIN–WITHIN DESIGN

Purpose of the test

The within–within design is a two-way factorial ANOVA where both factors are within-subjects effects. These within-subjects effects could be repeated measures or other conditions related to a single group of participants. The purpose of the test is to determine if each of the main effects has an influence on some numerical dependent variable when controlling for variance due to the other factor as well as determining if the combined effect of the two factors has an influence on the dependent variable of interest.

Assumptions

The assumptions of a factorial ANOVA test including within-subjects effects are similar to those for factorial ANOVA tests that use between-subjects effects. The residual values should be normally distributed but the assumption of independence does not apply because repeated measures will come from the same set of participants. Where within-subjects effects are measured at more than two levels, the data should satisfy the assumption of sphericity which was covered in Chapter 10.

Example: Fluid loss with and without a wetsuit

This example is an experiment to compare weight loss during a defined controlled period of running on a treadmill between performing the treadmill run while wearing a wet suit and while wearing normal running kit. In order to avoid individual participant effects on either condition, all 10 participants perform the treadmill run under both conditions. They are all weighed before and after each of the treadmill runs. The data in the example used here are fictitious and readers should be aware of the risk involved in fluid loss during such experiments. Body mass reductions during exercise of 2 per cent are associated with impaired performance, reductions of 5 per cent are associated with heat exhaustion, with more serious consequences being experienced such as hallucinations and heat stroke if body mass is reduced by 7 per cent during exercise (Rehner, 1994).

SPSS

As we have already seen with paired samples t-tests in Chapter 10 and repeated measures ANOVA tests in Chapter 11, repeated measures cause a mismatch between the conceptual variables of our study and the 'variables' that we create within our SPSS datasheet. In the current example, we conceptually have two independent variables ('condition' and 'time') and a single dependent variable (body mass). 'Condition' is measured at two levels: wearing a wetsuit or normal training kit. 'Time' is also measured at two levels: before the treadmill run or after the treadmill run. As we have seen before, the only independent factors that are included as variables within an SPSS datasheet are between-subjects factors whose values distinguish members of different independent groups. The data for the current example are found in the file 12-Wet_suit.SAV where there is no variable for 'condition' or 'time' and there is no single variable for body mass. Instead, we have a variable for body mass for each repeated measurement of it. In this case, we have two overlapping within-subjects factors in a 2 × 2 design. That is, each factor is measured at two levels and we need to know the value of the dependent variable for each combination of values of 'condition' and 'time'. In order to perform a two-way ANOVA test using the within-within design, we use **Analyse →** **General Linear Model → Repeated Measures**. Figure 12.7 shows that we identify and add

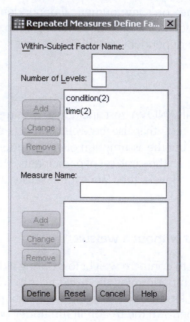

Figure 12.7 Defining a within–within design.

both within-subjects factors when defining the ANOVA test. It is necessary to advise SPSS that each of these factors is measured at two levels.

When we click on **Define**, note that each combination of 'condition' and 'time' needs to be represented by a 'variable' from the datasheet. The numbers 1 and 2 are used to represent the levels of each within-subjects factor. Therefore, we need to identify the level combinations of the independent factors that correspond to (1,1), (1,2), (2,1) and (2,2). In this example, our first factor is 'condition' as shown in Figure 12.8. Therefore, if (1,1) is used to represent the pre-test when wearing normal training kit ('ns_pre'), then (1,2) must be the post-test when wearing normal training kit ('ns_post'). This is because both (1,1) and (1,2) have the value 1 for 'condition' and so the 'variables' used must be based on the same level (or value) for 'condition'.

Once the 2 × 2 repeated measures have been transferred into the within subjects variables area, we click on **Save** so as we can use the pop-up window in Figure 12.9 to save residual values allowing us to test the assumptions of the test.

We use the **Options** button to activate the pop-up window in Figure 12.10 to allow us to obtain descriptive statistics, estimates of effect size and observed power.

When we click on **OK**, the test we have defined is executed by SPSS. Tables 12.9 and 12.10 show the most important output provided. Because each of the factors in the current

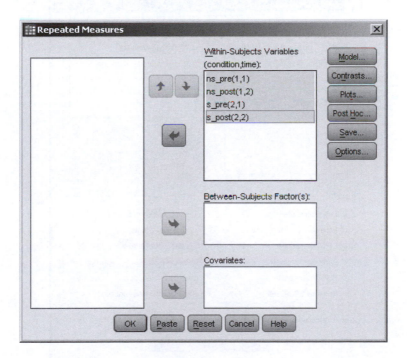

Figure 12.8 Repeated measures of our conceptual dependent variable are assigned to factor level combinations.

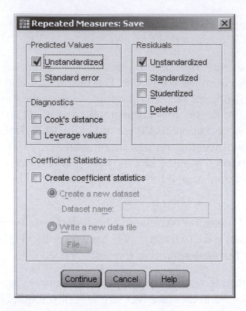

Figure 12.9 Saving predicted and residual values when performing an ANOVA test including two within-subjects effects.

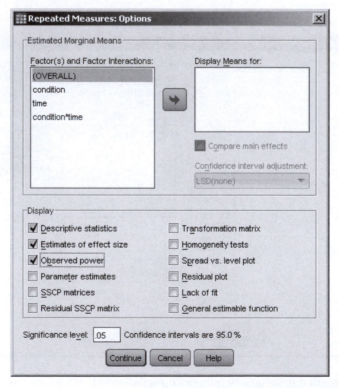

Figure 12.10 Selecting options within a repeated measures ANOVA test.

factorial ANOVA

Table 12.9 SPSS output for two repeated measures (Descriptive Statistics)

Descriptive Statistics			
	Mean	Std. Deviation	N
Without suit pre	73.9100	6.99483	10
Without suit post	72.9700	6.75591	10
With suit pre	73.9000	7.09726	10
With suit post	71.5300	6.96851	10

Table 12.10 SPSS output for two repeated measures (ANOVA results)

Tests of Within-Subjects Effects

Measure: MEASURE_1

Source	Type III Sum of Squares	df	Mean Square	F	Sig.	Partial Eta Squared	Noncent. Parameter	Observed Power[a]
condition	Sphericity Assumed	5.256	1	5.256	43.751	.000	.829	43.751	1.000
	Greenhouse-Geisser	5.256	1.000	5.256	43.751	.000	.829	43.751	1.000
	Huynh-Feldt	5.256	1.000	5.256	43.751	.000	.829	43.751	1.000
	Lower-bound	5.256	1.000	5.256	43.751	.000	.829	43.751	1.000
Error (condition)	Sphericity Assumed	1.081	9	.120					
	Greenhouse-Geisser	1.081	9.000	.120					
	Huynh-Feldt	1.081	9.000	.120					
	Lower-bound	1.081	9.000	.120					
time	Sphericity Assumed	27.390	1	27.390	94.188	.000	.913	94.188	1.000
	Greenhouse-Geisser	27.390	1.000	27.390	94.188	.000	.913	94.188	1.000
	Huynh-Feldt	27.390	1.000	27.390	94.188	.000	.913	94.188	1.000
	Lower-bound	27.390	1.000	27.390	94.188	.000	.913	94.188	1.000
Error(time)	Sphericity Assumed	2.617	9	.291					
	Greenhouse-Geisser	2.617	9.000	.291					
	Huynh-Feldt	2.617	9.000	.291					
	Lower-bound	2.617	9.000	.291					
condition * time	Sphericity Assumed	5.112	1	5.112	45.770	.000	.836	45.770	1.000
	Greenhouse-Geisser	5.112	1.000	5.112	45.770	.000	.836	45.770	1.000
	Huynh-Feldt	5.112	1.000	5.112	45.770	.000	.836	45.770	1.000
	Lower-bound	5.112	1.000	5.112	45.770	.000	.836	45.770	1.000
Error (condition* time)	Sphericity Assumed	1.005	9	.112					
	Greenhouse-Geisser	1.005	9.000	.112					
	Huynh-Feldt	1.005	9.000	.112					
	Lower-bound	1.005	9.000	.112					

a Computed using alpha = .05

example is measured at two levels, it is impossible for sphericity to be violated. Indeed, Mauchly's test of sphericity is not computed by SPSS for factors with only two levels. Furthermore, we do not require post hoc tests because each factor is only measured at two levels. If condition has a significant influence, then the difference must be between wearing a wetsuit and normal training kit. If time has a significant effect, then the difference must be between the pre-test body mass and post-test body mass.

The results of the Shapiro–Wilk test in Table 12.11 show that the residuals are not normally distributed. This was done using the Llewelyn approach of creating a single variable in SPSS with all 40 residual values (four values for each of the 10 participants). This was done in the file 12-Wet_suit-assumptions.SAV. This file also contains a dummy grouping variable allowing a one-way ANOVA to be done just so as we can see the results of Levene's test of equality of variances when applied to the residuals. Table 12.12 shows that we can assume equality of variances between the repeated measurements. Given that the maximum standard deviation for any factor level combination of 7.1 is less than twice the minimum of 6.8, the test can be considered robust enough to cope with the violation of normality.

REPORTING THE RESULTS

The results of the two-way ANOVA using within-subjects factors can be presented by using a summary table for descriptive statistics or a chart with inferential statistics presented in supporting text as follows:

Figure 12.11 shows that there was a greater pre/post-decrease in body mass when a wetsuit was worn than when normal training kit was worn. Condition ($F_{1,9} = 43.8$, $p < 0.001$, partial $\eta^2 = .829$), time ($F_{1,9} = 94.2$, $p < 0.001$, partial $\eta^2 = .903$) and the interaction of condition and time ($F_{1,9} = 45.8$, $p < 0.001$, partial $\eta^2 = .836$) all had a significant influence on body mass.

Table 12.11 Results of normality tests for repeated measures

	Tests of Normality					
	Kolmogorov–Smirnov[a]			Shapiro–Wilk		
	Statistic	df	Sig.	Statistic	df	Sig.
Residual for ns_pre	.162	40	.009	.900	40	.002

a Lilliefors Significance Correction

Table 12.12 Results of testing homogeneity of variances between repeated measures

Test of Homogeneity of Variances			
Residual for ns_pre			
Levene Statistic	df1	df2	Sig.
.005	3	36	1.000

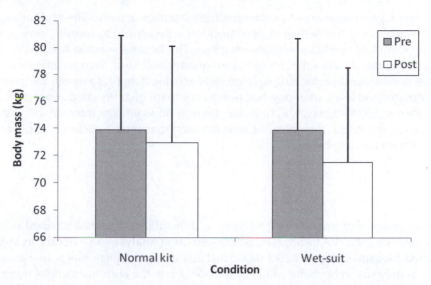

Figure 12.11 Body mass before and after treadmill running in different clothing.

THE MIXED DESIGN

Purpose of the test

The mixed design is a factorial ANOVA where at least one factor is a within-subjects effect (repeated measure) and at least one factor is a between-subjects effect. The purpose of the test is to determine if each of the main effects has an influence on some numerical dependent variable when controlling for variance due to the other factors as well as determining if the combined effect of the factors has an influence on the dependent variable of interest. This between-within design has been referred to as a mixed ANOVA (Tabachnick and Fidell, 1996: 45–4; Vincent, 1999: 201; Hinton, 2004: 181; Fallowfield et al., 2005: 256–7). Others use the term 'mixed effects' to describe statistical procedures that combine fixed and random factors (Newell et al., 2010, 325–56).

Assumptions

The assumptions described for the between–between design and the within–within design apply to the between–within design although repeated measures are dependent samples.

Example: Level and quarter effect on work-rate in netball

Recall the example used in Chapter 11 to illustrate the repeated measures ANOVA test. A set of 28 netball players who played all four quarters of a netball match were included and the repeated measures ANOVA test was used to determine if quarter of the match had an effect on the percentage of time spent performing high-intensity activity. The current example extends the analysis of the netball players' high-intensity activity to include level of player (international or club) as a between-subjects effect. The between–within ANOVA is used to test if any of quarter, level or the interaction of quarter and level have an influence on the percentage of time spent performing high-intensity activity. If there is a significant interaction between quarter and level, then post hoc tests to compare quarters are done separately for the two different levels of player. If, however, there is no significant interaction but there is a significant quarter effect, then post hoc tests are used to compare each pair of quarters for the set of players as a whole.

SPSS

Any univariate test that involves one or more within-subjects effects is defined using the repeated measures ANOVA facility. This is accessed using **Analyse → General Linear Model → Repeated Measures**. Figure 12.12 shows that any within-subjects effects are defined the same way as they would be in the within–within design. In the current example, there is just one within-subjects effect: 'Qtr'.

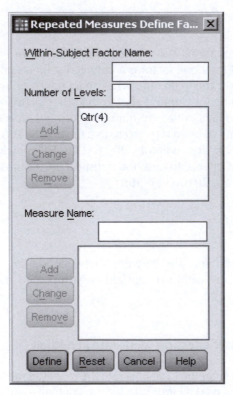

Figure 12.12 Identifying a within-subjects factor within a mixed ANOVA.

Once this within-subjects factor is added to the model, the test must be defined as shown in Figure 12.13. The 'variables' within the datasheet that represent the repeated measurements of our dependent variable are transferred into the *Within-subjects variable(s)* area. It should be noted at this point that there is no specific facility for a between–within ANOVA in SPSS and it is achieved using the repeated measures ANOVA facility by entering a grouping variable into the *Between-subjects factor(s)* area. In this case, the grouping variable is 'level'. Once a between-subjects factor is included in the pop-up window for a repeated measures ANOVA, the test becomes a between–within ANOVA test. Figure 12.14 shows that we can select **Options** including post hoc tests for a within-subjects effect ('Qtr'). Descriptive statistics, estimated effect sizes and observed power are also requested.

Tables 12.13 to 12.17 show the most important items produced as output for the between–within ANOVA test. Note that Mauchly's test revealed that sphericity could be assumed (p = 0.107). Furthermore, there was no significant interaction between quarter and level (p = 0.160) and, therefore, post hoc tests are applied to the sample of 28 players as a whole to compare each pair of quarters.

Reporting the results

With two factors being analysed, a clustered bar graph or table could be used to show the descriptive results with inferential statistics shown in supporting text as follows:

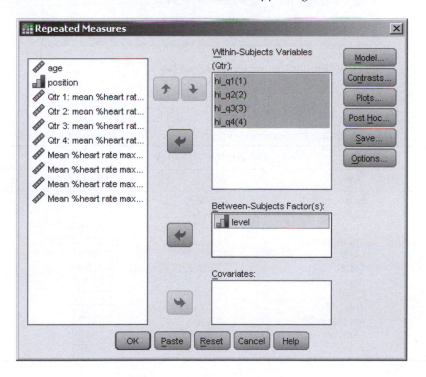

Figure 12.13 Defining a mixed ANOVA test.

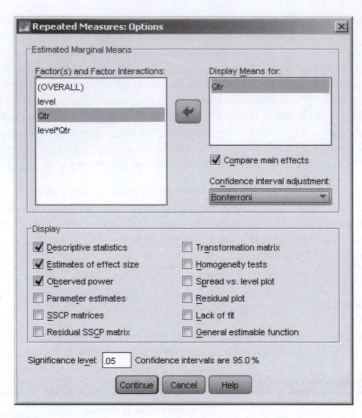

Figure 12.14 Selecting options in a mixed ANOVA test.

Table 12.13 SPSS output for a between-within ANOVA test (Descriptive Statistics)

Descriptive Statistics

	level	Mean	Std. Deviation	N
Qtr 1	International	23.6886	4.35665	14
	Club	22.4064	6.17029	14
	Total	23.0475	5.28167	28
Qtr 2	International	21.6921	5.11154	14
	Club	22.3243	5.81031	14
	Total	22.0082	5.37944	28
Qtr 3	International	21.7814	5.76409	14
	Club	19.5700	5.77903	14
	Total	20.6757	5.77453	28
Qtr 4	International	18.5986	5.54271	14
	Club	19.8443	5.39139	14
	Total	19.2214	5.40273	28

Table 12.14 SPSS output for a between-within ANOVA test (Mauchly's test)

Mauchly's Test of Sphericity[b]

Measure: MEASURE_1

Within Subjects Effect	Mauchly's W	Approx. Chi-Square	df	Sig.	Epsilon[a]		
					Greenhouse-Geisser	Huynh-Feldt	Lower-bound
Qtr	.693	9.067	5	.107	.794	.913	.333

Tests the null hypothesis that the error covariance matrix of the orthonormalized transformed dependent variables is proportional to an identity matrix.
a May be used to adjust the degrees of freedom for the averaged tests of significance. Corrected tests are displayed in the Tests of Within-Subjects Effects table.
b Design: Intercept + level
Within Subjects Design: Qtr

Table 12.15 SPSS output for a between-within ANOVA test (repeated measure and its interactions with the between-subjects effect)

Tests of Within-Subjects Effects

Measure: MEASURE_1

Source		Type III Sum of Squares	df	Mean Square	F	Sig.	Partial Eta Squared	Noncent. Parameter	Observed Power[a]
Qtr	Sphericity Assumed	231.007	3	77.002	7.444	.000	.223	22.333	.982
	Greenhouse-Geisser	231.007	2.382	96.962	7.444	.001	.223	17.735	.957
	Huynh-Feldt	231.007	2.740	84.314	7.444	.000	.223	20.396	.974
	Lower-bound	231.007	1.000	231.007	7.444	.011	.223	7.444	.747
Qtr * level	Sphericity Assumed	54.832	3	18.277	1.767	.160	.064	5.301	.444
	Greenhouse-Geisser	54.832	2.382	23.015	1.767	.173	.064	4.210	.390
	Huynh-Feldt	54.832	2.740	20.013	1.767	.166	.064	4.841	.422
	Lower-bound	54.832	1.000	54.832	1.767	.195	.064	1.767	.249
Error (Qtr)	Sphericity Assumed	806.827	78	10.344					
	Greenhouse-Geisser	806.827	61.943	13.025					
	Huynh-Feldt	806.827	71.236	11.326					
	Lower-bound	806.827	26.000	31.032					

a Computed using alpha = .05

Table 12.16 SPSS output for a between-within ANOVA test (between-subjects effects)

Tests of Between-Subjects Effects

Measure: MEASURE_1
Transformed Variable: Average

Source	Type III Sum of Squares	df	Mean Square	F	Sig.	Partial Eta Squared	Noncent. Parameter	Observed Power[a]
Intercept	50518.916	1	50518.916	557.335	.000	.955	557.335	1.000
level	4.568	1	4.568	.050	.824	.002	.050	.055
Error	2356.738	26	90.644					

a Computed using alpha = .05

Table 12.17 SPSS output for a between-within ANOVA test (Post hoc test results for the repeated measures)

Pairwise Comparisons

Measure: MEASURE_1

(I) Qtr	(J) Qtr	Mean Difference (I-J)	Std. Error	Sig.[a]	95% Confidence Interval for Difference[a]	
					Lower Bound	Upper Bound
1	2	1.039	.844	1.000	−1.372	3.450
	3	2.372	.931	.103	−.288	5.031
	4	3.826*	1.099	.011	.689	6.963
2	1	−1.039	.844	1.000	−3.450	1.372
	3	1.332	.755	.535	−.822	3.487
	4	2.787*	.751	.006	.641	4.933
3	1	−2.372	.931	.103	−5.031	.288
	2	−1.332	.755	.535	−3.487	.822
	4	1.454	.715	.314	−.588	3.497
4	1	−3.826*	1.099	.011	−6.963	−.689
	2	−2.787*	.751	.006	−4.933	−.641
	3	−1.454	.715	.314	−3.497	.588

Based on estimated marginal means
a Adjustment for multiple comparisons: Bonferroni.
* The mean difference is significant at the .05 level.

Figure 12.15 shows the percentage of high-intensity activity performed by international and club level netball players in the four quarters of a match. Quarter had a significant influence on the percentage of time spent performing high intensity activity ($F_{3,78} = 7.4$, $p < 0.001$, partial $\eta^2 = 0.223$) with Bonferroni-adjusted, post hoc tests revealing that players performed significantly less high intensity activity in the 4th quarter than in the 1st ($p = 0.011$) and 2nd quarters ($p = 0.006$). Neither level of the player ($F_{1,26} = 0.1$, $p = 0.824$, partial $\eta^2 = .002$) not the interaction of level and quarter ($F_{3,78} = 1.8$, $p = 0.160$, partial $\eta^2 = 0.064$) had a significant influence on the percentage of time spent performing high-intensity activity.

factorial ANOVA

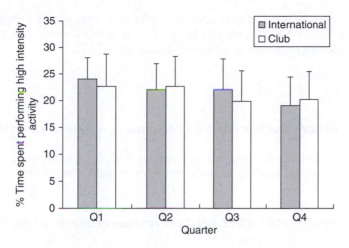

Figure 12.15 Percentage time spent performing high intensity activity by international and club level netball players in each quarter of a match.

ANOVA TESTS WITH MORE THAN TWO FACTORS

The three examples described so far in this chapter are all two-way ANOVA tests. The three-way factorial ANOVA includes three factors which may all be within-subjects effects, or which may all be between-subjects effect or which may be a combination of between-subjects effects and within-subjects effects. The purpose of the test is to determine if each of the three main effects has an influence on some numerical dependent variable when controlling for variance due to the other two factors as well as determining if the combined effect of any pair of factors or all three factors has an influence on the dependent variable of interest. These tests are performed in SPSS using the facilities already described in this chapter, the only difference being that three factors are included instead of two. On occasions, ANOVA tests may include more than three factors. For example, one study of score-line effect on the proportion of points won in tennis done by the author used two within-subjects factors, two between-subjects factors and two covariates (O'Donoghue, 2001). The within-subjects effects were point score within the game (measured at 16 levels – 0–0, 0–15, . . ., 40–40) and player within the match (measured at two levels – winning and losing player). Remember that although the winning and losing players are different players, their performances are related through opposition effect. Furthermore, the unit of analysis in this particular study was match rather than player performance. The two between-subjects effects were gender of the players (measured at two levels) and court surface on which the match was played (measured at four levels). The use of covariates makes this test a multifactor ANCOVA test. The two covariates were log transformed versions of the world rankings of the two players within the match.

SUMMARY

Factorial ANOVA tests are ANOVA tests where the effect of more than one factor on some dependent variable is analysed. These tests allow the effect of each factor to be tested controlling for other factors in the model. The effects of interactions between factors are also tested by factorial ANOVA tests. The factors included in a factorial ANOVA can include between-subjects effects as well as within-subjects effects.

EXERCISES

Exercise 12.1. Gender and age effect on playground activity at lunch time

Using the file ex12.1-Playground.SAV, repeat the analysis of playground activity described in this chapter except apply this to playground activity at lunch break rather than morning break.

Exercise 12.2. Gender and surface effect on inter-serve time in Grand Slam tennis

The file ex12.2-Inter-serve.SAV contains the mean time between first and second serves for second serve points in Grand Slam tennis matches played by men and women on four different court surfaces. Use a two-way factorial ANOVA to analyse the effect of gender and court surface and their interaction on mean inter-serve time.

Exercise 12.3. Venue and period effect on work-rate in professional soccer

A player tracking system is used by a professional soccer club to determine the percentage of time players spend moving at 4 m.s⁻¹ or faster. There are 21 players who have played at least three full 90-minute games at home and at least three full 90-minute games away from home. For each of these players, mean values are produced for home and away matches for the percentage of each of the six 15 minute sections of a match the player spent moving at 4 m.s⁻¹ or faster. Do an analysis of variances test including venue and period as within subjects factors measured at two and six levels respectively. The data are found in the file ex12.3-soccer-tracking.SAV.

Exercise 12.4. 400m hurdles performance

Perform a two-way ANOVA on the data in file ex12.4-hurdles.SAV. The conceptual variable of interest is time between touchdown of the lead foot for consecutive hurdles and the independent variables whose effect is to be tested are hurdle (H2 to H10), which is a within-subjects factor, and gender which is a between-subjects effect.

Exercise 12.5. Three-way ANOVA with the soccer player tracking data

Consider the soccer example used in Exercise 12.3 where a within–within design was used to test the effect of venue and 15-minute period within a set of 21 players who had played at least three full 90-minute matches at home and away from home. Repeat the analysis of this data, except on this occasion add positional role as a between-subjects effect measured at three levels (defender, midfielder and forward).

Exercise 12.6. Three-way ANOVA to analyse activity in the playground

Consider the example used in this chapter to illustrate the between–between design which was also used in Exercise 12.1. Add break time as a within-subjects factor measured at two levels (morning break and lunch break). Perform the three-way ANOVA including break time as a within-subjects effect as well as class and gender as between-subjects factors. Report the significance of all three main effects and all possible interaction effects.

PROJECT EXERCISE

Exercise 12.7. Training and competition hours done by athletes in different types of sport

Devise a data collection form to record training and competition hours performed in a typical training week by sportsmen and sportswomen. Include sections allowing the estimation of competition, tactical preparation, technical sessions, conditioning work and any other training activity requiring an investment of time. There are obvious limitations to any such data collection form, especially with training patterns varying between pre-season preparation and competition seasons in some sports. Add two final questions to the form for gender and type of sport (individual sport or team sport). Gather data from your classmates using the form and enter the total training hours, competition hours, gender and preferred sport type for each participant into an SPSS data sheet. Apply a two-way ANOVA test to investigate the effect of gender, type of sport and their interaction on training hours and then do the same for competition hours. Do the data satisfy the assumptions of the test.

CHAPTER 13

MULTIVARIATE ANOVA

INTRODUCTION

The ANOVA and ANCOVA tests covered in Chapter 12 are univariate tests because they only involve a single dependent variable. No matter how many independent factors are included, if there is a single dependent variable, the test is a univariate ANOVA or ANCOVA. This chapter covers multivariate ANOVA (MANOVA) and multivariate ANCOVA (MANCOVA) tests where there are two or more dependent variables. In a MANOVA test, all participants must be measured on the set of dependent variables of interest. Procedures have been proposed for replacing missing data in multivariate tests but there are no firm guidelines for dealing with this issue (Tabachnick and Fidell, 2007: 63). Therefore, it is recommended that any participant who has not been measured using the full set of dependent variables should be excluded from a study where a MANOVA test is being done. The MANOVA test applies the F test like univariate ANOVA tests do except to a single variable which is an optimal linear combination of the dependent variables of interest (Mardia *et al.*, 1994: 2; Thomas and Nelson, 1996: 180–1). Weights are used to produce this optimal linear combination. This single combined variable is optimized to maximize any variance that can be attributed to the independent variables(s). One text suggested that MANOVA tests also involve more than one independent variable (Thomas and Nelson, 1996: 180). The argument for this is that discriminant analysis can be used where there is a single discrete independent variable. The author of this book takes a different view that discriminant function analysis and MANOVA tests have different purposes as we will see in Chapter 16. In discriminant function analysis, we hypothesize that some discrete dependent variable is influenced by a set of continuous independent variables. A single factor MANOVA test, however, tests the hypothesis that the combined set of continuous variables is influenced by the discrete factor which is the independent variable in the analysis. Whether using MANOVA tests or discriminant tests, we should always consider the conceptual model that we are testing, understanding which variables may be influenced by other variables. This will help avoid putting the 'cart before the horse'. Other authors agreeing with this point of view include Ntoumanis (2001: 100–5) and Hinton *et al.* (2004: 241–50) who show how a one-way MANOVA tests can be done in SPSS.

Typically, we are interested in the individual dependent variables in their own right. So why don't we not just performance series of univariate ANOVA tests to analyse the effect of our independent variable(s) and any interaction effects on the dependent variables? Hinton

(2004: 308–9) explained the rationale for MANOVA tests from the point of view of Type I Error inflation. If we had a set of D dependent variables each using an α value of 0.05, the overall chance of a Type I Error being made in one or more univariate ANOVA tests would be $1 - 0.95^D$. This is because there is a 0.95 chance of avoiding a Type I Error each time we perform one of our univariate ANOVA tests and as we perform further univariate ANOVA tests, the overall chance of avoiding a Type I Error is reduced ($0.95 \times 0.95 \times \ldots \times 0.95$). If we have two dependent variables, the experimentwise Type I Error probability inflates to 0.098; if we have six dependent variables, it inflates to 0.265; and if we have 14 dependent variables, it inflates to 0.512 making a Type I Error more likely than not. The MANOVA test is a single test that can performed using an α value of 0.05 (for example). Therefore, MANOVA is to ANOVA what ANOVA is to post hoc tests. The MANOVA test can be used to contain Type I Error probability prior to performing individual univariate ANOVA tests on the dependent variables in the same way that a univariate ANOVA test can be used to contain Type I Error probability prior to comparing different samples distinguished by the independent variable. If the MANOVA test fails to find a significant influence of some independent factor, then there is no need to use univariate ANOVA tests to analyse the effect of this factor on the individual dependent variables. This is also the case for interaction effects so the typical process of using a MANOVA is:

1   Apply the MANOVA test to the set of dependent variables.
2   For any main or interaction effects where the MANOVA has revealed a significant influence, apply univariate ANOVA tests to the individual dependent variables.
3   For any dependent variables that are significantly influenced by any main effect measured at more than two levels, apply post hoc tests to compare individual pairs of samples.
4   For any dependent variables that are influenced by the interaction of two or more independent variables, compare the different levels of the independent variables the same way as when an interaction effect is found in a factorial ANOVA.

Some sport and exercise scientists wish to restrict the probability of a Type I Error to 0.05 (or some other value) at each stage of data analysis. For example, consider a study to compare the effect of socio-economic group (measured at six levels) on a set of five items measured using a well-being questionnaire instrument. We could initially perform a single one-way MANOVA test to analyse the effect of socio-economic group on the combined set of well-being items using an α value of 0.05. In the event of the MANOVA test revealing a significant socio-economic group effect, we could perform a series of five univariate ANOVA tests: one for each item of the well-being construct. Equation 13.1 shows how we would calculate the α value to be used with each of these univariate ANOVA tests to ensure the probability of a Type I Error occurring in one or more of these tests does not inflate beyond 0.05; D is the number of dependent variables, D = 5. In this example the critical p value to be used is 0.0102.

$$\alpha = 1 - e^{\ln(0.95)/D} \tag{13.1}$$

Now let us imagine that three of these items are significantly influenced by socio-economic group according to the univariate ANOVA tests. With six socio-economic groups, there will be 15 pairs of socio-economic groups to be compared for each of the three significant dependent variables. The figure of 15 is determined using equation 11.1 in Chapter 11 of this book. With a total of 45 pairwise comparisons of socio-economic group to be performed, restricting the overall Type I Error probability to 0.05 at this stage will require us to use an α value of 0.0011 according to equation 13.1. Note that in this use of equation 13.1 we are using D to represent the number of post hoc tests to be performed rather than the number of dependent variables.

There are a number of different ways to compute the F statistic used in a MANOVA test: SPSS reports Wilk's λ, Hotelling's T^2, Pillai's trace and Roy's Largest Root. In this book, Wilk's λ is used as it is the most commonly used method of doing a MANOVA test (Hinton, 2004: 311). Wilk's λ uses the determinants of matrices to express the within group (or error) variances as a proportion of the total variance. This gives a value between zero and one with lower values associated with F values that could be significant. The matrix algebra used in the calculation of Wilk's λ is beyond the scope of this book, but interested readers are referred to Tabachnick and Fidell, 2007: 253–61). This chapter covers single factor MANOVA tests as well as multifactor MANOVA tests using examples from fitness testing, exercise and health and sports psychology. Both between-subjects factors and within-subjects factors are covered as well as factorial MANOVA tests that include both between- and within-subjects factors. MANCOVA tests will be mentioned but without using an example for reasons explained later in the chapter.

SINGLE FACTOR MANOVA TESTS (BETWEEN-SUBJECTS EFFECTS)

Purpose of the test

We will refer to this test as a one-way MANOVA. The independent factor of interest to this test is the effect of some discrete grouping variable that allows independent samples to be identified. The purpose of the one-way MANOVA test is to determine if a single discrete grouping variable has a significant influence on a set of two or more interval or ratio scale dependent variables. If there is a significant influence, follow-up one-way ANOVA tests are applied to analyse the effect of the independent variable on each individual dependent variable.

Assumptions

The assumptions of the one-way MANOVA are the same as those of the one-way ANOVA test with some additional assumptions. The first of the additional assumptions is that there should be more cases than dependent variables for every level (group) in our between-subjects factor (Tabachnick and Fidell, 2007: 250–1).

This author recommends using at least three cases for each combination. Therefore, if we were comparing six socio-economic groups using a set of five items from a well-being construct, we would require at least 90 participants. The one-way MANOVA test also requires the variance-covariance matrices to be similar for the different samples being compared (Ntoumanis, 2001: 99–100, Hinton, 2004: 310). This can be tested using Box's M test where a p value of greater than 0.05 indicates that the assumption is satisfied (Hinton *et al.*, 2004: 247).

The dependent variables should not be highly correlated to each other (Ntoumanis, 2001: 100) and there should be no univariate or multivariate outliers. Univariate outliers can be identified by exploring the individual dependent variables. Multivariate outliers can be identified by exploring 'distance' variables that can be produced for the combined dependent variables. Ntoumanis (2001: 100) names the Mahalanobis distance criterion as a way of testing for multivariate outliers.

Example: Fitness testing of women's Gaelic footballers

A study was done to investigate differences between female Gaelic footballers of different positions for fitness test scores and anthropometric variables (Boyle *et al.*, 2002). There were 18 dependent variables which were hypothesized to be influenced by positional role. These were:

- Height (m)
- Body mass (kg)
- Age (years)
- %Body fat (Durnin and Wormersley, 1974)
- Estimated $\dot{V}O_2$ max (mL.kg^{-1}.min^{-1}) from a multistage fitness test (Ramsbottom *et al.*, 1988)
- Vertical jump performance made with the arms (cm) (Impellizzeri *et al.*, 2007)
- Vertical jump performance made without the use of the arms (cm) (Carling *et al.*, 2009: 143)
- Sit and reach test (cm) (YMCA of the USA, 2000: 158–60)
- Back hyperextension test (cm)
- The number of press ups that can be performed in a minute
- Grip strength with left arm (kgF)
- Grip strength with right arm (kgF)
- Back leg test strength (kgF)
- The number of sit ups that can be performed in a minute
- 10 m sprint performance (s) (best of three attempts with a running start timed using electronic speed gates)
- 20 m sprint performance (s) (best of three attempts with a running start timed using electronic speed gates)
- Fatigue (%)
- Hand pass test (Mulligan, 2001).

Readers interested in the precise methods of measuring these variables are referred to Mulligan's (2001) MSc dissertation. Positional role was classified into one of five different broad positional groups; the study included 41 players with the smallest group being goalkeepers (n = 4) with 9 full backs, 13 wing halves, 7 midfielders and 8 full forwards also being included in the study. Clearly, there is not enough data for the number of groups and dependent variables involved which is why Boyle *et al.* (2002) just reported univariate analysis results. However, we will use this data to illustrate the process of performing a MANOVA test.

A MANOVA test is used to determine if positional role has a significant influence on the set of fitness test scores and anthropometric variables as a whole. If there is a significant influence, then follow up one-way ANOVA tests and possibly Bonferroni-adjusted, post hoc tests are used in the analysis of individual fitness test scores and anthropometric variables.

SPSS

The data for the ladies Gaelic football fitness tests are found in the file 13-l_gaelic.SAV. In SPSS we use **Analyse → General Linear Model → Multivariate** in order to perform a MANOVA test involving exclusively between subjects factors (the one-way MANOVA falls into this category of MANOVA tests). The pop-up window shown in Figure 13.1 appears and we transfer 'position' into the *Fixed factor(s)* area and all the remaining variables except 'BMI' (body mass index) into the *Dependent variables* area. BMI is functionally dependent on height and body mass that are already included in the dependent variables to be analysed.

We can use **Options** to request descriptive statistics, effect size, observed power and homogeneity of variances tests as shown in Figure 13.2. Figure 13.3 shows the **Post Hoc** tests

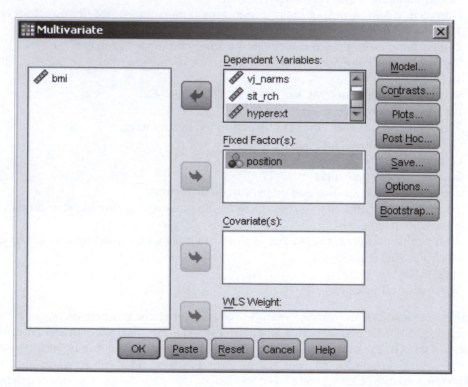

Figure 13.1 Multivariate pop-up window in SPSS.

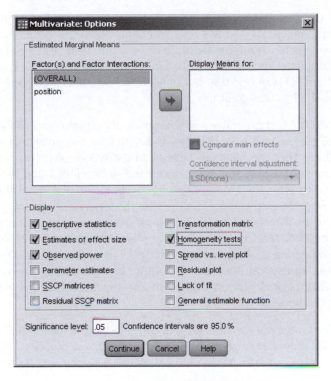

Figure 13.2 Options pop-up window within the Multivariate facility of SPSS.

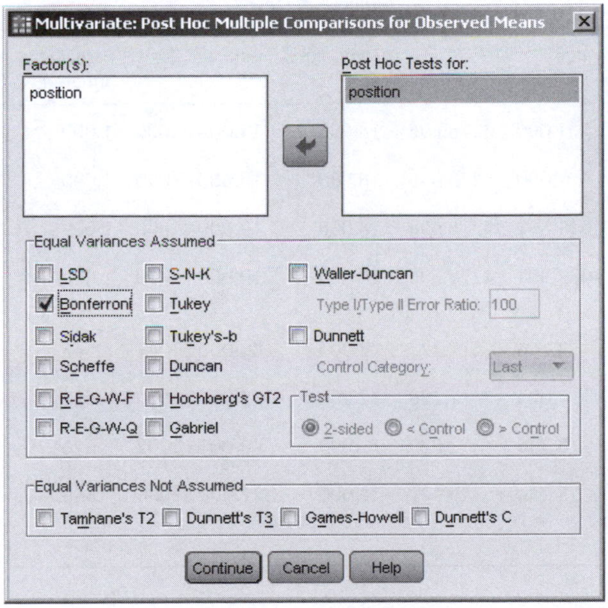

Figure 13.3 Post hoc test selection for MANOVA tests.

pop-up window used to select a post hoc procedure to be applied in the event of the MANOVA and any individual ANOVA tests revealing significant positional effects. In this example we have selected Bonferonni-adjusted, post hoc tests. If we did wish to apply a different α value using the approach described in the introduction to this chapter, we would actually choose the LSD post hoc test, but only consider p values below the α value computed in equation 13.1 to be significant.

The one-way MANOVA test generates a large amount of output as you will see when you try this example in SPSS. First, we have descriptive statistics for five positional groups for all 18 dependent variables. Then we have our table of MANOVA results. This is followed by the results for 18 separate one-way ANOVA tests for the dependent variables individually. All these are produced whether the MANOVA reveals a significant positional effect or not. Then we have 20 pairwise comparisons for each of the 18 dependent variables: equation 11.1 indicates that there are 10 pairs of positional roles if we have five positional roles. However, SPSS provided a post hoc test for each pair of samples both ways round. That is, we see post hoc test results comparing goalkeepers and full backs as well as comparing full backs and goalkeepers. If we request post hoc tests, this output is produced for each dependent variable no matter whether or not the MANOVA and/or any individual one-way ANOVA tests reveal significant positional effects.

Table 13.1 shows the Multivariate Tests table of the SPSS output for the one-way MANOVA. We are interested in the effect of position using the Wilk's λ row of results. We can see that λ = 0.023, the F ratio is 1.726, the p value is 0.010 which is significant. The partial η^2 value of

Table 13.1 One-way MANOVA results produced by SPSS

				Multivariate Tests[d]					
Effect		Value	F	Hypothesis df	Error df	Sig.	Partial Eta Squared	Noncent. Parameter	Observed Power[b]
Intercept	Pillai's Trace	1.000	4278.696[a]	18.000	19.000	.000	1.000	77016.521	1.000
	Wilks' Lambda	.000	4278.696[a]	18.000	19.000	.000	1.000	77016.521	1.000
	Hotelling's Trace	4053.501	4278.696[a]	18.000	19.000	.000	1.000	77016.521	1.000
	Roy's Largest Root	4053.501	4278.696[a]	18.000	19.000	.000	1.000	77016.521	1.000
position	Pillai's Trace	2.186	1.472	72.000	88.000	.042	.546	106.013	.996
	Wilks' Lambda	.023	1.726	72.000	77.069	.010	.611	121.043	.998
	Hotelling's Trace	8.346	2.029	72.000	70.000	.002	.676	146.052	1.000
	Roy's Largest Root	5.254	6.421c	18.000	22.000	.000	.840	115.584	1.000

a Exact statistic
b Computed using alpha = .05
c The statistic is an upper bound on F that yields a lower bound on the significance level.
d Design: Intercept + position

0.611 indicates a meaningful positional effect and the observed power probability of 0.998 means that if there is a genuine positional effect in the wider population of women's Gaelic football players, we are most likely to show it with this test if we have used a random sample.

Reporting results

The MANOVA results would be reported in text as follows with a table of descriptive results also showing follow-up univariate ANOVA results and using symbols to show significant post hoc test results.

A one-way MANOVA test revealed that positional role had a significant influence on the set of 18 fitness test scores and anthropometric variables ($\lambda = 0.023$, $F_{72, 77.069} = 1.726$, $p = 0.010$, partial $\eta^2 = 0.611$). Table 13.2 shows that there were six of the 18 fitness test scores and anthropometric variables that were significantly influenced by positional role.

Table 13.2 Summary of analysis

| Variable | Positional role | | | | | ANOVA |
	GK (n = 4)	FB (n = 9)	WH (n = 13)	MF (n = 7)	FF (n = 8)	$F_{4,36}$
Height	1.64 ± 0.06	1.64 ± 0.08	1.63 ± 0.03	1.66 ± 0.04	1.63 ± 0.05	0.6
Weight	70.5 ± 7.1	64.4 ± 9.6	59.0 ± 5.4 ^	66.1 ± 3.2	64.5 ± 4.6	3.3 *
Age	24.0 ± 8.0	19.4 ± 5	22.3 ± 7.5	25.1 ± 6.1	21.8 ± 3.8	0.9
Body fat	31.8 ± 5.2	29.0 ± 4.7	28.1 ± 3.3	29.7 ± 3.7	29.3 ± 1.9	0.8
Est $\dot{V}O_2$max	31.3 ± 1.8	37.4 ± 4.5	38.1 ± 4.4	42.7 ± 4.0 ^	40.8 ± 6.0 ^	4.6 **
Vert jump (arms)	46.0 ± 2.4	48.0 ± 9.7	45.8 ± 6.2	54.3 ± 5.4	49.8 ± 9.8	1.6
Vert jump (no arms)	39.8 ± 1.9	40.2 ± 6.8	38.8 ± 7.3	46.9 ± 4.3	42.1 ± 9.2	1.7
Sit and reach	26.8 ± 2.2	26.9 ± 5.4	27.6 ± 5.7	27.9 ± 8.2	27.6 ± 7.2	0.0
Babk hyper-ext	33.9 ± 1.9	35.6 ± 7.7	32.8 ± 6.6	29.8 ± 8.1	32.5 ± 8.9	0.7
Grip str (L)	30.0 ± 3.0	30.9 ± 4.4	29.2 ± 4.2	34.7 ± 3.3	28.4 ± 4.6	2.8 *
Grip str (R)	29.5 ± 4.0	62.3 ± 90.2	29.8 ± 6.8	36.1 ± 3.7	29.4 ± 5.1	1.0
Back leg test	111.5 ± 13.8	96.3 ± 15.4 #	99.5 ± 27.2 #	138.3 ± 20.3	110.9 ± 35.2	3.5 *
Press ups	32.5 ± 7.9	41.7 ± 15.5	44.8 ± 15.8	39.6 ± 9.6	36.9 ± 14.9	0.8
Sit ups	44.0 ± 2.7	51.8 ± 10.5	59.0 ± 12.3	59.6 ± 14.1	59.3 ± 6.5	2.2
10m sprint	2.1 ± 0.2	1.8 ± 0.2	1.8 ± 0.1	1.8 ± 0.1	1.8 ± 0.1	2.6
20m sprint	4.2 ± 0.2	3.9 ± 0.5	4.0 ± 0.3 #	3.4 ± 0.6	3.4 ± 0.5	4.4 **
Fatigue	20.4 ± 8.9	5.5 ± 3.2 ^	5.1 ± 2.4 ^	5.7 ± 3.3 ^	6.1 ± 3.2 ^	13.9 ***
Hand pass	27.3+8.5	28.1+4.5	29.9+4.5	27+7.1	28.4+6.9	0.4

Univariate ANOVA results: * $p < 0.05$, ** $p < 0.01$, *** $p < 0.001$.
Bonferonni adjusted post hoc tests results:
^ Significantly different to GK
# Significantly different to MF

The Bonferroni-adjusted, post hoc test is a conservative post hoc test and so we occasionally find a situation such as our result for grip strength (left), which is that there is a significant positional effect but the post hoc tests could not identify any pairs of positions that significantly differed for grip strength (left). Examining the post hoc tests also revealed a significant difference in the performance of the 10m sprint between goalkeepers and full backs. This is not reported in Table 13.2, because the univariate ANOVA test for this dependent variable was not significant.

FACTORIAL MANOVA TESTS (BETWEEN-SUBJECTS EFFECTS)

Purpose

As well as being used with a single between-subjects effect, MANOVA tests can be used with two or more between-subjects effects. A factorial MANOVA can test the effect of two or more independent factors as well as the various interactions of factors on a set of two or more dependent variables. The stages of performing a factorial MANOVA test are similar to those used to perform a one-way MANOVA test except we also need to address interaction effects as well as main effects. Consider a two-way MANOVA to analyse the effect of gender and socio-economic group on a set of five items of a well-being construct. Gender is measured at two levels and socio-economic group is measured at six levels. The steps of the analysis are as follows:

1   Perform the two-way MANOVA test which produces p values for the effects of gender, socio-economic group and their interactions. If neither of the main effects nor their interaction have a significant influence on the set of dependent variables, then we stop the inferential testing at this point.
2   For each of the main effects that were significant according to the MANOVA test, inspect the results of the univariate tests of those main effects on individual dependent variables. Gender is only measured at two levels so we do not have to consider whether or not to employ post hoc tests to compare male and female participants. Socio-economic group, however, is measured at six levels so post hoc tests may need to be produced.
3   If the interaction of gender and socio-economic group has a significant influence on the combined set of dependent variables according to the MANOVA test, then inspect the results for the interaction effect within the univariate ANOVA tests of the individual dependent variables.
4   It is possible, but unlikely, that we may have one or more significant effect from the MANOVA test but no significant effects arising from any of the univariate ANOVA tests. If this does happen, the inferential analysis can stop at this point.
5   If we reach this point, there may be some dependent variables that are influenced by main effects only, some influenced by main effects and interaction effects and some influenced by interaction effects only. If there are any significant interaction effects according to the univariate ANOVA tests, then post hoc tests should be used to compare pairs of socio-economic groups for the male and female participants separately. If socio-economic group has a significant effect on any dependent variable but there is no significant interaction of gender and socio-economic group on that dependent variable, then post hoc tests should be applied to each pair of socio-economic groups for the sample as a whole.

250

A MANOVA test involving two or more between subjects effects is performed in SPSS using **Analyse → General Linear Model → Multivariate** as before except transferring all independent variables of interest into the *Fixed factors* area in the multivariate pop-up window shown in Figure 13.1. The default multivariate MANOVA will analyse each main effect and all possible interaction effects. This can result in the same proliferation of interaction effects discussed in Chapter 12. A four-way MANOVA test is much more complex than a four-way univariate ANOVA described in Chapter 12. The **Model** button in Figure 13.1 allows us to specify which main effects and interaction effects we wish to analyse within a more focused MANOVA test.

Assumptions

The assumptions of the one-way MANOVA test also apply to the factorial MANOVA test. Note that there should be at least one participant for each dependent variable for each combination of factor levels. An example with sufficient data to give a chance of satisfying the assumptions will be used to illustrate the factorial MANOVA and so the SPSS output for Box's M test will be covered.

Example: Gender and sport type effect on burnout

A preliminary version of the Athlete Burnout Questionnaire instrument (Raedeke and Smith, 2001) produces three dimensions of burnout or potential for burnout that are measured on a 5 to 25 scale. Each dimension score was made up of the sum of five question responses measured on a five-point Likert scale: 1 (almost never), 2 (rarely), 3 (sometimes), 4 (frequently) and 5 (almost always). These dimensions are: reduced sense of accomplishment (RSA), emotional and physical exhaustion (EE) and devaluation (DEV). The questionnaire is completed by 93 male student athletes and 82 female student athletes. The 93 males consist of 29 who participate in individual sports and 64 who participate in team sports, while the 82 females consist of 28 who participate in individual sports and 54 who participate in team sports. We wish to determine if gender, type of sport and/or their interaction have an influence on any dimension of athlete burnout. A two-way MANOVA can be applied to the three dimensions including gender and type of sport as between-subjects effects.

SPSS

This example uses data for the 175 fictitious participants from the file 13-burnout.SAV. We use **Analyse → General Linear Model → Multivariate** which activates the pop-up window shown in Figure 13.1. The variables 'RSA', 'EE' and 'DEV' are transferred to the *Dependent variables* area and 'Gender' and 'Sport' are transferred into *Fixed factor(s)* area. Because both of our independent variables are measured at two levels, we do not require post hoc tests. The **Options** facility is still used to request descriptive statistics, homogeneity tests, effect size estimates and power calculations. Table 13.3 shows that the assumption of homogeneity of covariance matrices is satisfied because the p value from Box's M test is greater than 0.05 (p = 0.761).

Table 13.3 Result of Box's M test to evaluate the assumption of homogeneity of covariance matrices

Box's Test of Equality of Covariance Matrices[a]

Box's M	14.026
F	.750
df1	18
df2	45228.580
Sig.	.761

Tests the null hypothesis that the observed covariance matrices of the dependent variables are equal across groups.
a Design: Intercept + Gender + Sport + Gender * Sport

The results of the MANOVA test are shown in Table 13.4. Using the Wilk's λ row of results, we can see that neither gender nor sport has a significant influence on athlete burnout and that there is no significant interaction of gender and sport. Therefore, we do not need to report on the univariate ANOVA results for RSA, EE or DEV individually.

Reporting results

In this example, the descriptive results should still show the means and standard deviations for the 2 × 2 (gender × type of sport) groups even though there were no significant differences. Some may prefer to show the descriptive statistics for the sample of 175 student athletes as a single group in a situation like this. However, showing the separate groups helps reinforce any similarities that the researcher wishes to discuss. Table 13.5 shows the descriptive results which would be accompanied by the following text.

> Table 13.5 shows that the scores for athlete burnout were similar between the different groups of student athletes. There was no significant influence of gender ($\lambda = 0.992$, $F_{3,169} = 0.5$, $p = 0.718$), type of sport ($\lambda = 0.986$, $F_{3,169} = 0.8$, $p = 0.511$) or their interaction ($\lambda = 0.973$, $F_{3,169} = 1.6$, $p = 0.197$).

REPEATED MEASURES MANOVA TESTS

Purpose

The repeated measures MANOVA test analyses the effect of one or more within-subjects factors on a set of dependent variables. The repeated measures MANOVA tests of concern to this section of the chapter are those without any between-subjects factors (grouping variables) being included. Where a between-subjects factor is included in a repeated measures MANOVA, the current textbook refers to the test as a mixed MANOVA. In a repeated measures MANOVA test, each dependent variable is measured for each participant for each level of the independent variable. The levels of the independent variable represent different conditions under which data are collected for all participants. For example, in a diurnal variation we may be interested in the effect of time of day on a set of vigilance and reaction

multivariate ANOVA

Table 13.4 Results of a factorial MANOVA results with two between subjects effects

Multivariate Tests[c]

Effect		Value	F	Hypothesis df	Error df	Sig.	Partial Eta Squared	Noncent. Parameter	Observed Power[b]
Intercept	Pillai's Trace	.934	801.471[a]	3.000	169.000	.000	.934	2404.413	1.000
	Wilks' Lambda	.066	801.471[a]	3.000	169.000	.000	.934	2404.413	1.000
	Hotelling's Trace	14.227	801.471[a]	3.000	169.000	.000	.934	2404.413	1.000
	Roy's Largest Root	14.227	801.471[a]	3.000	169.000	.000	.934	2404.413	1.000
Gender	Pillai's Trace	.008	.450[a]	3.000	169.000	.718	.008	1.350	.139
	Wilks' Lambda	.992	.450[a]	3.000	169.000	.718	.008	1.350	.139
	Hotelling's Trace	.008	.450[a]	3.000	169.000	.718	.008	1.350	.139
	Roy's Largest Root	.008	.450[a]	3.000	169.000	.718	.008	1.350	.139
Sport	Pillai's Trace	.014	.772[a]	3.000	169.000	.511	.014	2.317	.214
	Wilks' Lambda	.986	.772[a]	3.000	169.000	.511	.014	2.317	.214
	Hotelling's Trace	.014	.772[a]	3.000	169.000	.511	.014	2.317	.214
	Roy's Largest Root	.014	.772[a]	3.000	169.000	.511	.014	2.317	.214
Gender * Sport	Pillai's Trace	.027	1.576[a]	3.000	169.000	.197	.027	4.728	.410
	Wilks' Lambda	.973	1.576[a]	3.000	169.000	.197	.027	4.728	.410
	Hotelling's Trace	.028	1.576[a]	3.000	169.000	.197	.027	4.728	.410
	Roy's Largest Root	.028	1.576[a]	3.000	169.000	.197	.027	4.728	.410

a Exact statistic

b Computed using alpha = .05

c Design: Intercept + Gender + Sport + Gender * Sport

Table 13.5 Athlete burnout in student athletes (mean+SD)

| Dimension | Male | | Female | |
	Individual Sport (n = 29)	Team Sport (n = 64)	Individual Sport (n = 28)	Team Sport (n = 44)
RSA	13.0 ± 4.0	13.5 ± 3.7	13.6 ± 3.6	12.6 ± 3.6
EE	14.1 ± 4.9	12.3 ± 4.7	12.4 ± 4.0	12.6 ± 3.1
DEV	12.7 ± 6.4	11.5 ± 5.1	12.6 ± 4.1	12.3 ± 5.1

variables. All participants in the study would be tested at 04:00, 08:00, 12:00, 16:00, 20:00 and 00:00. Participants would be measured in each of these conditions a week apart in a Latin square design to counter fatigue and familiarity effects.

It is also possible to perform a multifactor MANOVA test involving exclusively within-subjects factors as independent variables. This is done by defining more than one within-subject factor as shown in the within–within design Chapter 12 but setting up measure variables as will be illustrated in the current section of this chapter.

Assumptions

Repeated measures MANOVA tests make the same assumptions as MANOVA tests with between-subjects effects. In addition, the assumption of sphericity should be checked when performing any follow-up repeated measures ANOVA tests on individual dependent variables.

Example: Comparing anxiety before training and competitive matches

The purpose of this example is to compare different dimensions of a modified version of the Sport Anxiety Scale 2 (SAS2) between training and competitive matches. The SAS2 instrument used (Smith et al., 2006) has had a directional scale added to it (Jones and Swain, 1992). This determines whether participants view their anxiety as facilitative or debilitative. The participants in this study would have been instructed to complete the questionnaire in relation to a training session or a forthcoming competition. The same 175 fictitious participants used in the athlete burnout example participate in the current example. The participants complete the modified SAS2 questionnaire prior to a training match and prior to a competitive match. There are three main components of the modified SAS2 instrument: worry, somatic anxiety and concentration disruption. Each of these is represented by an intensity score between 5 and 20 and a direction score between −15 and +15. This is because the SAS2 instrument allows for the possibility of anxiety being viewed as facilitative or debilitative by different athletes. This gives a total of six values that make up the SAS2 instrument. A repeated measures MANOVA test can be used to compare the 175 student athletes between the two types of match in terms of anxiety as a whole, with follow-up repeated measures ANOVA tests being applied to the three intensity scores and three direction scores to be compared between the match types individually if necessary.

SPSS

This example uses the data in the file 13-SAS2.SAV. Hinton *et al.* (2004: 250–9) described how a repeated measures MANOVA test can be accomplished in SPSS. Instead of using **Analyse → General Linear Model → Multivariate** as we did for the MANOVA tests involving exclusively between subjects factors, we use **Analyse → General Linear Model → Repeated Measures** which activates the pop-up window shown in Figure 13.4. As with a univariate repeated measures ANOVA test, our SPSS datasheet will not contain a column for the within subjects factor (match type) so we need to enter 'type' into the *Within-subject factor name* area and '2' into the *Number of levels* area. We enter '2' because there are two match types: a training match and a competitive match. The fundamental difference in the use of this pop-up window between univariate and multivariate versions of the repeated measures ANOVA test is that we enter measure names. The SPSS datasheet will have a column for each of our six variables under training and competitive match conditions. We wish to enter the variable names of interest irrespective of their representation for training and competitive matches in the datasheet. Therefore, we enter six variable names in turn into the *Measure name* area clicking on **Add** to transfer each into the variables area below *Measure name*. As Figure 13.4

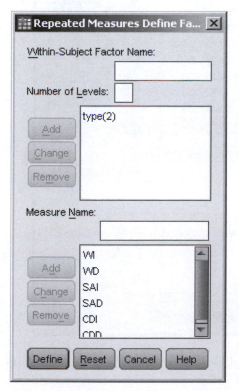

Figure 13.4 Use of the repeated measures ANOVA facility in SPSS to perform a repeated measures MANOVA.

shows, we have called the variables WI (worry intensity), WD (worry direction), SAI (somatic anxiety intensity), SAD (somatic anxiety direction), CDI (concentration disruption intensity) and CDD (concentration disruption direction).

Once we have added all six conceptual dependent variables, we click on **Define** which activates the pop-up window used to identify the repeated measures of each conceptual variable within the SPSS datasheet. This is illustrated in Figure 13.5. For example 'SAI_TRAIN' represents SAI before a training match, while 'SAI_COMP' represents SAI before a competitive match. These two variables are therefore intentionally transferred into the *Within-subjects variables* area for the repeated measures of the SAI measure labelled '(1,SAI)' and '(2,SAI)' respectively. We have chosen to use 1 to represent training matches and 2 to represent competitive matches. We could have done this the other way round as long as what we do is done consistently for all six dependent variables.

The **Options** facility allows us to request descriptive statistics, effect size estimates and power calculations. We do not need to request post hoc tests because match type is only measured at two levels so any difference would have to be between those two types of

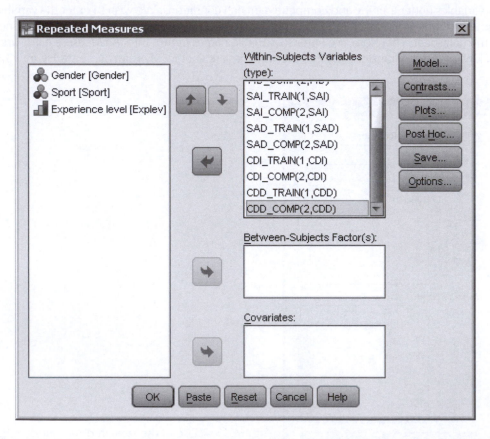

Figure 13.5 Linking variables (columns) in the SPSS datasheet to repeated measures of conceptual dependent variables.

multivariate ANOVA

match. It is worth reminding readers that if it was necessary to perform post hoc tests on any within-subjects factor, this has to be done through **Options** rather than **Post Hoc** as explained in Chapter 11. Once we click on **Continue** to close down the Options pop-up window and **OK** to close down the repeated measures pop-up window, SPSS performs the repeated measures MANOVA test on our data producing the output shown in Tables 13.6 and 13.7. Because we have two repeated measures, we automatically satisfy the assumption of sphericity. Using the Wilk's λ row of results in Table 13.6 reveals a significant effect of match type with a p value (Sig.) of less than 0.001. In Table 13.7, we can use the Sphericity Assumed row of results for each dependent variable revealing that each of three intensity scores and each of the three direction scores is significantly influenced by match type. Note that in each case, there is one degree of freedom for the main effect and 174 degrees of freedom for the error effect.

Reporting the results

A key part of using SPSS effectively is to be able to extract the most important information from the SPSS output. The results for this example would typically be represented in a table such as Table 13.8 that allows a direct comparison between training and competitive matches. The right-most two columns are used to augment the univariate inferential results to the descriptive results which helps cut down on the number of words required in the results section of a paper of thesis. The table would be accompanied by the following text which would provide the multivariate results:

> Match type had a significant influence on the overall construct of anxiety ($\lambda = 0.039$, $F_{6,169} = 701.8$, $p < 0.001$, partial $\eta^2 = 0.961$). Table 13.8 shows that each intensity and direction score was significantly influenced by match type with raised intensity values and a negative direction of scores during competitive matches.

MIXED FACTORIAL MANOVA TESTS

Purpose of the test

There are occasions where we are interested in the effect of some repeated measure(s) and some between-subjects factors on the set of dependent variables. The mixed MANOVA is the best test to perform in such a situation. The test will analyse the effect of each factor and the effect of each interaction of factors on the set of dependent variables. If any are significant, follow-up repeated measures, ANOVA tests and factorial ANOVA tests can be applied to the individual dependent variables.

Assumptions

The assumptions of a mixed MANOVA test combine the assumptions of MANOVA tests with between and within-subjects effects.

Table 13.6 Results of a repeated measures MANOVA

Multivariate Tests[c]

Effect			Value	F	Hypothesis df	Error df	Sig.	Partial Eta Squared	Noncent. Parameter	Observed Power[b]
Between Subjects	Intercept	Pillai's Trace	.961	690.305[a]	6.000	169.000	.000	.961	4141.830	1.000
		Wilks' Lambda	.039	690.305[a]	6.000	169.000	.000	.961	4141.830	1.000
		Hotelling's Trace	24.508	690.305[a]	6.000	169.000	.000	.961	4141.830	1.000
		Roy's Largest Root	24.508	690.305[a]	6.000	169.000	.000	.961	4141.830	1.000
Within Subjects	type	Pillai's Trace	.961	701.833[a]	6.000	169.000	.000	.961	4210.999	1.000
		Wilks' Lambda	.039	701.833[a]	6.000	169.000	.000	.961	4210.999	1.000
		Hotelling's Trace	24.917	701.833[a]	6.000	169.000	.000	.961	4210.999	1.000
		Roy's Largest Root	24.917	701.833[a]	6.000	169.000	.000	.961	4210.999	1.000

a Exact statistic
b Computed using alpha = .05
c Design: Intercept
Within Subjects Design: type

Table 13.7 Follow-up repeated measures ANOVA tests for individual dependent variables (only the Sphericity Assumed results from the SPSS output are shown here)

Univariate Tests

Source	Measure		Type III Sum of Squares	Df	Mean Square	F	Sig.	Partial Eta Squared	Noncent. Parameter	Observed Power[a]
type	WI	Sphericity Assumed	124.803	1	124.803	50.187	.000	.224	50.187	1.000
	WD	Sphericity Assumed	915.303	1	915.303	690.354	.000	.799	690.354	1.000
	SAI	Sphericity Assumed	31.500	1	31.500	25.375	.000	.127	25.375	.999
	SAD	Sphericity Assumed	1992.071	1	1992.071	3196.763	.000	.948	3196.763	1.000
	CDI	Sphericity Assumed	89.511	1	89.511	85.582	.000	.330	85.582	1.000
	CDD	Sphericity Assumed	1716.071	1	1716.071	513.557	.000	.747	513.557	1.000
Error (type)	WI	Sphericity Assumed	432.697	174	2.487					
	WD	Sphericity Assumed	230.697	174	1.326					
	SAI	Sphericity Assumed	216.000	174	1.241					
	SAD	Sphericity Assumed	108.429	174	.623					
	CDI	Sphericity Assumed	181.989	174	1.046					
	CDD Sphericity Assumed		581.429	174	3.342					

a Computed using alpha = .05

Table 13.8 Scores for intensity and direction of different dimensions of a modified SAS2 instrument

Dimension	Match type		Univariate ANOVA results	
	Training match	Competitive match	$F_{1,74}$	p
WI	12.3 ± 3.7	13.5 ± 3.7	50.2	< 0.001
WD	−2.1 ± 7.4	−5.4 ± 7.1	690.4	< 0.001
SAI	10.1 ± 3.3	10.7 ± 8.5	25.4	< 0.001
SAD	+0.8 ± 6.4	−4.0 ± 6.0	3196.8	< 0.001
CDI	7.9 ± 2.4	8.9 ± 2.7	85.6	< 0.001
CDD	−0.6 ± 7.7	−5.1 ± 7.3	513.6	< 0.001

Example: Comparing anxiety before training and competitive matches

The mixed MANOVA test will be illustrated using the previous example of modified SAS2 data except including gender as a between-subjects effect. This will allow the influence on anxiety of gender, type of match and the interaction of gender and type of match to be tested.

SPSS

The file 13-SAS2.SAV provides the data for 175 fictitious participants used in this example. The repeated measures facility is used as shown in Figure 13.5 except 'Gender' is now transferred into the *Between-subjects factors* area. Once again, our repeated measure only has two levels and so we have automatically satisfied Mauchly's test of sphericity. Table 13.9 shows that the mixed MANOVA tests the significance of Wilk's λ for gender, type and their interaction. Both main effects are significant but there is no significant interaction. Table 13.10 provides four rows of results for type of sport and the interaction of gender and type of sport for each dependent variable. We are only interested in the Sphericiy Assumed row of results in this example. Table 13.11 shows the effect of gender on each of the dependent variables.

Reporting the results

The descriptive results for female and male participants are shown separately in Table 13.12 with the inferential results being placed in a separate table of ANOVA results (Table 13.13) due to the F-ratios applying to the whole sample rather than individual genders. Note that because the MANOVA test did not reveal a significant interaction for anxiety as a whole, interaction results are not reported for the individual dependent variables. The following text could be used to support the Tables 13.12 and 13.13.

> Both gender ($\lambda = 0.893$, $F_{6,168} = 3.4$, $p = 0.004$, partial $\eta^2 = 0.107$) and type of match ($\lambda = 0.039$, $F_{6,168} = 697.7$, $p < 0.001$, partial $\eta^2 = 0.961$) had a significant influence on anxiety. However, there was no significant influence of the

Table 13.9 SPSS output for mixed MANOVA test results (Multivariate tests)

Multivariate Tests[c]

Effect			Value	F	Hypothesis df	Error df	Sig.	Partial Eta Squared	Noncent. Parameter	Observed Power[b]
Between Subjects	Intercept	Pillai's Trace	.964	758.306[a]	6.000	168.000	.000	.964	4549.835	1.000
		Wilks' Lambda	.036	758.306[a]	6.000	168.000	.000	.964	4549.835	1.000
		Hotelling's Trace	27.082	758.306[a]	6.000	168.000	.000	.964	4549.835	1.000
		Roy's Largest Root	27.082	758.306[a]	6.000	168.000	.000	.964	4549.835	1.000
	Gender	Pillai's Trace	.107	3.361[a]	6.000	168.000	.004	.107	20.168	.933
		Wilks' Lambda	.893	3.361[a]	6.000	168.000	.004	.107	20.168	.933
		Hotelling's Trace	.120	3.361[a]	6.000	168.000	.004	.107	20.168	.933
		Roy's Largest Root	.120	3.361[a]	6.000	168.000	.004	.107	20.168	.933
Within Subjects	type	Pillai's Trace	.961	697.730[a]	6.000	168.000	.000	.961	4186.377	1.000
		Wilks' Lambda	.039	697.730[a]	6.000	168.000	.000	.961	4186.377	1.000
		Hotelling's Trace	24.919	697.730[a]	6.000	168.000	.000	.961	4186.377	1.000
		Roy's Largest Root	24.919	697.730[a]	6.000	168.000	.000	.961	4186.377	1.000
	type * Gender	Pillai's Trace	.031	.885[a]	6.000	168.000	.507	.031	5.312	.344
		Wilks' Lambda	.969	.885[a]	6.000	168.000	.507	.031	5.312	.344
		Hotelling's Trace	.032	.885[a]	6.000	168.000	.507	.031	5.312	.344
		Roy's Largest Root	.032	.885[a]	6.000	168.000	.507	.031	5.312	.344

a Exact statistic
b Computed using alpha = .05
c Design: Intercept + Gender
Within Subjects Design: type

Table 13.10 SPSS output for mixed MANOVA test results (Univariate tests for repeated measure and its interaction with between-subjects effect) (only the Sphericity Assumed results from the SPSS output are shown here)

Univariate Tests

Source	Measure		Type III Sum of Squares	Df	Mean Square	F	Sig.	Partial Eta Squrd	Noncent. Parameter	Observed Power[a]
type	WI	Sphericity Ass	122.376	1	122.376	49.146	.000	.221	49.146	1.000
	WD	Sphericity Ass	917.301	1	917.301	694.452	.000	.801	694.452	1.000
	SAI	Sphericity Ass	31.587	1	31.587	25.309	.000	.128	25.309	.999
	SAD	Sphericity Ass	1985.846	1	1985.846	3170.983	.000	.948	3170.983	1.000
	CDI	Sphericity Ass	87.899	1	87.899	84.081	.000	.327	84.081	1.000
	CDD	Sphericity Ass	1719.806	1	1719.806	515.332	.000	.749	515.332	1.000
type * Gender	WI	Sphericity Ass	1.919	1	1.919	.771	.381	.004	.771	.141
	WD	Sphericity Ass	2.181	1	2.181	1.652	.200	.009	1.652	.248
	SAI	Sphericity Ass	.090	1	.090	.072	.789	.000	.072	.058
	SAD	Sphericity Ass	.086	1	.086	.138	.711	.001	.138	.066
	CDI	Sphericity Ass	1.133	1	1.133	1.084	.299	.006	1.084	.179
	CDD	Sphericity Ass	4.080	1	4.080	1.223	.270	.007	1.223	.196
Error(type)	WI	Sphericity Ass	430.778	173	2.490					
	WD	Sphericity Ass	228.516	173	1.321					
	SAI	Sphericity Ass	215.910	173	1.248					
	SAD	Sphericity Ass	108.342	173	.626					
	CDI	Sphericity Ass	180.856	173	1.045					
	CDD	Sphericity Ass	577.349	173	3.337					

a Computed using alpha = .05

Table 13.11 SPSS output for mixed MANOVA test results (Univariate results for between-subjects effect)

Tests of Between-Subjects Effects

Transformed Variable:Average

Source	Measure		Type III Sum of Squares	df	Mean Square	F	Sig.	Partial Eta Squared	Noncent. Parameter	Observed Power[a]
Intercept	dimension 2	WI	58561.154	1	58561.154	2539.307	.000	.936	2539.307	1.000
		WD	4959.422	1	4959.422	47.963	.000	.217	47.963	1.000
		SAI	37715.083	1	37715.083	1738.724	.000	.910	1738.724	1.000
		SAD	882.962	1	882.962	11.513	.001	.062	11.513	.921
		CDI	24536.142	1	24536.142	2108.202	.000	.924	2108.202	1.000
		CDD	2793.184	1	2793.184	25.530	.000	.129	25.530	.999
Gender	dimension 2	WI	389.885	1	389.885	16.906	.000	.089	16.906	.983
		WD	47.342	1	47.342	.458	.500	.003	.458	.103
		SAI	63.883	1	63.883	2.945	.088	.017	2.945	.400
		SAD	2.916	1	2.916	.038	.846	.000	.038	.054
		CDI	61.970	1	61.970	5.325	.022	.030	5.325	.631
		CDD	.064	1	.064	.001	.981	.000	.001	.050
Error	dimension 2	WI	3989.703	173	23.062					
		WD	17888.532	173	103.402					
		SAI	3752.585	173	21.691					
		SAD	13267.512	173	76.691					
		CDI	2013.447	173	11.638					
		CDD	18927.491	173	109.407					

a Computed using alpha = .05

interaction of gender and type of match on anxiety ($\lambda = 0.969$, $F_{6,168} = 0.9$, $p = 0.507$, partial $\eta^2 = 0.031$). Table 13.12 shows that females had higher intensity scores for all three dimensions of anxiety before training matches and competitive matches. However, Table 13.13 shows that these gender differences were only significant for WI and CDI. There was no significant gender effect on any of the direction scores. Both female and male participants showed an increase in all three intensity scores before competitive matches. All direction scores became more debilitative before competitive matches for both female and male participants.

MANCOVA TESTS

A MANCOVA test is a MANOVA test where one or more covariates have been included. These are rarely used and often there is little rationale to adjust an optimal linear composite

Table 13.12 Scores for intensity and direction of different dimensions of a modified SAS2 instrument

Dimension	Match type	
	Training match	Competitive match
Females		
WI	13.5 ± 3.4	14.5 ± 3.6
WD	−2.4 ± 7.6	−5.8 ± 7.2
SAI	10.5 ± 3.6	11.2 ± 3.7
SAD	+0.7 ± 7.0	−4.1 ± 6.6
CDI	8.4 ± 2.6	9.3 ± 2.9
CDD	−0.5 ± 7.3	−5.2 ± 7.0
Males		
WI	11.2 ± 3.6	12.6 ± 3.6
WD	−1.9 ± 7.2	−5.0 ± 7.0
SAI	9.7 ± 3.0	10.3 ± 3.2
SAD	+0.9 ± 5.9	−3.9 ± 5.5
CDI	7.4 ± 2.2	8.5 ± 2.5
CDD	−0.7 ± 8.0	−5.0 ± 7.5

Table 13.13 ANOVA results for individual dimensions of modified SAS2

Dimension	Gender		Type of Match	
	$F_{1,137}$	p	$F_{1,137}$	p
WI	16.9	<0.001	49.1	<0.001
WD	0.5	0.500	649.5	<0.001
SAI	2.9	0.088	25.3	<0.001
SAD	0.0	0.848	3171.0	<0.001
CDI	5.3	0.022	84.1	<0.001
CDD	0.0	0.981	515.3	<0.001

of dependent variables with a covariate (Thomas and Nelson, 1996: 181–2). In those situations where the researcher has a good rationale for using a MANCOVA, covariates can simply be included by transferring them into the *Covariate* area of the Multivariate pop-up window shown in Figure 13.1 or the *Covariate* area of the Repeated Measures Definition pop-up window shown in Figure 13.5.

SUMMARY

MANOVA tests analyse the effect of independent variables on a set of two or more dependent variables. There are single factor MANOVA tests and multifactor MANOVA tests, with many different combinations of between-subjects factors and within-subjects factors being included. Where a MANOVA test finds any factor to have a significant effect on the set of dependent variables, follow-up univariate ANOVA tests can be applied to the individual dependent variables. Similarly, if there is an interaction effect of two or more factors according to the MANOVA test, follow-up univariate ANOVA tests can test for the same interaction effect on the individual dependent variables. The MANOVA test allows us to restrict the probability of making a Type I Error by performing a single test using the chosen α value. If the MANOVA test finds no significant differences, the inferential results of a study can be presented very concisely because follow-up univariate tests will not be required.

EXERCISES

Exercise 13.1. Effect of exercise participation on wellbeing

A total of 123 students complete a wellbeing questionnaire instrument (Corbin *et al.*, 2000: 17–18) that produces scores from 3 to 12 for emotional wellbeing, intellectual wellbeing, physical wellbeing, social wellbeing and physical wellbeing. These all add up to a total wellbeing score between 15 and 60. Each participant classifies themselves as inactive, semi-active, a regular participant in exercise or a competitive sports person according to definitions provided on an attached sheet to the questionnaire. The data for this exercise are found in the file ex13.1-wellbeing.SAV.

a) Perform a one-way MANOVA test to determine the effect of activity level on wellbeing. Use emotional, intellectual, physical, social and spiritual wellbeing as dependent variables within the MANOVA test. Do not include the total wellbeing score because it is functionally dependent on the five individual types of wellbeing. Use follow-up univarite ANOVA tests and post hoc tests if necessary. Show how you would present the descriptive and inferential results.

b) Apply a one-way ANOVA to the total wellbeing score using activity level as the independent variable. Is the conclusion of his test consistent with that of the MANOVA test in part (a)?

Exercise 13.2. Gender and type of sport effect on behavioural regulation in sport

The file ex13.2-BRSQ.SAV contains data for 175 fictitious student athletes who have completed a behavioural regulation in sport questionnaire (Lonsdale *et al.*, 2008). The sample consists of 93 males and 82 females who classify themselves as either participants in individual sports or team sports. The behavioural

regulation in sport instrument (BRSQ) is composed of nine subscales that constitute the concept of self-determination theory with four of the subscales being concerned with intrinsic motivation. The nine subscales are scored on a 4 to 28 scale and are listed below:

- Intrinsic Motivation
- Intrinsic Motivation for Accomplishment
- Intrinsic Motivation to Gain Knowledge
- Intrinsic Motivation to Experience Stimulation
- Integration
- Identified Regulation
- Introjected Regulation
- External Regulation
- Amotivation.

Use a factorial MANOVA to determine if gender, type of sport preferred and/or their interaction have a significant influence on self-determination. Apply follow-up univariate analysis tests to the nine individual subscales if necessary. How would you present the results?

Exercise 13.3. The effect of gender, type of sport and type of match on anxiety (modified SAS2)

This exercise uses the same data as the example used for the mixed MANOVA test. This exercise requires you to add an additional between-subjects effect (type of sport preferred). Therefore, you will be applying a three-way MANOVA test including type of match as a within-subjects effect and gender and type of sport preferred as between-subjects effects. The dependent variables will still be WI, WD, SAI, SAD, CDI and CDD. The data are found in the file ex13.3-SAS2.SAV. Perform the MANOVA test, applying any follow-up ANOVA tests on individual dependent variables if necessary. How would you report the results?

PROJECT EXERCISE

Exercise 13.4. Training and competition hours done by male and female athletes

This is a variant of Exercise 12.7 from Chapter 12. Therefore, if Exercise 12.7 was done, you do not need to repeat the form design and data collection activities. Devise a data collection form to record training and competition hours performed in a typical training week by sportsmen and sportswomen. Include sections allowing the estimation of competition, tactical preparation, technical sessions, conditioning work and any other training activity requiring an investment of time. Add a final question to the form for gender. Gather data from your classmates using the form and enter the total training hours, competition hours and gender for each participant into an SPSS datasheet. Apply a one-way MANOVA test to investigate the effect of gender on training and competition hours.

CHAPTER 14

NON-PARAMETRIC TESTS

INTRODUCTION

In Chapter 10, the independent samples t-test and paired samples t-test were described. These tests are used for comparing numerical scale dependent variables between two independent and two related samples respectively. Chapter 11 described the one-way ANOVA test and the repeated measures ANOVA test which are used for comparing numerical scale dependent variables between three or more independent and three or more related samples respectively. These t-tests and ANOVA tests are parametric tests that are calculated using the measured values. The tests have assumptions that should be satisfied by the data they are applied to. The dependent variables should be interval or ratio scale variables, they should be normally distributed and there should be similar variances for the dependent variable between samples. The repeated measures ANOVA test has the additional assumption of sphericity which is homogeneity of variances and covariances between samples. The scale of measurement of the dependent variable can be checked by examining the definition of the variable. Normality is tested using either the Shapiro–Wilk test or the Kolmogorov–Smirnov test as described in Chapter 6. Homogeneity of variances can be tested using Levene's test as described in Chapter 10 while sphericity is tested using Mauchly's test as described in Chapter 11. Where the data fail to satisfy the assumptions of the given test and there are no transformations of the data that will result in the assumptions being satisfied, it may be necessary to use an alternative non-parametric test to compare the samples. Table 14.1 shows the alternative non-parametric tests that should be used in different comparisons.

Non-parametric tests are distribution-free tests as they have no assumptions about the distribution of the dependent variables. The tests compute test statistics just as t-tests produce t-scores and ANOVA tests produce F-ratios. Non-parametric test statistics are used to deter-

Table 14.1 Parametric and alternative nonparametric tests

Comparison	Parametric Test	Non-Parametric Test
2 independent samples	Independent t-test	Mann-Whitney U test
2 related samples	Paired t-test	Wilcoxon Signed Ranks test
3+ independent samples	One-way ANOVA	Kruskal-Wallis H test
3+ related samples	Repeated measures ANOVA	Friedman test

mine p values just as they are in the parametric tests. A major difference between parametric and non-parametric tests is that parametric tests compute the test statistics using the values of the dependent variables whereas non-parametric tests use ranks instead of the actual values. This means that there is information loss when using non-parametric tests and so they are considered to be not as powerful as parametric tests (Thomas and Nelson, 1996: 102; Diamantopoulos and Schlegelmilch, 1997: 142). When one considers the equation 10.2 of the independent samples t-test, we can see that the equation is in terms of the values within the samples being compared. This is because the means and standard deviations are ultimately calculated using all the values in the samples. This is not the case with non-parametric procedures where ranks are used. Consider an international track athletics meeting where two teams of four athletes take part in an 800m race. The result is shown in Table 14.2.

If mean time was used to determine which team won the race, Great Britain would be declared the winner with a mean time of 1 minute 53.3s compared to the USA's 1 minute 56.2s. If finishing position was used to determine the winning team, the USA would be declared the winner with a mean finishing position of 3.5 compared with Great Britain's 5.5. This example shows how the use of ranks and values can lead to different conclusions from exactly the same data. It is, therefore, important to use the correct test (parametric or non-parametric) when analysing data.

This chapter will describe four non-parametric tests that can be used to compare samples where numerical scale dependent variables fail to satisfy the assumptions of parametric tests. The non-parametric tests can also be used with ordinal dependent variables because ordinal variables can also be ranked.

MANN–WHITNEY U TEST

Purpose of the test

The Mann–Whitney U test is used to compare two independent samples in terms of some ordinal, interval or ratio scale variable. For example, we might wish to compare reported level of agreement ('strongly disagree', 'disagree', 'undecided', 'agree', 'strongly agree') from a questionnaire survey between female and male respondents.

Table 14.2 Result of a fictitious 800m team race

Position	Athlete	Team	Time
1	Adam	Great Britain	1:44.0
2	Brian	USA	1:56.1
3	Colin	USA	1:56.2
4	David	USA	1:56.2
5	Edward	USA	1:56.3
6	Frank	Great Britain	1:56.3
7	George	Great Britain	1:56.4
8	Henry	Great Britain	1:56.5

Example: Comparing rally lengths between women's and men's Grand Slam singles tennis

We will use a performance analysis example to describe the use of the Mann–Whitney U test. Rallies are timed in men's and women's singles matches with the mean rally duration within each match being determined. There are 252 matches within the data: 116 are women's singles matches and 136 are men's singles matches. The mean rally duration for each of the 252 matches has been determined and we wish to compare this performance indicator between men's and women's matches to see if gender has a significant influence on mean rally duration or not.

SPSS

The data for the tennis example are found in the file 14-tennis.SAV. In version 18.0 of SPSS, the Mann–Whitney U test is accessed with **Analyse → Non-parametric → Legacy Dialogs → 2 Independent Samples**. In versions of SPSS prior to version 18, the Legacy Dialogs menu was not provided and the Mann–Whitney U test was accessed using **Analyse → Non-parametric → 2 Independent Samples**. In the pop-up window that appears (Figure 14.1), 'gender' is transferred into the *Grouping Variable* area; this is our independent variable. It is worth noting that the independent variable needs to be numerically coded for it to be used as a potential grouping variable within a Mann–Whitney U test. Therefore, we use the values 1 and 2 rather than 'Female' and 'Male'. These are set up when we click on the **Define** button within the Two Independent Samples Tests pop-up window. Of course, the values 1 and 2 can be labelled with 'Female' and 'Male'. There is another good reason for using 1 and 2 rather than 'Female' and 'Male' which is that it is easier to enter 1 and 2 when preparing the datasheet and also 'Female' and 'female' would count as different values if we tried to type in full names of the genders. If we have entered 'Female' and 'Male' as text string values, there is still a way to use the data within the Mann–Whitney U test. This is done by automatically recoding the variable in SPSS (**Transform → Automatic Recode**) which will produce a new version of the variable with numerical codes that are labelled with the original strings. 'Rally Length' is transferred into the *Test Variable List*; this is our dependent variable. The default test is the Mann–Whitney U test as shown in Figure 14.1.

The output of the test is shown in Tables 14.3 and 14.4. With 252 matches, if there was no difference between the rally lengths of men's and women's matches, then we would expect the mean rank of each group to be 126.5. As we can see from the Table 14.3, the women's matches have a higher mean rank while the men's matches have a lower mean rank. The Sig. value in the Table 14.4 is the p value which in this case is less than 0.001.

We also need descriptive statistics to present with the results of the Mann–Whitney U test. We use **Analyse → Compare Means → Means** to produce means or medians for the two groups. If medians are required, then the **Options** button is used to allow the median to be included in the descriptive statistics. 'Rally length' is transferred into the dependent variable list and gender is transferred into the *Independent Variable List*. This will tell us that the median for mean rally length is 6.6s in women's singles and 5.0s in men's singles. Table 14.5 shows the output.

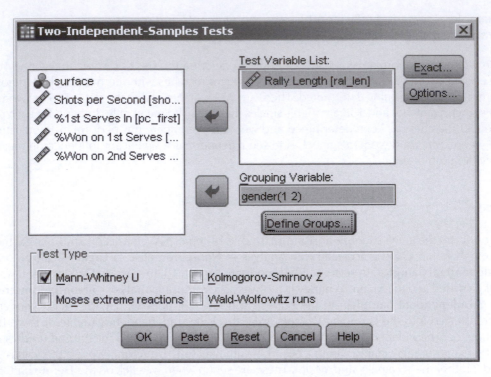

Figure 14.1 The pop-up window to compare two independent samples using a non-parametric test.

Table 14.3 Mean ranks of two independent groups for a dependent variable

	Ranks			
	gender	*N*	*Mean Rank*	*Sum of Ranks*
Rally Length	female	116	160.18	18581.00
	male	136	97.77	13297.00
	Total	252		

Table 14.4 SPSS output for the Mann-Whitney U test

Test Statistics[a]	
	Rally Length
Mann-Whitney U	3981.000
Wilcoxon W	13297.000
Z	−6.774
Asymp. Sig. (2-tailed)	.000

a Grouping Variable: gender

Table 14.5 SPSS Output for Compare means

Report

Rally Length

gender	Mean	N	Std. Deviation	Median
female	7.223	116	2.3217	6.620
Male	5.242	136	1.7464	5.049
Total	6.154	252	2.2559	5.934

Presentation of results

Figure 14.2 shows how the mean and standard deviation could be presented in the form of a bar chart. If, instead, we wished to use medians, then the results could be reported as follows:

> The median for mean rally duration in women's singles matches was 6.6s, which was significantly longer than the 5.0s in men's singles matches ($z = -6.8$, $p < 0.001$).

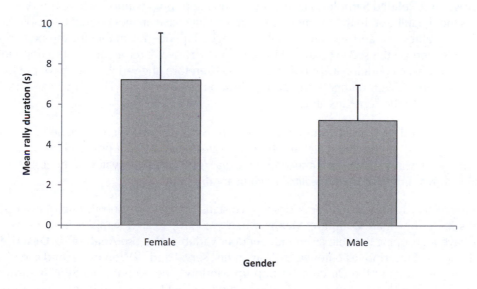

Figure 14.2 Mean rally length (s) in Grand Slam singles tennis.

WILCOXON SIGNED RANKS TEST

Purpose of the test

The Wilcoxon signed ranks test is the non-parametric alternative to the paired samples t-test. The test is used to compare two samples related to the same group in terms of some ordinal, interval or ratio scale dependent variable. For example, the Wilcoxon signed ranks test could be used to compare resting heart rate at 6am and 6pm within a group of participants.

Example: Percentage of points won in tennis when the first serve is in and when a second serve is required

We will illustrate the Wilcoxon signed ranks test using the same tennis study that was used to illustrate the paired samples t-test in Chapter 10. For each of the 252 matches, the percentage of points won when the first serve was in and the percentage of points won when a second serve was required have been calculated. We wish to determine if the percentage of points won is similar between points where the first serve was in and points when a second serve was required. Therefore, we have two samples related to the same set of 252 matches.

SPSS

The Wilcoxon signed ranks test is done using **Analyse → Non-parametric Tests → Legacy Dialogs → 2 Related Samples**. Figure 14.3 shows the pop-up window where the Wilcoxon test is the default test. In this example, we conceptually have an independent variable 'service' with values first and second as well as a single dependent variable for the percentage of points won by the serving player. However, in SPSS we enter a copy of the dependent variable for each condition it is measured under. Therefore, the variables we have put in our datasheet are '%Won on first serve' and '%Won on second serve'. These two variables are transferred into the *Test Pairs* area.

Tables 14.6 and 14.7 are the output produced by SPSS for the Wilcoxon signed ranks test. There were 237 of the 252 matches where a greater percentage of points were won when the first serve was in than when a second serve was required. The p value of the test (Asymp. Sig.) is shown in Table 14.7 revealing a significant difference.

It is necessary to determine some descriptive statistics. This is done through **Descriptive Statistics** rather than **Compare Means** when we have related samples because we do not have a grouping variable as an independent variable. We use **Analyse → Descriptive Statistics → Descriptives** entering '%Won on first serve' and '%Won on second serve' into the *Variables* area of the Descriptives pop-up window. The output from SPSS is shown in Table 14.8. This does not provide us with the medians but these could be obtained through the **Explore** facility.

272

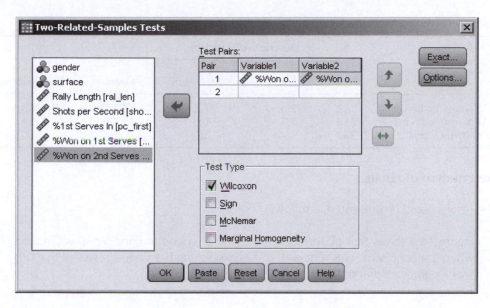

Figure 14.3 Pop-up window for two related samples tests to be done non-parametrically.

Table 14.6 Mean positive and negative ranks when comparing two related samples

Ranks

		N	Mean Rank	Sum of Ranks
% Won on 2nd Serves – % Won on 1st Serves	Negative Ranks	237[a]	130.48	30924.00
	Positive Ranks	15[b]	63.60	954.00
	Ties	0[c]		
	Total	252		

a % Won on 2nd Serves < % Won on 1st Serves
b % Won on 2nd Serves > % Won on 1st Serves
c % Won on 2nd Serves = % Won on 1st Serves

Table 14.7 Wilcoxon Signed ranks test results

Test Statistics[b]

	%Won on 2nd Serves – %Won on 1st Serves
Z	−12.938[a]
Asymp. Sig. (2-tailed)	.000

a Based on positive ranks.
b Wilcoxon Signed Ranks Test

Table 14.8 Descriptive statistics produced by SPSS

	N	Minimum	Maximum	Mean	Std. Deviation
			Descriptive Statistics		
% Won on 1st Serves	252	37.8	88.5	67.382	9.1782
% Won on 2nd Serves	252	7.7	100.0	48.286	10.5399
Valid N (listwise)	252				

Presentation of results

The results could be presented as follows:

> Figure 14.4 shows that players won more points when their first serve was in than
> when a second serve was required. A Wilcoxon signed ranks test revealed that this
> was a significant difference ($z = -12.9$, $p < 0.001$).

KRUSKAL–WALLIS H TEST

Purpose of the test

The Kruskal–Wallis H test is used to compare three or more independent samples in terms
of some dependent ordinal, interval or ratio scale dependent variable. For example, we
might wish to compare how often people of different socio-economic groups experience
given situations ('never', 'seldom', 'sometimes', 'often', 'always') using a questionnaire survey. Where a significant difference is found, we can either use follow-up Mann–Whitney U
tests to compare individual pairs of socio-economic groups or use the post hoc procedure
described by Thomas and Nelson (1996: 204–2) to perform post hoc tests.

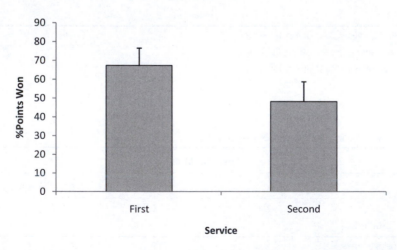

Figure 14.4 Percentage of points won by the serving player in Grand Slam singles tennis.

Assumptions

The Kruskal–Wallis H test does have an assumption which is that there are at least five values in each independent sample being compared. The H statistic produced by the Kruskal–Wallis H test is approximately chi square distributed.

Example: Surface effect on rally duration in Grand Slam tournaments

Before the introduction of the Type 1 and Type 3 balls and surface grading in professional tennis in 2002, the same type of tennis balls was used at all four Grand Slam tournaments. Furthermore, between 1979 and 2007 all four Grand Slam tennis tournaments were played on different surfaces. In 2008, the Australian Open changed to using the same surface as the US Open. This example uses data from Grand Slam tennis matches played between 1997 and 1999 and, therefore, any differences between tournaments can be considered to be due to surface effects. The mean rally duration has been determined for 252 matches (62 matches played on synthetic hard courts at the Australian Open, 57 matches played on clay courts at the French Open, 66 matches played on grass at Wimbledon and 67 matches played on hard courts at the US Open). The matches involve different pairs of players with a few exceptions and can be considered as independent.

SPSS

The data for the 252 matches are found in the file 14-tennis.SAV. We use **Analyse → Non-parametric tests → Legacy Dialogs → K Independent Samples**; note that the Kruskal–Wallis H test is the default non-parametric test to compare three or more independent samples. We transfer the variable 'Surface' into the *Grouping Variable* area of the pop-up window shown in Figure 14.5. We have to **Define** the range of values as 1 to 4 which we have used to represent the four surfaces. Again, we need to use a numeric coding to represent the values of the independent variable to all to allow the variable to be used as a grouping variable in the non-parametric test. If we have failed to do this, we can use the automatic recode facility of SPSS in order to produce a coded version of the variable. 'Rally length' is transferred into the *Test Variable List* and then we can click on **OK**.

Tables 14.9 and 14.10 show the SPSS output for the Kruskal–Wallis H test. With there being 252 matches ranked from 1 to 252, if there was absolutely no surface effect, we would expect the mean rank for each surface to be 126.5, but as the Ranks table in Table 14.9 shows, some mean ranks are much higher than this and some are much lower. The chi square value shown in the Table 14.10 is actually the H statistic which approximates to the chi square distribution with K – 1 degrees of if there are five or more values in each sample being compared: K is the number of independent samples. The p value (Asymp. Sig.) shown in Table 14.10 is less than 0.05 which would be the typical value below which we would indicate statistical significance.

The mean ranks are relative to the data in the sample and do not tell us whether the mean rally durations are short, medium length or long. Therefore we use compare means

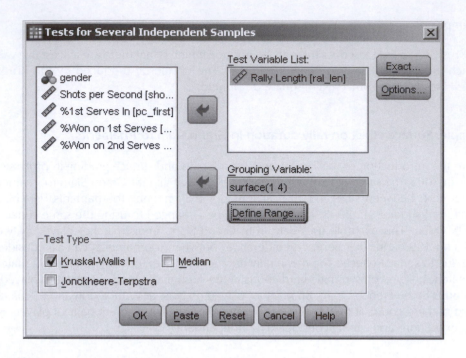

Figure 14.5 Pop-up window for the Kruskal–Wallis H test.

Table 14.9 Mean ranks for four independent samples

	surface	N	Mean Rank
Ranks			
Rally Length	Synthetic	62	135.31
	Clay	57	188.65
	Grass	66	72.92
	Cement	67	118.25
	Total	252	

Table 14.10 SPSS output for the Kruskal–Wallis H test

Test Statistics[a,b]	Rally Length
Chi-Square	78.858
Df	3
Asymp. Sig.	.000

a Kruskal Wallis Test
b Grouping Variable: surface

non-parametric tests

(**Analyse** → **Compare Means** → **Means** entering 'surface' into the *Independent List* and 'Rally length' into the *Dependent List*) to determine the essential descriptive statistics for mean rally length on the different surfaces. Table 14.11 shows the output obtained from SPSS. We could also obtain medians through the **Options** button within the Compare Means pop-up window if we wished.

Given that the Kruskal–Wallis H test has revealed a significant surface effect on rally length, we wish to determine which pairs of surfaces are different by using pairwise comparisons. SPSS does not provide a post hoc test option for the Kruskal–Wallis H test in the way that it does for the one-way ANOVA. Therefore, we need to choose a test to apply to each of the pairs of surfaces. Given that we are using the Kruskal–Wallis H test, we are treating the matches played on different surfaces as being independent. The use of the Kruskal–Wallis H test also means that we have made a decision to use non-parametric procedures. Therefore, the most appropriate test to use in order to compare pairs of court surfaces is the Mann–Whitney U test.

If there are K samples, then the number of pairs of samples is K(K–1)/2. In this tennis example there are four court surfaces and therefore there are six pairs of court surfaces. The α value used with these post hoc tests is adjusted to avoid Type I Error inflation. Using equation 11.7 in Chapter 11, α for the follow-up Mann–Whitney U tests will be 0.0085 when we have four groups of matches and six pairs of groups. Doing these adjusted Mann–Whitney U tests reveals that each pair of surfaces have significantly different rally lengths except the synthetic grass surface used at the Australian Open and the hard court surface used at the US Open.

Thomas and Nelson (1996: 204–5) described a process to perform pairwise comparisons in the event of a significant Kruskal–Wallis H test result. Equation 14.1 is used to determine the standard error of any difference in mean rank between two groups where n is the total number of values used in the study and n_1 and n_2 are the numbers of values in the two particular groups being compared in a given post hoc test.

$$SE = \sqrt{\left(\frac{(n(n-1))}{12}\right)\left(\frac{1}{n_1}+\frac{1}{n_2}\right)} \tag{14.1}$$

This is multiplied by a z-score in order to determine a confidence interval for rank difference outside which the difference between two samples would be considered significantly differ-

Table 14.11 Descriptive statistics

Report			
Rally Length			
surface	Mean	N	Std. Deviation
Synthetic	6.288	62	1.5852
Clay	8.156	57	2.4088
Grass	4.632	66	1.5999
Cement	5.828	67	1.9193
Total	6.154	252	2.2559

ent. Based on equation 11.7, we use an α value of 0.0085 instead of 0.05. Because we are using a two-tailed post hoc test, we use a p value of 0.0043 for which the critical z-values are ± 2.63. The SEs for the different pairs are multiplied by this $z_{\alpha/2}$ value to get a required confidence interval for the two samples being equal. This is shown in the fourth column of Table 14.12. The mean ranks from SPSS are used to determine rank differences and compare these to critical confidence interval. Here we see that all pairs are significantly different except Synthetic vs Hard court. The 14-KWH-pairwise-comparisons.XLS spreadsheet file implements Thomas and Nelson's (1996: 204–5) procedure and users simply need to enter mean ranks, sample sizes, α level and number of groups into the blue areas.

Presentation of results

The results could be presented as follows:

> Figure 14.6 shows the mean rally duration of singles matches played on the four different court surfaces used in Grand Slam tennis. Surface had a significant influence on mean rally length ($H_3 = 78.9$, p < 0.001) with each pair of surfaces having significantly different rally lengths (p < 0.0085) except for the synthetic grass surface used at the Australian Open and the hard court surface used at the US Open.

Table 14.12 Results of Thomas and Nelson's (1996: 204–5) procedure for post hoc comparisons for the Kruskal–Wallis H test

Group 1	Group 2	SE from (14.1)	$\pm z_{\alpha/2}SE$	Rank diff	Significant
Synthetic	Clay	13.38	± 35.19	53.34	Yes
Synthetic	Grass	12.89	± 33.92	62.39	Yes
Synthetic	Hard	12.84	± 33.80	17.06	No
Clay	Grass	13.18	± 34.68	115.73	Yes
Clay	Hard	13.13	± 34.56	70.40	Yes
Grass	Hard	12.64	± 33.26	45.33	Yes

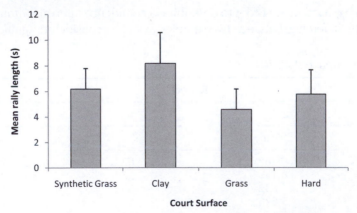

Figure 14.6 Mean rally duration at Grand Slam tennis tournaments.

278

non-parametric tests

FRIEDMAN TEST

Purpose of the test

The Friedman test is used to compare three or more samples related to the same group of participants using some dependent ordinal, interval or ratio scale dependent variable. For example, a questionnaire for soccer spectators may ask about the convenience of travelling to home matches when those matches are played on (a) Saturday afternoons, (b) Sunday afternoons and (c) Monday evenings. There are three questions, one for each match time, each of which requires a response of 'very convenient' 'convenient', 'inconvenient', 'very inconvenient' and 'not possible' to attend. The responses to these three questions can be compared for the responding spectators using a Friedman test. Where there is a significant influence of match time, we can use follow-up Wilcoxon signed ranks tests to compare individual pairs of match times.

Example: High-intensity activity performed during the four quarters of a netball match

An observational study is conducted of club level netball players where the percentage of time they spend performing high-intensity activity is recorded by an observer using a computerized timing system. We will assume that the method has been demonstrated to be valid and reliable. The data for those participating players who competed in all four quarters of the match where they were observed are included in the study. The percentage of time spent performing high-intensity activity is a performance analysis measure that typically fails to satisfy the assumptions of the repeated measures ANOVA test and, therefore, the Friedman test is used to compare this performance indicator between the four different quarters of a netball match.

SPSS

The high intensity activity data for 28 club level netball players are found in the file 14-netball.SAV. We perform the Friedman test in SPSS using **Analyse → Non-parametric Tests → Legacy Dialogs → K Related Samples** which causes the pop-up window in Figure 14.7 to appear with the Friedman test as the default test. The four variables 'Qtr 1: %time spent performing high intensity activity' through to 'Qtr 4: %time spent performing high intensity activity' are transferred into the *Test Variable List*. It is worth noting that we conceptually have an independent variable (quarter) and a dependent variable (%time spent performing high-intensity activity). However, in SPSS we do not have a variable quarter because all player performances included in the study involved all four quarters. Therefore, we have four variables in SPSS representing the four repeated measurements of our conceptual dependent variable.

When we click on **OK**, we see the output shown in Tables 14.13 and 14.14. The Friedman test takes dependent variable's values and sorts them into ascending order before assigning ranks according to the number of related samples in the test. In this example, the lowest

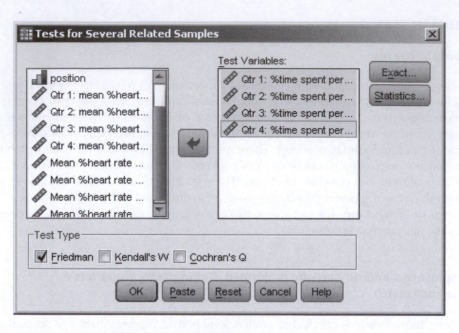

Figure 14.7 Pop-up window for the Friedman test in SPSS.

Table 14.13 Mean ranks used with the Friedman test

Ranks	
	Mean Rank
Qtr 1	2.82
Qtr 2	2.96
Qtr 3	2.39
Qtr 4	1.82

quarter of values are given a rank of 1, the next lowest quarter of values are given a rank of 2, and so on. If quarter of a netball match had nothing to do with the percentage of time a player spent performing high-intensity activity, then we would expect to see a mean rank of 2.5 for each of the four quarters. The mean rank of each sample is shown in Table 14.13. The Friedman test determines whether the ranks are significantly different to the expected distribution of ranks. Table 14.14 shows that there is a significant influence of quarter; the p value (Asymp. Sig.) is less than 0.05.

We also need to determine descriptive statistics (means and standard deviations or medians) for the percentage of time spent performing high-intensity activity in the four quarters. This can be done using **Analyse → Descriptive Statistics → Descriptives** if we do not need to determine the median or **Analyse → Descriptive Statistics → Explore** if we do need to determine the median. Table 14.15 shows the means and standard deviations determined by SPSS.

Table 14.14 SPSS output for the Friedman test

Test Statistics[a]	
N	28
Chi-Square	13.286
Df	3
Asymp. Sig.	.004

a Friedman Test

Table 14.15 Descriptive statistics

Descriptive Statistics					
	N	Minimum	Maximum	Mean	Std. Deviation
Qtr 1	28	15.07	33.17	23.0475	5.28167
Qtr 2	28	12.01	32.33	22.0082	5.37944
Qtr 3	28	10.57	33.28	20.6757	5.77453
Qtr 4	28	11.31	32.08	19.2214	5.40273
Valid N (listwise)	28				

Because the percentage of time spent performing high-intensity activity is significantly influenced by quarter, we use a series of six Wilcoxon signed ranks tests to compare each pair of quarters. Unlike when Mann–Whitney U tests are used to perform pairwise comparisons following a Kruskal–Wallis H test, SPSS allows all six pairs of Wilcoxon tests to be entered into the two Related Samples pop-up window in one go. Table 14.16 shows the output of these tests. It is necessary to adjust the α level to 0.0085 in order to restrict the probability of a significant difference between any of the pairs of quarters to a maximum of 0.05 (based on equation 11.2). Therefore, there are only significant differences between quarters 1 and 4 ($p < 0.0085$) and quarters 2 and 4 ($p < 0.0085$).

The results would be expressed as follows:

> Figure 14.8 shows the percentage of time that players spent performing high-intensity activity in the four quarters of a netball match. Quarter had a significant influence on the percentage of time spent performing high-intensity activity ($\chi^2_3 = 13.3$,

Table 14.16 Follow up Wilcoxon signed ranks tests

Test Statistics[b]						
	Qtr 2–Qtr 1	Qtr 3–Qtr 1	Qtr 4–Qtr 1	Qtr 3–Qtr 2	Qtr 4–Qtr 2	Qtr 4–Qtr 3
Z	−1.093[a]	−2.186[a]	−2.756[a]	−1.822[a]	−3.189[a]	−1.936[a]
Asymp. Sig. (2-tailed)	.274	.029	.006	.068	.001	.053

a Based on positive ranks.
b Wilcoxon Signed Ranks Test

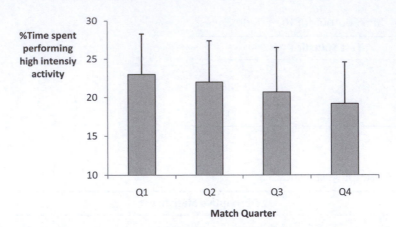

Figure 14.8 Percentage of time spent performing high-intensity activity in netball.

p = 0.004) with a significantly lower percentage of the 4th quarter being spent performing high-intensity activity than the 1st quarter (p < 0.0085) and 2nd quarter (p < 0.0085).

SUMMARY

The tests covered in this chapter are non-parametric alternatives to the t-tests and ANOVA tests covered in Chapters 10 and 11. These non-parametric tests can be used if the dependent variable is measured on an ordinal, interval or ratio scale. The tests can be used for interval and ratio scale data where the data fail to satisfy the assumptions of the t-tests and ANOVA tests. Non-parametric tests use ranks to compute test statistics and are, therefore, not as powerful as the parametric tests. However, the non-parametric tests are also distribution-free tests that can be used with any ordinal or numerical scale data irrespective of the distribution of data values.

EXERCISES

Exercise 14.1. Service dominance in women's and men's singles tennis

The file ex14.1-tennis.SAV contains data for a sample of 252 tennis matches. The variables are the number of shots played per second, the percentage of points where the first serve is in, the percentage of points won when the first serve is in and the percentage of points won when a second serve is required. Use appropriate descriptive and inferential statistics to test if any of these four variables are influenced by gender.

Exercise 14.2. Burnout potential of individual sport and team sport athletes

This exercise is a sports psychology example. The example uses data for 187 fictitious student athletes that can be found in the file ex14.2-burnout.SAV. The file contains nominal variables for gender, type of sport played and level at which the respondent participates in the sport. The purpose of this exercise is to test whether individual sports performers and team sports performers (the variable is 'Sport') differ with respect to burnout. The specific variables to be compared between individual and team sports performers are:

- Reduced sense of accomplishment [RSA]
- Emotional–Physical exhaustion [EE]
- Devaluation [DEV].

Exercise 14.3. Comparing worry and somatic anxiety within student athletes

This example uses data for the same 187 fictitious student athletes that were used in Exercise 14.2. These data can be found in the file ex14.3-SAS2.SAV. Competitive anxiety is composed of worry and somatic anxiety. The purpose of this exercise is to test whether student sports performers experience similar levels of worry intensity (WI_COMP) and somatic anxiety intensity (SAI_COMP) when preparing for competition; these are measured on comparable scales. Use appropriate descriptive and inferential statistics to compare the two types of anxiety and report findings from the data.

Exercise 14.4. Surface effect in men's and women's singles tennis

Using the tennis data in the file ex14.4-tennis.SAV, determine for men's and women's singles matches separately whether there is a significant surface effect on rally length, shots played per second, the percentage of points where the first serve is played in, the percentage of points won when the first serve is in and the percentage of points won when a second serve is required. Note that **Data → Split File** will allow you to logically split the file on gender so as any analysis done will be applied to men's and women's singles matches separately.

Exercise 14.5. Heart rate response during the four different quarters of club level netball matches

The file ex14.5-netball.SAV contains the mean heart rate response (age related %heart rate max) for 28 players who competed in all four quarters of the netball matches where their heart rates were successfully recorded. Use a repeated measures ANOVA test and a Friedman test to determine whether quarter had an influence on heart rate response. Comment on the findings of the two different tests? Which test would be the best to use in this situation and why?

PROJECT EXERCISE

Exercise 14.6. Home advantage in sport

Find a season's final league table for a sport of your choice that includes the number of won, drawn and lost matches played at home as well as away from home. Enter these six values for each team that participated in the league into an SPSS datasheet. Compare the number of wins achieved at home with the number of wins achieved away from home using a Wilcoxon signed ranks test. Do the same for the number of matches drawn at home and away as well as the number of matches lost at home and away from home. In the particular league you looked at, is there evidence of home advantage?

Exercise 14.7. Surface effect on double faults played in Grand Slam singles tennis

The official tournament websites for the four Grand Slam tennis tournaments provide match statistics for most of the matches played under the Draws area.

- www.ausopen.org
- www.frenchopen.org
- www.wimbledon.org
- www.usopen.org

Using completed matches from the 3rd round of the women's singles, determine the percentage of all points that are double faults. There will be a maximum of 16 completed 3rd round matches per tournament where these data are provided. Enter this data into SPSS along with a grouping variable that represents the tournament where the match was played. Use a Kruskal–Wallis H test with follow-up Bonferroni-adjusted Mann Whitney U tests to compare the percentage of points that are double faults between the four tournaments. Repeat this exercise for 3rd round men's singles matches.

Exercise 14.8. Preferred equipment for cardio-vascular exercise in the gym

Develop a questionnaire that asks the below questions to your classmates. For each question, the response is one of the following:

- Never use
- Rarely use
- Sometimes use
- Often use
- Very often use.

Q1. How often do you exercise on a treadmill?
Q2. How often do you exercise on an exercise bike?
Q3. How often do you exercise on a stepping machine?
Q4. How often do you exercise on a rowing ergometer?
Q5. How often do you exercise on a cross-training machine?

Which items of equipment are used most often and least often? Use a Friedman test to see if the frequency of use of these items of gym equipment significantly differs for the respondents. If there are significant differences, which pairs of equipment types are significantly different in terms of how often they are used?

CHAPTER 15

CHI SQUARE

INTRODUCTION

Chi square tests are typically used with nominal variables but they are also used with ordinal variables of few values and numerical variables that can be formed into subranges of values. There are two types of chi square test: the chi square goodness of fit test and the chi square test of independence. The chi square goodness of fit test is used to compare the distribution of cases among the values of some categorical variable with a theoretically expected distribution. For example, a gym may have 1,300 male members and 700 female members. The management of the gym would like to know if the 35 per cent of their members who are female is similar enough to the proportion of females in the wider population (assumed to be 50 per cent). The chi square goodness of fit test compares the observed distribution of the gym's membership (1,300 males and 700 females) with a theoretically expected distribution (1,000 males and 1,000 females) to determine if the difference in these two distributions is down to chance or whether there is a significantly different gender distribution than expected.

The main difference between the chi square goodness of fit test and the chi square test of independence is that the chi square goodness of fit test is concerned with a single categorical variable while the chi square test of independence is concerned with two categorical variables. For example, a survey may be conducted where one of the questions asks whether the respondent is a member of a gym and another question asks for the respondent's gender. Imagine that 80 out of 180 male respondents (44.4 per cent) indicate that they have gym membership compared to 40 out of 220 female respondents (18.2 per cent). We might wish to determine if the differing proportion of people with gym membership is down to chance or if there is a gender effect. This can be determined using a chi square test of independence. A second difference between the chi square test of independence and the chi square goodness of fit test is that the chi square test of independence does not use some theoretically expected distribution. Instead, the figure of 120 out of 400 respondents of the whole sample who are members of a gym (30 per cent) is used to determine the expected distribution. That is, if gender had absolutely no impact on whether or not people were members of a gym, we would expect the 30 per cent of the sample who are members of a gym to apply to the male and female respondents equally. Therefore, we would expect 54 out of the 180 males and 66 out of the 220 females to be members of a gym. The chi square test of independence will then compare the observed frequencies with

285

chi square

these expected frequencies to determine if gender and gym membership are independent variables or if there is a gender effect.

THE CHI SQUARE GOODNESS OF FIT TEST

Purpose of the test

Sometimes we have a theoretical expected frequency distribution for a single categorical variable. An example of a categorical variable is gender which has two values. The purpose of the chi squared goodness of fit test is to compare the observed frequency distribution with this theoretical expected frequency distribution. The theoretically expected distribution could come from national statistics, census data or other sources. As mentioned in Chapter 6, there are many chi square distributions that are distinguished by their degrees of freedom. The number of degrees of freedom is one less than the number of values in the given categorical variable. For example, gender has two values (female and male) and so there is one degree of freedom in a chi square test comparing observed and expected frequency distributions of gender.

Assumptions

The minimum number of values within a categorical variable being tested with a chi square goodness of fit is two. The test is valid if we have an expected frequency of at least five for at least 80 per cent of the values of the categorical variable. If this assumption is violated, we can merge some categorical values for the purpose of performing the test. The individual values can still be displayed in descriptive statistics.

Example: Relative age in women's Grand Slam singles tennis

In junior sport, participants are grouped into cohorts separated by cut-off dates; for example, the International Tennis Federation (ITF) has used a cut-off date of 1 January for the ITF junior tour. Players born in the first six months of the year have a relative age advantage over those born in the second six months. This may lead to attrition of late born players who may be perceived by themselves and others as having less talent than they actually have. This in turn may lead to a different birth month distribution of senior players to the general birth month distribution of the global population, with more senior players having been born in the first half of the year than the second half of the year. Table 15.1 shows the birth month distribution of the female and male players who competed in the singles events at one or more Grand Slam tennis tournament in 2009. The sources of these data are the official websites of the four Grand Slam tournaments in 2009 (www.ausopen.org, www.frenchopen. org, www.wimbledon.org and www.usopen.org all accessed on 2 September 2009).

We wish to determine if the birth month distribution of the female players fits the theoretically expected distribution. In the absence of global population birth month data, we will create an expected birth month distribution that assumes an equal number of people are

Table 15.1 Month of birth of players who competed in the first round of the senior singles event at any Grand Slam tournament in 2009

Gender						Month of Birth							
	J	F	M	A	M	J	J	A	S	O	N	D	Total
Female	23	14	16	15	21	18	14	16	17	9	9	11	183
Male	17	17	12	18	19	17	22	17	12	14	12	16	193
Total	40	31	28	33	40	35	36	33	29	23	21	27	376

born on each day of the calendar year. Therefore, a fraction of 31 out of 365¼ of the population would be expected to be born in January as well as March, May, July, August, October and December, a fraction of 28¼ out of 365¼ of the population would be expected to be born in February with a fraction of 30 out of 365¼ of the population being expected to be born in each of the remaining months. With a total number of 183 females participating in senior singles events at Grand Slam tennis tournaments in 2009, we would therefore expect 15.53 (183 × 31 / 365.25) to be born in each the months January, March, May, July, August, October and December, 14.15 (183 × 28.25 / 365.25) to be born in February with 15.03 (183 × 30 / 365.25) to be born in each of the remaining months. The chi square goodness of fit test compares the observed birth month distribution with this theoretically expected birth month distribution. We will use a p value of 0.05 or less to determine if there is a significant difference between the observed and expected frequency distributions.

SPSS

The months of birth of the 183 players are found in the file 15-womens-singles-2009-grand-slam-tennis-senior.SAV. We can inspect the frequency distribution to ensure it is the same as shown in Table 15.1 by using **Analyse → Descriptive Statistics → Frequencies** and transferring the variable 'month of birth' into the *Variables* area of the pop-up window. To perform the chi square goodness of fit test we use **Analyse → Nonparametric Tests → Legacy Dialogs → Chi Square** and transfer the variable 'month of birth' into the test *Test Variable List* of the chi square pop-up window as shown in Figure 15.1. In the *Expected Values* area of the pop-up window, we click on the 'Values' radio button because we are not using equal expected values for each month. We enter the expected frequency for each month in turn using the **Add** button to include it in the list of expected frequencies. The particular values we enter here are 15.53, 14.15, 15.53, 15.03, 15.53, 15.03, 15.53, 15.53, 15.03, 15.53, 15.03 and 15.53. We then click on the **OK** button and the output of the test is produced as shown in Tables 15.2 and 15.3.

Here, we conclude that there was no significant difference between the observed month of birth distribution for the female players and an expected distribution where there is an equal chance of being born on each day of the calendar year ($\chi^2_{11} = 13.0$, p = 0.291). A limitation of the method used in this example is the assumption of a consistent birth rate throughout the calendar year. A more sophisticated approach is possible where national statistics provide birth month distributions for the wider population (Joll and O'Donoghue, 2007).

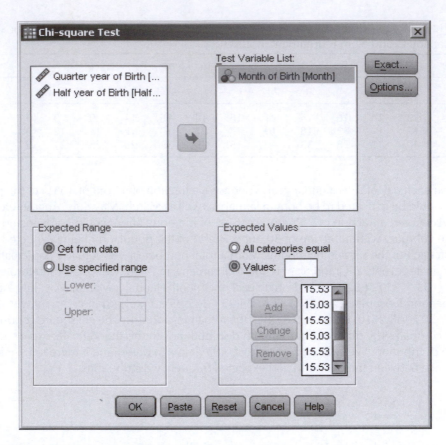

Figure 15.1 Chi Square pop-up window.

Table 15.2 Observed and expected frequencies

Month of Birth			
	Observed N	*Expected N*	*Residual*
January	23	15.5	7.5
February	14	14.2	−.2
March	16	15.5	.5
April	15	15.0	.0
May	21	15.5	5.5
June	18	15.0	3.0
July	14	15.5	−1.5
August	16	15.5	.5
September	17	15.0	2.0
October	9	15.5	−6.5
November	9	15.0	−6.0
December	11	15.5	−4.5
Total	183		

Table 15.3 SPSS output for the chi square goodness of fit test

Test Statistics	
	Month of Birth
Chi-Square	13.031[a]
Df	11
Asymp. Sig.	.291

a 0 cells (.0%) have expected frequencies less than 5. The minimum expected
cell frequency is 14.2.

Presentation of results

The observed and expected frequencies can be expressed in graphical form (as in Figure
15.2) or tabular form. Key chi square results can be shown in text as follows:

> Figure 15.2 shows the observed and expected distribution of female tennis players
> participating in Grand Slam singles tennis in 2009. The observed birth month distri-
> bution was not significantly different to an expected distribution assuming an equal
> chance of being born on each day of the calendar year (χ^2_{11} = 13.0, p = 0.291).

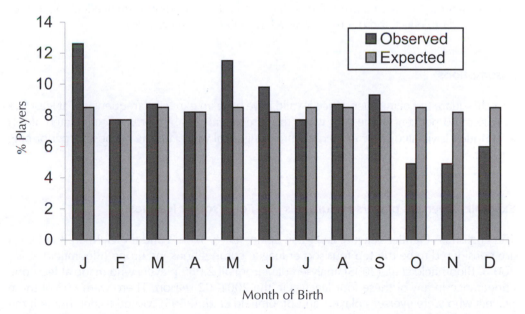

Figure 15.2 Birth month distribution of female tennis players participating in Grand Slam
singles tennis in 2009.

CHI SQUARE TEST OF INDEPENDENCE

Purpose of the test

The chi square test of independence is used to test if two categorical variables are related or whether they are independent. Consider one of the categorical variables consisting of two values (say female and male) and the other also consisting of two categories (say preference for individual sports or preference for team sports). If gender and type of sport preferred were completely independent, then we would expect the proportion of our overall sample that preferred individual sports to apply exactly equally to the female subset of the sample and to the male subset of the sample. The chi square test of independence uses the overall sample proportions to test whether the two variables are independent or not. The two variables being compared are cross-tabulated to show the frequency of cases for each gender value and sports preference. The number of degrees of freedom in a chi square test of independence is the product of the number of values in one categorical variable less one and the number of values in the other categorical variable less one. That is, the number of degrees of freedom is the number of rows less one in the cross-tabulation multiplied by the number of columns less one. The minimum number of degrees of freedom in a chi square test of independence is one and, therefore, a 2 × 2 cross-tabulation is the smallest that a chi square test of independence can be applied to. Indeed, the 2 × 2 cross-tabulation is the easiest one in which to obtain a significant association and so Fisher's exact test is used in this case to reduce the chance of making a Type I Error.

The chi square test of independence can be used with a measure of effect size. Where a 2 × 2 cross-tabulation is done, phi (ϕ) can be reported as the effect size. In all other situations, Cramer's V should be used. For both ϕ and V, 0.1 represents a small effect, 0.3 represents a medium effect and 0.5 represents a large effect (Cramer, 1999).

Assumptions

The chi square test of independence is valid if we have an expected frequency of at least five in at least 80 per cent of the cells when the categorical variables are cross-tabulated. If this assumption is violated, we can merge some categorical values for the purpose of performing the test.

Example: Overseas players in Europe's 'Big Four' soccer leagues

The English FA Premier League, La Liga in Spain, Italy's Serie A and the German Bundesliga are considered to be the top four soccer league competitions in Europe (Bloomfield et al., 2005). Bloomfield et al. (2005) analysed data from all 2,085 players who made at least one appearance in any of these four leagues in the 2001–02 season. There were 802 of these players who were overseas players and Bloomfield et al. (2005) wished to determine if the geographical origin of those players was similar between the four leagues. Table 15.4 shows the frequency and percentage profile of overseas players in the four leagues. If geographical

Table 15.4 Geographical origin of foreign players Europe's four top soccer leagues 2001–02 (Bloomfield et al., 2007)

Origin	English FA Premier League	Spanish La Liga	Italian Serie A	German Bundesliga	Total
Western Europe	66 (30.0%)	22 (12.4%)	29 (17.5%)	41 (17.2%)	158 (19.7%)
Southern Europe	29 (13.2%)	33 (18.6%)	32 (19.3%)	44 (18.4%)	138 (17.2%)
Northern Europe	51 (23.2%)	7 (4.0%)	7 (4.2%)	18 (7.5%)	83 (10.3%)
Eastern Europe	14 (6.4%)	11 (6.2%)	14 (8.4%)	62 (25.9%)	101 (12.6%)
North & Central America	14 (6.4%)	4 (2.3%)	0 (0.0%)	4 (1.7%)	22 (2.7%)
South America	16 (7.3%)	95 (53.7%)	72 (43.4%)	29 (12.1%)	212 (26.4%)
Africa	17 (7.7%)	4 (2.3%)	9 (5.4%)	32 (13.4%)	62 (7.7%)
Asia & Oceania	13 (5.9%)	1 (0.6%)	3 (1.8%)	9 (3.8%)	26 (3.2%)
Total	220 (100%)	177 (100%)	166 (100%)	239 (100%)	802 (100%)

Western Europe – UK, Ireland, Germany, Belgium, the Netherlands, Luxembourg, France, Switzerland, Austria and Liechtenstein.
Southern Europe – Italy, Spain, Portugal, Greece, Turkey, Yugoslavia, Croatia, Slovenia, Bosnia, Macedonia, Albania, Kosova.
Northern Europe – Sweden, Norway, Denmark, Finland, Iceland, Latvia, Lithuania, Estonia.
Eastern Europe – Bulgaria, Romania, Czech Republic, Slovakia, Russia, Ukraine, Moldova, Belarus, Poland, Hungary.

origin of these players was completely independent of the league that they played in, then we would expect the same percentages to apply to each league. Specifically, we would expect the percentages shown in the 'Total' column to apply to each individual league. There are differences and the chi square test of independence can determine whether those differences are significant. We will use a p value of 0.05 or less to identify a significant difference between the observed and expected frequency distributions.

SPSS

The league and geographical origin of the 802 players are found in the file 15-overseas-players-2001–02.SAV. We cross-tabulate the frequencies to ensure they are the same as shown in Table 15.4 by using **Analyse → Descriptive Statistics → Crosstabs** and transfer the variable 'League' into the *Column(s)* and 'Geographical origin' into the *Row(s)* areas of the pop-up window. Figure 15.3 shows the Crosstabs pop-up window. The cells button allows us to indicate that we wish to view both the observed and expected frequencies and that we wish to also express the frequencies as percentages of the columns they are in. We then click on the **Continue** button to clear the cell pop-up window. The chi square test of independence is not performed as a default action of the cross-tabulation and so we must click on the **Statistics** button and indicate that we wish to perform a chi square test and have Cramer's V and phi (φ) reported. This is shown in Figure 15.4. We then click on the **Continue** button to clear the Statistics pop-up window and then click on the **OK** button so that the output of the cross-tabulation and chi square test of independence is produced as shown in Tables 15.5 to 15.7.

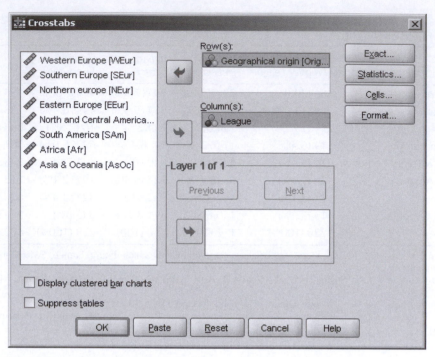

Figure 15.3 Cross-tabulation pop-up window.

Figure 15.4 Statistics pop-up window
for cross-tabulations.

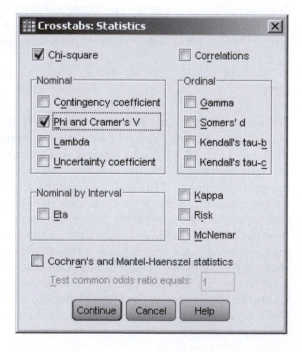

Table 15.5 Cross-tabulation of frequencies output by SPSS

Geographical origin * League Cross-tabulation

			League				Total
			FA Premier League	La Liga	Serie A	Bundeslige	
Geog origin	Western Europe	Count	66	22	29	41	158
		Expected Count	43.3	34.9	32.7	47.1	158.0
		% within League	30.0%	12.4%	17.5%	17.2%	19.7%
	Southern Europe	Count	29	33	32	44	138
		Expected Count	37.9	30.5	28.6	41.1	138.0
		% within League	13.2%	18.6%	19.3%	18.4%	17.2%
	Northern Europe	Count	51	7	7	18	83
		Expected Count	22.8	18.3	17.2	24.7	83.0
		% within League	23.2%	4.0%	4.2%	7.5%	10.3%
	Eastern Europe	Count	14	11	14	62	101
		Expected Count	27.7	22.3	20.9	30.1	101.0
		% within League	6.4%	6.2%	8.4%	25.9%	12.6%
	North & Central America	Count	14	4	0	4	22
		Expected Count	6.0	4.9	4.6	6.6	22.0
		% within League	6.4%	2.3%	.0%	1.7%	2.7%
	South America	Count	16	95	72	29	212
		Expected Count	58.2	46.8	43.9	63.2	212.0
		% within League	7.3%	53.7%	43.4%	12.1%	26.4%
	Africa	Count	17	4	9	32	62
		Expected Count	17.0	13.7	12.8	18.5	62.0
		% within League	7.7%	2.3%	5.4%	13.4%	7.7%
	Asia & Oceania	Count	13	1	3	9	26
		Expected Count	7.1	5.7	5.4	7.7	26.0
		% within League	5.9%	.6%	1.8%	3.8%	3.2%
Total		Count	220	177	166	239	802
		Expected Count	220.0	177.0	166.0	239.0	802.0
		% within League	100.0%	100.0%	100.0%	100.0%	100.0%

Table 15.6 Results for the chi square test of independence

Chi-Square Tests			
	Value	df	Asymp. Sig. (2-sided)
Pearson Chi-Square	280.008a	21	.000
Likelihood Ratio	276.191	21	.000
Linear-by-Linear Association	4.972	1	.026
N of Valid Cases	802		

a 2 cells (6.3%) have expected count less than 5. The minimum expected count is 4.55.

Table 15.7 Effect size results for the chi square test of independence

Symmetric Measures		Value	Approx. Sig.
Nominal by Nominal	Phi	.591	.000
	Cramer's V	.341	.000
N of Valid Cases		802	

The Pearson's Chi Square row of Table 15.6 provides the results of interest. The note 'a' beneath the Chi Square Tests table shows that there were two of the 32 cells in the cross-tabulation of frequencies where there was an expected count of less than five. This is less than 20 per cent of the cells and therefore the data satisfy the assumption of the chi square test that at least 80 per cent of the cells have an expected count of five or greater. The 'Value' is the chi square test statistic and 'Asymp. Sig.' is the p value. There were four columns and eight rows in the cross-tabulation of frequencies and, therefore, the number of degrees of freedom used in the test was $(4 - 1) \times (8 - 1) = 21$. We conclude that there was a significant association between league and the geographical origin of players recruited by clubs within the leagues ($\chi^2_{21} = 280.0$, $p < 0.001$). We can also see in Table 15.7 that the size of the effect is medium (Cramer's V = 0.341). An inspection of the observed and expected frequencies shows the biggest differences between observed and expected frequencies are that there were more overseas players from Western Europe, Northern Europe and North and Central America than expected in the English FA Premier League than expected, more South American players than expected in La Liga and Serie A and more overseas players from Eastern Europe and Africa than expected in the Bundesliga.

If we wished to rigorously confirm the significance of league on any of the geographical regions, we can perform follow-up post hoc chi square tests of independence using dichotomous (yes or no) variables for each of the eight broad regions. You will notice in the file that 15-overseas-players-2001–02.SAV that there are eight such variables that simply have values 'yes' or 'no' indicating whether or not the player is from the given region. We can apply post hoc chi square tests of independence to these variables to determine if the proportion of players from any given region significantly differs between the leagues. Because eight of these tests are being performed, a Bonferroni adjustment is applied to the p value required

294

chi square

for significance to avoid inflating the probability of making a Type I Error. Therefore, only p values of 0.006 or less (from equation 11.7) will be considered significant. Performing these post hoc chi square tests reveals that there were significant differences with small to medium effects between the leagues for the proportion of overseas players who were from Western Europe ($p < 0.001$, $\phi = 0.166$), Northern Europe ($p < 0.001$, $\phi = 0.263$), Eastern Europe ($p < 0.001$, $\phi = 0.263$), North and Central America ($p = 0.001$, $\phi = 0.144$), South America ($p < 0.001$, $\phi = 0.456$) and Africa ($p < 0.001$, $\phi = 0.133$).

Where a chi square test of independence is applied to a 2×2 cross-tabulation, an additional row of results is produced in the Chi Square Test results table in the SPSS output. This is a row for Fisher's Exact Test. This test is only performed on a 2×2 cross-tabulation and this is the only situation when its results will be shown in the SPSS output. When using a 2×2 cross-tabulation, we should report the two-tailed p value (Asymp. Sig.) from Fisher's Exact test rather than the p value from Pearson's Chi Square.

Reporting results

A table such as Table 15.4 could be used to present the results with a possible additional row to show the post hoc chi square test results. The table could be supported by the following text:

> Table 15.4 shows the number of overseas players in each league from different geographical areas. There was a significant association between league and the geographical origin of players recruited by clubs within the leagues ($\chi^2_{21} = 280.0$, $p < 0.001$, $V = 0.341$) with more overseas players from Western Europe ($p < 0.001$, $\phi = 0.166$), Northern Europe ($p < 0.001$, $\phi = 0.263$) and North and Central America ($p = 0.001$, $\phi = 0.144$) in the English FA Premier League than expected, more South American players than expected ($p < 0.001$, $\phi = 0.456$) in La Liga and Serie A and more overseas players from Eastern Europe ($p < 0.001$, $\phi = 0.263$) and Africa ($p < 0.001$, $\phi = 0.133$) than expected in the Bundesliga.

SUMMARY

There are two types of chi square test. The chi square goodness of fit test is used to compare the frequency distribution of a categorical variable with a theoretically expected frequency distribution. This requires us to produce the theoretically expected distribution using some method that we can justify. The chi square test of independence tests whether two categorical variables are associated with each other or whether they are independent. This test does not require us to produce expected frequencies because these can be determined from the data themselves. Where we are performing a chi square test on a 2×2 frequency cross-tabulation, we should use Fisher's exact test to help reduce the chance of making a Type I error. For any chi square test to be valid, at least 80 per cent of the expected frequencies must be five or greater.

EXERCISES

Exercise 15.1. Relative age in women's tennis

In the ex15.1-womens-singles-2009-grand-slam-tennis-senior.SAV file, there are two additional variables: 'quarter year of birth' and 'half year of birth'. Quarter year groups the players into four quarters based on their month of birth (January–March, April–June, July–September and October–December). Half year of birth simply classifies the players as H1 if they were born in the first six months of the year or H2 if they were born in the second half of the year. Using the same approach as before to determining an expected frequency distribution, do the following:

a)   Determine if the distribution of quarter year of birth is significantly different to the expected distribution based on an equal chance of being born on each day of the calendar year.
b)   Determine if the distribution of half year of birth is significantly different to the expected distribution based on an equal chance of being born on each day of the calendar year.

Exercise 15.2. Relative age in men's tennis

The ex15.2-mens-singles-2009-grand-slam-tennis-senior.SAV file contains the 'month of birth', 'quarter year of birth' and 'half year of birth' for the 193 male players who participated in the first round of the senior men's singles events at one or more of the four Grand Slam tennis tournaments in 2009. Using the same approach as before to determining an expected frequency distribution, do the following:

a)   Determine if the distribution of month of birth is significantly different to the expected distribution based on an equal chance of being born on each day of the calendar year.
b)   Determine if the distribution of quarter year of birth is significantly different to the expected distribution based on an equal chance of being born on each day of the calendar year.
c)   Determine if the distribution of half year of birth is significantly different to the expected distribution based on an equal chance of being born on each day of the calendar year.

Exercise 15.3. Injuries in women's netball

The file ex15.3-netball-injuries-survey.SAV contains the responses to the following six questions within a questionnaire survey administered to 106 netball players (possible responses are shown in parentheses):

1   What level of netball do you play? (Elite or Non-elite)
2   Have you suffered an injury in the last 12 months that caused you to miss at least one match or training session? (Yes or No)
     If the answer to Question 2 was no, go to Question 6.
3   Did the most recent injury occur during a match or training? (Match or Training)
4   How was the most recent injury treated? (Not-treated, Sports massage, Physiotherapy, Surgery or Other)
5   What was the rehabilitation process for the most recent injury? (None, Sports massage, Occupational therapy, Physiotherapy or Other)
6   What equipment do you use to prevent injury? (None, Strapping, Splints, Braces or Other).

Use cross-tabulation of frequencies and chi square tests of independence to answer the following questions:

a) Is there an association between level of player and whether the player suffered an injury in the last 12 months?

b) Is there an association between level of player and any of the following for those players who suffered injuries: (i) whether the most recent injury occurred during a match or training, (ii) treatment option taken for the most recent injury and (iii) rehabilitation process for the most recent injury.

c) Is there an association between level of player and type of equipment used to prevent injury? Note there are 13 respondents who did not answer this question and we can neither assume that they used equipment or not. Therefore, only the 93 who responded should be analysed.

PROJECT EXERCISE

Exercise 15.4. Relative age of athletes

Find a squad of 20 to 30 players in a sport of your choice who are willing to provide you with their months of birth. This could be a squad of athletes you know or it could be a squad of players who have published birth date details of their players on the official club internet site. Find out the cut-off date for the junior competition year in the sport in the given country; this might be 1 January in some sports but might coincide with the school year in other sports. Indeed, the squad you are analysing might be a school squad. Classify each player as being born in the first or second six months of the junior competition year. Note that if the junior competition year has a cut-off date of 1 September (for example), the first six months of the year will be September to December as well as January and February. Use a chi square goodness of fit test to determine if the distribution of half year of birth observed for the squad fits a theoretically expected distribution. Whether it does or not, try to list factors that might explain the result of the project.

Exercise 15.5. Proportion of unseeded players in the 3rd round of Grand Slam tennis tournaments

For each of the four Grand Slam tennis tournaments played this year, examine the results of the 3rd round matches of the men's and women's singles events. These can be found in the 'Draws' section of the tournaments' official internet sites (www.ausopen.org, www.frenchopen.org, www.wimbledon. org and www.usopen.org). Note the number of seeded and unseeded players participating in the 3rd round. Use four chi square tests of independence to determine if the proportion of unseeded players making the 3rd round significantly differs between men's and women's singles at any of the four tournaments.

CHAPTER 16

STATISTICAL CLASSIFICATION

INTRODUCTION

Statistical classification techniques can be seen as the opposite of analysis of variances (ANOVA) tests. ANOVA tests have categorical independent variables which are hypothesized to influence the values of some numerical scale dependent variables. In statistical classification, however, we produce a predictive model of categorical group membership in terms of some numerical scale independent variables. Therefore, the grouping variable is the dependent variable. Two statistical classification techniques are described using prediction of match outcomes in the 2010 FIFA World Cup as an example. Discriminant function analysis is used to predict membership of two or more groups. This technique is used to produce a predictive model of the pool stage matches of the World Cup which could be wins, draws or losses for the higher ranked team. The second technique is binary logistic regression which predicts membership of some dichotomous grouping variable. The technique is used to predict the outcomes of the knockout stage matches of the World Cup which are classified as being won by the higher ranked team or as upsets. Even if a knockout match goes to extra time and a penalty shoot out, one of the two teams involved will still be eliminated from the tournament.

DISCRIMINANT FUNCTION ANALYSIS

Purpose of the test

Discriminant function analysis is a technique used to predict group membership using a set of numerical predictor variables. Analysis of existing cases with known group membership produces a predictive model that can be applied to new cases to predict their group memberships. The predictive model consists of a territorial map onto which new cases can be plotted using canonical discriminant functions in order to predict their group membership.

Assumptions

There are a number of assumptions that should be satisfied by the previous case data used to produce the predictive model. If the means of the predictor variables are similar for

all groups, then discriminant function analysis is unsuitable (Mardia *et al.*, 1994: 318). The variables used to predict group membership should be normally distributed and the within-group covariance matrices should similar for all groups (Manly, 2005: 205). A covariance matrix is similar to a correlation matrix in that the matrix is a square with the same number of rows and columns and the same set of variables form the rows and columns. A correlation matrix shows the correlation coefficients (Pearson's r) between each pair of variables with the top left to bottom right diagonal showing perfect correlations between the same variables (r = 1.000) and the remainder of the correlation matrix being symmetrical about the top left to bottom right diagonal. This is illustrated in Table 16.1(a). Table 16.1(b) is the covariance matrix for the same variables. In the covariance matrix, the top left to bottom right diagonal contains the variances of the three variables A, B and C. The variance is the square of the standard deviation. The remaining cells contain the covariances for each pair of different variables. The covariance between the variables A and B is given by equation 16.1 where n is the number of cases, $\bar{A}$ and $\bar{B}$ are the means for variables A and B respectively. The covariance matrices for the different groups within the previous case data must be sufficiently similar to allow discriminant function analysis to be performed validly. Box's M test is used to test equality of covariances between groups.

$$\text{Co var} = \frac{\sum_{i=1}^{n}(A_i - \bar{A})(B_i - \bar{B})}{n-1} \tag{16.1}$$

Tabachnick and Fidell (2007: 381–3) listed additional assumptions that the predictor values should not include significant outliers, each pair of predictor variables must have a linear relationship within each group and that predictor variables should not be highly correlated (absence of multicollinearity) or functionally dependent (absence of singularity).

Example: Predicting outcomes of pool matches of the 2010 FIFA World Cup

The FIFA World Cup is arranged into a pool stage followed by a knockout tournament. During the pool stage, the teams are arranged into eight pools of four teams with each team in a pool playing each other pool member once. Therefore, there are 48 pool matches in total (8 pools x 6 matches per pool). In order to predict the outcomes of these matches (win, draw

Table 16.1 (a) Correlation matrix, (b) Covariance matrix. The variables A, B and C are the difference variables form FIFA ranking points, distance travelled and recovery days from previous match for the two teams in 153 soccer matches. These variables can be found in the file 16.1-pool-matches-2007-2010.SAV ('RP_Diff', 'Dist_Diff' and 'Rec_Diff')

(a) Correlation Matrix				(b) Covariance Matrix		
	A	B	C		A	B	C
A	1.000	0.101	0.002	A	58817.6	70194.8	0.4
B	0.101	1.000	−0.149	B	70194.8	8172500.3	−354.4
C	0.002	−0.149	1.000	C	0.4	−354.4	0.7

or lose), discriminant function analysis is applied to 153 pool matches from international soccer tournaments played since the 2006 World Cup when FIFA changed the way in which it ranks international teams. The two teams participating in a match are termed the superior team and the inferior team based on their FIFA World rankings at the time the match was played. The superior team is the higher ranked of the two teams and the inferior team is the lower ranked of the two. This is a crude prediction technique based on three hypothesized predictor variables:

1   FIFA World ranking points which are published by FIFA with the values for any previous month being provided (www.fifa.com accessed 26 May 2010).
2   Distance between a team's capital city and the capital city of the host nation of the tournament (km) according to a tourist website (www.indo.com/distance accessed 26 May 2010).
3   Recovery days from the previous match played.

These variables together with the match outcome are known for each of the two teams participating in the 153 previous matches. Because each match involved two teams, the difference between the two teams' values (superior team – inferior team) was used for each variable. Therefore, the three variables are effectively:

1   How many more FIFA world ranking points the superior team has than the inferior team.
2   How much further the superior team's capital city is to the host nation's capital city than the inferior team's capital city is (km).
3   How many more recovery days did the superior team have from their previous match than the inferior team.

The outcome variable classified matches as wins, draws or losses for the superior team. The number of canonical discriminant functions produced is either the number of predictor variables or one less than the number of groups, whichever is less. Therefore, in the current example discriminant function analysis produces a pair of canonical discriminant functions. The territorial map produced is a two-dimensional space because we have two canonical discriminant functions. The territorial map has three areas: one for each match outcome class (win, draw or loss for the superior team). The three predictor values for each of the pool matches of the 2010 World Cup were known before the matches were played because we knew the tournament was hosted in South Africa, the schedule of matches was known since the World Cup draw was made in December 2009 and the FIFA world ranking points of the 32 teams was known at the beginning of the tournament. The outcomes of the matches were unknown at the beginning of the tournament. Applying the two canonical discriminant functions to the known values of the predictor variables for a 2010 World Cup pool match produces co-ordinates that can be plotted on the territorial map in order to classify the predicted match outcome.

SPSS and Excel

The data for the 153 pool matches from previous international soccer tournaments are found in the file 16.1-pool-matches-2007–2010.SAV. The three predictor variables are 'RP_Diff'

(difference in ranking points), 'Dist_Diff' (how much further the superior team travelled to the tournament than the inferior team) and 'Rec_Diff' (how many more recovery days from the previous match the superior team had than the inferior team). However, when we explore these variables (**Analyse** → **Descriptive Statistics** → **Explore** asking for normality plots with tests), the Kolmogorov–Smirnov tests reveal that none are normally distributed ($p \leq 0.002$). Therefore, the file also contains three additional variables that are produced by mapping functions. There are a number of ways of transforming variables to cope with violations of the assumptions of statistical procedures (Nevill, 2000; Taqbachnick and Fidell, 2007: 86–8). These transformations do not always result in the data satisfying the necessary assumptions. Therefore, O'Donoghue (2006) proposed the use of mapping functions that mapped any variable onto the standard normal distribution (mean of zero and standard deviation of one). Microsoft Excel can be used to produce these mapping functions using the function NORMSINV. The 153 variables of a value are arranged in ascending order. The z-scores are given by NORMSINV((i – 0.5) / 153) for each value of i from 1 to 153. Where two or more matches have the same value for a given variable, it is necessary to use the mean of the z-scores for these matches as each value of the original variable can only be mapped onto a single z-score. The Microsoft Excel spreadsheet 16-Previous-From-2007.XLS contains a sheet 'Mapping Functions' that shows the mappings from original variable values onto z-scores. This is done for pool matches as well as knockout matches. It should be noted that the values of the original variables were not unique and, therefore, the mapping functions all contain less than 153 value mappings. The sheet 'z-score workings' illustrates how this was done for to map 'RP_Diff' onto 'zRP_Diff'. The z-transformed variables 'zRP_Diff' and 'zDist_Diff' are sufficiently normal according to Kolmogorov–Smirnov tests ($p > 0.05$). However, 'zRec_Diff' is not normally distributed despite the transformation ($p < 0.001$) and is, therefore, excluded from the analysis.

There were two outliers for 'zRP_Diff': the 2009 Confederations Cup matches between Spain and New Zealand and between Italy and Brazil. Italy and Brazil were ranked 4th and 5th in the world at the time and there were only four ranking points between them. Spain were ranked 1st with 1,761 ranking points while New Zealand were ranked 82nd with 431 ranking points. At this point we need to ask whether we should remove these outliers from the data set or not. It is possible for teams to have the same number of ranking points without there being any measurement error. Should New Zealand's matches be removed from the data? Is the Confederations Cup fundamentally different to other international soccer tournaments? Can the best team from Oceania qualify for the World Cup? The author decided not to exclude these outliers because New Zealand had actually qualified for the World Cup and closely ranked teams (such as Italy and Brazil were in 2009) could meet in the 2010 World Cup.

There were also two outliers in the 'zDist_Diff' variable: South Africa vs New Zealand and Spain vs South Africa from the 2009 Confederations Cup. We are attempting to predict the outcomes of the 2010 World Cup which is a tournament involving teams from different confederations (FIFA's continental zones). Because FIFA changed its world ranking system in 2006, the only matches included in the previous case data that involve teams from different confederations are Confederations Cup matches. These matches involve the greatest amount of travel in the data set and similar distances would have to be travelled to South Africa for the 2010 World Cup. It was, therefore, the view of the author that excluding Confederations Cup matches would risk a danger of extrapolating the effect of distance travelled within intra-confederations tournaments (such as Copa America and Euro 2008) beyond the distances used to create the predictive model. Therefore, these two outliers were not removed from the data set.

The correlation between the two predictor variables was not high (r = 0.119) and there was no evidence of a non-linear relationship between the variables when the scatter plot was inspected (**Graphs → Chart Builder** choosing scatter plot and placing 'zRP_Diff' and 'zDist_Diff' within the axes).

To perform the discriminant function analysis in SPSS, use **Analyse → Classify → Discriminant** and transfer the variables 'zRP_Diff' and 'zDist_Diff' into the *Independents* list of the pop-up window shown in Figure 16.1. 'Outcome' is transferred into the *Grouping Variable* with the range of values being defined as –1 to 1 (–1 indicates a loss for the superior team, 0 indicates a draw and +1 indicates a win). In order to check that the data satisfy the assumptions of the technique, click on the **Statistics** button and request Means, Box's M test, unstandardized function coefficients, within-groups covariance and separate-groups covariance. Click on **Continue**. The number of wins (77), draws (36) and losses (40) are not equal and so it is necessary to advise SPSS that the proportions should be derived from the data provided. Use the **Classify** button and request that prior probabilities are computed from group sizes and request a territorial map. While in the **Classify** pop-up window, also request SPSS to display a summary table. Click on **Continue**. Click on **OK**.

The **Save** button within the Discriminant Analysis pop-up window could be used to save predicted group membership (predicted match outcome) for the previous cases as well as the canonical discriminant function values. These are not necessary to produce the predictive model but may be of interest for checking purposes.

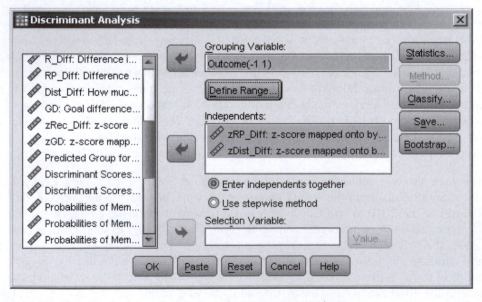

Figure 16.1 The discriminant function analysis pop-up window.

statistical classification

Tables 16.2 to 16.7 show the most important tables of the SPSS output and Figure 16.2 shows the territorial map produced. Box's M test was not significant (p = 0.480) and, therefore, the data satisfied the assumption of similar covariance matrices between matches of different outcomes. We use the unstandardized canonical discriminant functions in Table 16.7 which are also presented in equations 16.2 and 16.3.

$$F1 = 1.064 \text{ zRD\_Diff} - 0.096 \text{ zDist\_Diff} \tag{16.2}$$

$$F2 = -0.029 \text{ zRD\_Diff} + 1.003 \text{ zDist\_Diff} \tag{16.3}$$

Table 16.2 SPSS output for Box's M test

Test Results		
Box's M		5.632
F	Approx.	.918
	df1	6
	df2	149148.495
	Sig.	.480

Tests null hypothesis of equal population covariance matrices.

Table 16.3 Eigenvalues for canonical discriminant functions

Eigenvalues				
Function	Eigenvalue	% of Variance	Cumulative %	Canonical Correlation
1	.130[a]	91.3	91.3	.339
2	.012[a]	8.7	100.0	.111

a First 2 canonical discriminant functions were used in the analysis.

Table 16.4 Significance of different models

Wilks' Lambda				
Test of Function(s)	Wilks' Lambda	Chi-square	df	Sig.
1 through 2	.874	20.067	4	.000
2	.988	1.837	1	.175

Table 16.5 Standardized coefficients for the two discriminant functions

Standardized Canonical Discriminant Function Coefficients		
	Function	
	1	2
zRP_Diff	1.007	-.027
zDist_Diff	-.096	1.003

Table 16.6 Structure matrix

	Structure Matrix	
	Function	
	1	2
zRP_Diff	.995*	.095
zDist_Diff	.027	1.000*

Pooled within-groups correlations between discriminating variables and standardized
canonical discriminant functions
Variables ordered by absolute size of correlation within function.
* Largest absolute correlation between each variable and any discriminant function

Table 16.7 Coefficients of the two canonical discriminant functions

Canonical Discriminant Function Coefficients		
	Function	
	1	2
zRP_Diff	1.064	−.029
zDist_Diff	−.096	1.003
(Constant)	.000	.000

Unstandardized coefficients

Wilk's λ shown in Table 16.4 revealed that the pair of functions F1 and F2 together form a
significant predictive model (p < 0.001). Table 16.8 reveals that the predictive model correctly classified 58.8 per cent of the 153 previous matches.

The Excel spreadsheet 16-2010-matches.XLS contains details of the matches actually played
in the 2010 FIFA World Cup. Consider Pool A contested by France, South Africa, Mexico

Table 16.8 Classification results comparing observed and predicted group membership

			Classification Results[a]			
		Outcome	Predicted Group Membership			
			−1	0	1	Total
Original	Count	−1	11	2	27	40
		0	3	10	23	36
		1	5	3	69	77
	%	−1	27.5	5.0	67.5	100.0
		0	8.3	27.8	63.9	100.0
		1	6.5	3.9	89.6	100.0

a 58.8% of original grouped cases correctly classified.

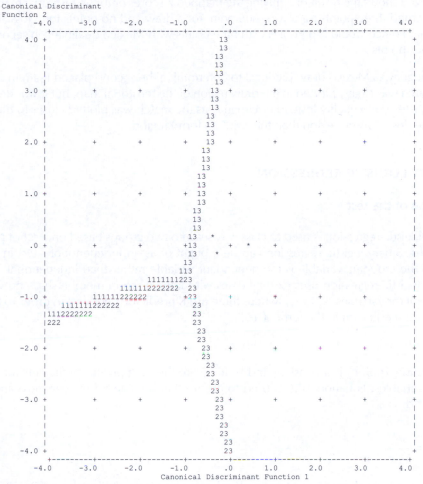

Figure 16.2 Territorial Map (−1 is loss, 0 is draw and +1 is win).

and Uruguay. The mapping functions are used to transform the original difference variables onto the z-score transformed variables. Where an original value x lies between two values of the mapping function x1 and x2 that are mapped onto z1 and z2 respectively, we use interpolation to determine the z-score as shown in equation 16.4.

$$z = z1 + (z2 - z1) \times (x - x1) / (x2 - x1)$$ (16.4)

305

statistical classification

Table 16.9 shows the result of applying the mapped z-scores onto the territorial map. Teams are awarded three points for a win, one point for a draw and no points for a loss. This predicts that France would top Pool A with nine points with Mexico qualifying in second place with four points.

The Uruguay vs Mexico draw is critical to determining the second placed team in the pool. The match was clearly plotted in the draw region of the territorial map, but it was decided to predict Mexico to qualify instead of Uruguay as the match was plotted closer to the border with the 'lose' (upset) region than the 'win' (to form) region.

BINARY LOGISTIC REGRESSION

Purpose of the test

Binary logistic regression is used to classify cases into two groups based on a set of predictor variables. Binary logistic regression can be thought of as an independent t-test in reverse; that is, the grouping variable is the dependent variable rather than independent variable. Binary logistic regression uses existing cases with known group membership and values of the predictor variables, x_1 to x_K, (where there are K predictor variables) in order to produce a regression equation in the form of 16.5.

$$y = b_0 + b_1.x_1 + b_2.x_2 + \ldots\ldots + b_K.x_K \tag{16.5}$$

In equation 16.5, b_0 is a constant and b_1 to b_K are the multiplying coefficients for the predictor variables. Equation 16.6 is used to ensure that a value π between zero and one is produced.

$$\pi = \frac{e^y}{1+e^y} \tag{16.6}$$

Values of π of less than 0.5 predict membership of one group, while values of 0.5 and greater predict membership of the other. Once the coefficients b_0 to b_K are determined by binary logistic regression, equations 16.5 and 16.6 can be used to predict future cases where the values of the predictor variables are known but group membership is unknown. The odds

Table 16.9 Pool A predictions

Superior team	Inferior team	RP_Diff	Dist_Diff	zRP_Diff	zDist_Diff	F1	F2	Predicted outcome
Mexico	S.Africa	503	14588	0.975	3.000	0.749	2.981	Win
France	Uruguay	145	740	−0.589	0.376	−0.663	0.394	Win
Uruguay	S.Africa	507	7919	1.040	2.405	0.875	2.382	Win
France	Mexico	149	−5929	−0.556	−2.217	−0.378	−2.207	Win
Uruguay	Mexico	4	−6669	−2.720	−2.320	−2.671	−2.248	Draw
France	S.Africa	652	8659	1.526	3.000	1.335	2.965	Win

statistical classification

ratio is $\pi / (1 - \pi)$ and is deetermined by changing the subject of equation 16.6 to e^y, taking the natural logarithm and substituting equation 16.5 for y, we obtain the expression of the logit (log of the odds membership of the second group) shown in equation 16.7.

$$\ln\left(\frac{\pi}{1-\pi}\right) = b_0 + b_1.x_1 + b_2.x_2 + ... + b_K.x_K \qquad (16.7)$$

Assumptions

Logistic regression does not have any assumptions relating to the distribution of the predictor variables; indeed, the predictor variables can be a combination of continuous or dichotomous variables (Tabachnick and Fidell, 2007: 437). However, there are three assumptions that need to be satisfied by the data used to form the predictive model (Tabachnick and Fidell, 2007: 442–4). First, the logit (equation 16.7) should have a linear relationship with any continuous predictor variables. Second, there should be no pair of predictor variables that are highly correlated (absence of multicollinearity). Third, there should be no outliers in the predicted group memberships. This can be checked by examining the difference (residual) between actual group membership signified by zero or one and the probability values (π) produced by logistic regression.

Example: Predicting outcomes of the knockout matches of the 2010 FIFA World Cup

Binary logistic regression is illustrated here using the knockout stages of the 2010 FIFA World Cup. The knockout tournament commences with round two consisting of 16 teams, then the quarter-finals, semi-finals, a third place play-off and the final. Binary logistic regression is applied to 54 knockout stage matches from international soccer tournaments played since the 2006 World Cup when FIFA changed the way in which it ranks international teams. As in the example used to illustrate discriminant function analysis, the two teams participating in a match are termed the superior team and the inferior team based on their FIFA world rankings at the time the match was played. The same predictor variables that were used to predict pool stage match outcomes with discriminant function analysis are used here:

1   FIFA world ranking points which are published by FIFA with the values for any previous month being provided (www.fifa.com accessed 26 May 2010).
2   Distance between a team's capital city and the capital city of the host nation of the tournament (km) according to a tourist website (www.indo.com/distance accessed 26 May 2010).
3   Recovery days from the previous match played.

The main difference to pool stage matches is that knockout stage matches only have two outcomes: a win for the higher ranked team or a win for the lower ranked team. Even if the match goes to extra time and a penalty shoot out, only one team can progress to the next round of the tournament. As before, the difference between the two teams' values (superior team – inferior team) is used for each variable. Therefore, the three variables are effectively:

1   How many more FIFA world ranking points the superior team has than the inferior team.
2   How much further the superior team's capital city is to the host nation's capital city than the inferior team's capital city (km).
3   How many more recovery days did the superior team have from their previous match than the inferior team.

The coefficients b_0 to b_K to be used in equation 16.5 and 16.6 are produced by binary logistic regression. The three predictor values for each of the knockout matches of the 2010 World Cup were known before the knockout stage of the tournament commenced. The outcomes of the matches were unknown. Applying the predictive model (π) to the known values of the predictor variables for a 2010 World Cup knockout match produces the probability of the match being won by the superior team; if this is greater than or equal to 0.5, then the superior team is predicted to win otherwise an upset is predicted.

SPSS and Excel

The data for the 54 knockout matches from previous international soccer tournaments are found in the file 16-ko-matches-2007–2010.SAV. The three predictor variables are 'RP_Diff' (difference in ranking points), 'Dist_Diff' (how much further the superior team had to travel to the tournament than the inferior team) and 'Rec_Diff' (how many more recovery days from the previous match the superior team had than the inferior team).

In order to test if the data satisfy the assumptions of binary logistic regression, we need to perform a binary logistic regression in order to obtain the logit values and the residuals between predicted and actual match outcomes.

To perform a binary logistic in SPSS, use **Analyse** → **Regression** → **Binary Logistic** and transfer the variables 'RP_Diff', 'Dist_Diff' and 'Rec_Diff' into the *Covariates* list of the pop-up window shown in Figure 16.3. 'Outcome' is transferred into the *Dependent* variable (0 is used to represent a win for the inferior team and 1 is used to represent a win for the superior team).

The **Categorical** button is used to identify any categorical predictor variables while the **Options** button is used to set a cut-off probability to classify groups of 0.5. The **Save** button provides a pop-up window that allows us to save those values that we need to test the assumptions of the test: the unstandardized residuals and the logit of the residual. Unfortunately, the logit value itself is not produced automatically and so we have to use **Transform** → **Compute** to create it. This allows inspection of the relationship between the logit and the predictor variables which needs to be linear. The predictor variables are not highly correlated with ($-0.037 \leq r \leq 0.274$). Furthermore, there are no outliers or extreme values when we **Explore** the residuals (differences between observed and expected probabilities of wins by the superior team). Therefore, we can use the original variables to produce a predictive model. Tables 16.10 to 16.12 show the main output for binary logistic regression.

Table 16.11 reveals that the model is incorrectly classifying 20 of the 54 previous tournament matches. The B column of Table 16.12 shows the multiplying coefficients which are used in equation 16.8. Where the number of decimal places is not sufficient to show the

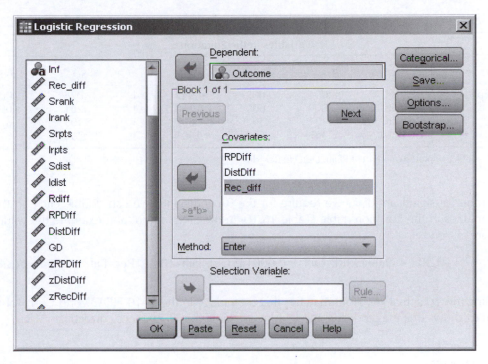

Figure 16.3 The pop-up window for binary logistic regression.

Table 16.10 Significance of independent variables in the model

			Score	df	Sig.
		Variables not in the Equation			
Step 0	Variables	RPDiff	5.328	1	.021
		DistDiff	1.855	1	.173
		Rec_diff	.927	1	.336
	Overall Statistics		7.727	3	.052

Table 16.11 Prediction accuracy based on historical data

			Predicted		
			Outcome		
	Observed		.00	1.00	Percentage Correct
			Classification Table[a]		
Step 1	Outcome	.00	7	12	36.8
		1.00	8	27	77.1
	Overall Percentage				63.0

a The cut value is .500

Table 16.12 Coefficients for the model and their significance

		B	S.E.	Wald	df	Sig.	Exp(B)
	Variables in the Equation						
Step 1[a]	RPDiff	.004	.002	4.942	1	.026	1.004
	DistDiff	.000	.000	1.160	1	.281	1.000
	Rec_diff	−.391	.300	1.697	1	.193	.676
	Constant	−.389	.502	.601	1	.438	.678

a Variable(s) entered on step 1: RPDiff, DistDiff, Rec_diff.

number of significant digits we require for the B coefficients, we can double click on the 'Variables in the Equation' table (Table 16.10) in the SPSS output and examine the values more closely.

$$y = -0.389 + 0.00433\text{RP\_Diff} + 0.000145\text{Dist\_Diff} - 0.391\text{Rec\_Diff} \tag{16.8}$$

Using the 16 actual knockout matches that took place in the 2010 FIFA World Cup, Table 16.13 shows that the predictive model forecast 10 of the 16 matches correctly.

SUMMARY

Discriminant function analysis and binary logistic regression are statistical techniques used to produce models of group membership using independent variables that are numerical scale variables although binary logistic regression can include dichotomous variables as well. Discriminant function analysis is used when membership of one of two or more groups is

Table 16.13 Match predictions

Superior Team	Inferior Team	RP_Diff	Dist_Diff	Rec_Diff	y	π	Prediction	Correct
Uruguay	S.Korea	267	−4553	0	0.107	0.527	Win	Yes
USA	Ghana	157	8392	0	1.508	0.819	Win	No
Holland	Slovakia	454	684	0	1.676	0.842	Win	Yes
Brazil	Chile	745	−443	0	2.773	0.941	Win	Yes
Argentina	Mexico	181	−6448	0	−.540	0.368	Lose	No
Germany	England	14	−202	0	−.358	0.412	Lose	No
Paraguay	Japan	138	−5112	0	−.533	0.370	Lose	No
Spain	Portugal	316	−85	0	0.967	0.725	Win	Yes
Uruguay	Ghana	99	3297	0	0.518	0.627	Win	Yes
Brazil	Holland	380	−1074	0	1.101	0.750	Win	No
Germany	Argentina	6	658	0	−.268	0.433	Lose	No
Spain	Paraguay	745	−377	0	2.782	0.942	Win	Yes
Holland	Uruguay	332	1032	0	1.198	0.768	Win	Yes
Spain	Germany	483	−760	0	1.592	0.831	Win	Yes
Germany	Uruguay	183	879	−1	0.922	0.715	Win	Yes
Spain	Holland	334	−913	−1	1.316	0.788	Win	Yes

statistical classification

being modelled while binary logistic regression can be used when membership of one of two groups is being modelled. Where models successfully predict group membership in previous cases, they may be used to predict group membership in future cases.

EXERCISES

Exercise 16.1. FIFA World Cup 2010

Use the predictive model produced by discriminant function analysis to predict the outcomes of the 42 matches of the remaining seven pools of the 2010 FIFA World Cup (B to H). The match data are found in the 'Exercise 16.1' sheet of the ex16.1-2010-Matches.XLS spreadsheet.

Exercise 16.2. Violating the assumptions

This exercise is going to investigate what happens when the data used violate the assumptions of discriminant function analysis.

a) Using the original difference variables ('RP_Diff', 'Dist_Diff', 'Rec_Diff'), rather than the z-transformed versions, produce a predictive model of the 2010 FIFA World Cup. The model needs to include the canonical discriminant functions and the territorial map. How many of the previous 153 cases would this model correctly predict? Use the file ex16.2-pool-matches-2007–2010.SAV.
b) Use this model to predict the outcomes of the 48 pool stage matches of the 2010 FIFA World Cup. The data for the 2010 pool matches are found in the 'Exercise 16.2(b)' sheet of the ex16.2-2010-Matches.XLS spreadsheet. Compare the results with those of Exercise 16.1 and what actually happened in the 2010 World Cup to consider the impact of violating the assumptions.

Exercise 16.3. Satisfying assumption of no outliers in predictor variables

Once the z-transformed variables were produced, there were still four matches with outlier values (two for 'zRP_Diff' and two for 'zDist_Diff'). These matches were all Confederations Cup matches. What happens to the predictive model when they are removed from the previous case data?

a) Use the SPSS datasheet ex16.3-pool-matches-2007–2010-without-outliers.SAV to produce the discriminant functions and territorial map.
b) Use the 'Exercise 16.3(b)' sheet of the ex16.3-2010-Matches.XLS spreadsheet to predict the outcomes of the 2010 pool stage matches.

Exercise 16.4. Pre-tournament forecast

Continue the prediction of the knockout stages using the teams that were predicted to be in the second round according to your solution to Exercise 16.1. Use the model in equations 16.8 and 16.6 during this exercise. Note the knockout structure in Table 16.14; the only matches where there are differences in the teams' recovery days from their previous matches are the third place play off and the final. Note that the teams in each predicted knockout match need to be considered as difference variables subtracting the values of the inferior team (according to FIFA's world rankings published on 26 May 2010) from those of the superior team.

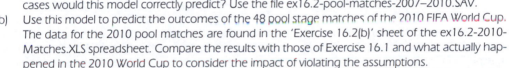

311

Table 16.14 Knock out matches of the 2010 FIFA World Cup

Match	Round	Date	Teams
1	2nd Round	26/6/10	Winner pool A v Runner Up pool B
2	2nd Round	26/6/10	Winner pool C v Runner Up pool D
3	2nd Round	28/6/10	Winner pool E v Runner Up pool F
4	2nd Round	28/6/10	Winner pool G v Runner Up pool H
5	2nd Round	27/6/10	Winner pool B v Runner Up pool A
6	2nd Round	27/6/10	Winner pool D v Runner Up pool C
7	2nd Round	29/6/10	Winner pool F v Runner Up pool E
8	2nd Round	29/6/10	Winner pool H v Runner Up pool G
9	Quarter-final	2/7/10	Winner match 1 v Winner match 2
10	Quarter-final	2/7/10	Winner match 3 v Winner match 4
11	Quarter-final	3/7/10	Winner match 5 v Winner match 6
12	Quarter-final	3/7/10	Winner match 7 v Winner match 8
13	Semi-final	6/7/10	Winner match 9 v Winner match 10
14	Semi-final	7/7/10	Winner match 11 v Winner match 12
15	3rd place play off	10/7/10	Loser match 13 v Loser match 14
16	Final	11/7/10	Winner match 13 v Winner match 14

PROJECT EXERCISE

Exercise 16.5. Predicting outcomes of soccer matches

Use a forthcoming international soccer tournament for which the draw has already been made and use discriminant function analysis to predict the outcomes of the pool matches. Use the same variables as were used in this chapter but include more previous case data such as the 2010 FIFA World Cup matches.

Exercise 16.6. Predicting the outcomes of international rugby union matches

Use binary logistic regression to predict the outcomes of matches of the Rugby World Cup 2011. Assume that there will be no draws and don't include any draws in the previous case data. Use the rugby world rankings (www.irb.com) and the distance between capital city of a country and capital city of the tournament host (www.indo.com/distance).

CHAPTER 17

CLUSTER ANALYSIS

INTRODUCTION

Traditionally, research in sport and exercise has used methods based on the assumptions of the normative or the interpretive paradigm (Cohen *et al.*, 2007: 7–26). The human nature assumptions of these paradigms motivate different approaches to data analysis. The normative paradigm assumes that human behaviour is deterministic and can be characterized by an average human being. The interpretive paradigm, however, assumes that human behaviour is voluntaristic with individual responses to situations. There are occasions where neither of these approaches is satisfactory due to the nature of the behaviour being investigated. There may be too many participants to analyse as individuals, but there may be different types of individual that we wish to identify and analyse rather than imposing a model of an average person. Sometimes different types of participant can be identified from existing variables such as positional role, gender, socio-economic group, level of participation and age category. However, there are other classifications based on combinations of variables collected during a study. These classifications may be based on behaviour, attitudes or other variables. Cluster analysis is a technique that allows such non-obvious groupings of participants to be identified.

Once different clusters have been identified, a clustering variable can be treated the same way as any other grouping variable. The full potential of cluster analysis is often realized by the follow-up analysis that is done using the clusters. They could be compared using some numerical variables of interest or they could be cross-tabulated with other grouping variables to determine if those variables are independent of cluster or not. Therefore, this chapter provides two examples of the use of cluster analysis. The first is a performance analysis example to identify clusters of tennis players whose strategy is effected by score-line in different ways. Follow-up ANOVA tests are used to investigate the score-line effects within the different clusters. The other example uses cluster analysis to identify different types of people with respect to preference for flavouring within energy drinks. These clusters are cross-tabulated with demographic variables to give a valuable market sector analysis.

HIERARCHICAL CLUSTER ANALYSIS

Purpose of cluster analysis

Cluster analysis is used to inspect data and identify groups of participants that would not have been known at the beginning of the investigation. There are many different types of cluster analysis with various techniques used to decide which participants belong in which clusters. This chapter will use hierarchical cluster analysis where clusters are developed through division or agglomeration. Division is done within a top-down approach that commences with the full sample as a single cluster and optimally partitions clusters (Mardia *et al.*, 1994: 369) until the base of the hierarchy is formed of individual participants. Agglomeration is a bottom-up approach commencing with individual participants who are formed into clusters by identifying 'nearest neighbours' within the data. Ultimately a hierarchical model is formed allowing the researcher to choose how many clusters to recognize. The decision on the number of clusters to use is based on the inherent structure of the cluster hierarchy, the need to have a manageable number of clusters for research purposes and the need for clusters to have sufficient participant numbers to be generalized.

Assumptions

Cluster analysis is not an inferential test and makes no attempt to estimate population parameters. However, there are some assumptions of the technique that need to be recognized. The most important assumption is that there are clusters of participants. Evidence of such clusters can be uncovered when exploring variables using scatter plots. However, on other occasions there may be a continuum of participants according to some variable rather than clear clusterings of participants. There are many algorithms for deciding on partitioning or agglomeration. Different algorithms are suitable in particular situations and making a good choice of algorithm is important to identifying appropriate clusters (Manly, 2005: 132–3).

First example: Score-line effect on net strategy in tennis

An example of clustering is how the strategy of tennis players is influenced by score-line (O'Donoghue, 2003). The percentage of points where a player attacks the net is an indicator of strategy which may be influenced by score-line. However, the strategy of different players may be influenced by score-line changes in different ways. For example, some players may apply a similar strategy irrespective of the score-line while others may attack the net more or less frequently when they are ahead or behind on service breaks than they do when the score is level on service breaks. This means that the assumption that there is such a person as the average tennis player with respect to score-line effects can be challenged. Consequently, this also challenges the assumption that conclusions based on a sample can be generalized to the relevant population of tennis players. Regarding the interpretive paradigm, the assumption that all players respond in individual ways would promote analysis of individual performances with respect to score-line effect. This might be feasible with a small purposive sample but not with a large sample of players. Therefore, what is needed is a means of identifying different player types with respect to how score-line effects net strategy in tennis. These sub-groups of players can then be analysed using quantitative techniques. Statistical techniques are routinely used to compare samples of players based on some independent grouping variable. For example independent t-tests, one-way ANOVA tests and their equivalent non-parametric tests are used to compare the performances of players of different genders and levels. The problem with classifying players by how their net strategy is influenced by score-line is that this characteristic of a player is not as obvious as a player's gender or level. It is necessary to examine the dependent variable(s) of interest (in this case the percentage of net points played) in order identify different types of player with respect to score-line effect. The data for this example come from 43 players (20 females and 23 males) who competed in Grand Slam tournaments in 2007; these are listed in Table 17.1. These players are included because their data came from more than one match and the data for each player included at least 25 points when the player was trailing on service breaks, at least 25 points where the player was level on service breaks and at least 25 points where the player was leading on service breaks. To control for opposition and surface effects in the different matches played by each player, the percentage of points the players went to the net when leading and trailing on service breaks was related to the percentage of points where they went to the net when the score was level. In O'Donoghue's (2003) investigation, the data was heteroscedastistic with positive associations between absolute differences in the percentage of net points in different score-lines and mean percentage of net points played across different score-lines. Therefore, the percentage of net points played when leading and trailing on service breaks were expressed as ratios of the percentage of net points played when level on service breaks. The data displayed in Table 17.1 is homoscedastic with no 'shotgun' effect when considering the percentage of net points played when level and behind on service breaks ($r = -0.09$) or when considering the percentage of net points played when level and behind on service breaks ($r = 0.14$). It was, therefore, decided to express the percentage of points the players went to the net when trailing and leading on service breaks as differences to the percentage of net points played when level on service breaks. These differences are shown in the final two columns of Table 17.1. Using differences rather than ratios has an added advantage of not losing the data for Kudryavtseva and Paszek due to division by zero problems when determining ratios.

SPSS

The cluster analysis facility in SPSS is typically executed at least twice during the analysis of data such as those in Table 17.1. There may be several executions in an exploratory way and one execution to save a range of candidate solutions, one of which will be chosen. The file 17-cluster-diff2007.SAV contains the tennis data shown in Table 17.1. In SPSS, hierarchical cluster analysis is done using **Analyse → Classify → Hierarchical Cluster**. The pop-up window shown in Figure 17.1 is activated.

The **Method** pop-up window provides a choice of different clustering algorithms and distance measures used to make decisions about cluster division or agglomeration depending on the method chosen. In the tennis example, we use the default 'between groups linkage' method and 'squared euclidian distance' interval. However, several different combinations of method and interval could be explored involving subjective researcher choice in the overall process of cluster analysis (Manly, 2005: 125–30). When exploring the data, the

Table 17.1 The percentage of points where players went to the net in different score-line states

Player	Gender	Game State (Service Breaks)			Difference between score-lines	
		Behind	Level	Ahead	Behind – Level	Ahead – Level
Bartoli	Female	6.30	8.85	7.02	−2.55	−1.83
Chakvetadze	Female	7.84	4.13	4.85	3.71	0.72
Clijsters	Female	7.69	3.83	6.52	3.87	2.70
Hantuchova	Female	8.54	6.16	3.03	2.38	−3.13
Henin	Female	11.76	10.94	12.26	0.82	1.32
Hingis	Female	11.70	6.21	5.04	5.48	−1.17
Ivanovic	Female	6.49	6.93	6.74	−0.45	−0.20
Jankovic	Female	7.09	8.79	4.32	−1.70	−4.47
Kudyavtseva	Female	7.14	0.00	2.63	7.14	2.63
Kuznetsova	Female	5.86	8.63	6.67	−2.78	−1.97
Li	Female	1.59	4.76	5.45	−3.17	0.69
Paszek	Female	0.00	0.00	4.00	0.00	4.00
Peer	Female	1.79	6.99	5.52	−5.21	−1.48
Petrova	Female	9.80	12.95	8.54	−3.15	−4.41
Sharapova	Female	5.43	8.81	6.58	−3.37	−2.23
Sugiyama	Female	0.00	11.32	7.41	−11.32	−3.91
Vaidisova	Female	6.38	7.25	5.17	−0.86	−2.07
WilliamsS	Female	8.61	5.64	8.08	2.98	2.44
WilliamsV	Female	15.69	16.30	14.18	−0.61	−2.12
Zvonereva	Female	4.00	4.05	5.41	−0.05	1.35
Berdych	Male	3.85	5.07	3.13	−1.22	−1.94
Blake	Male	10.14	12.57	8.40	−2.43	−4.17
Chela	Male	8.00	8.09	5.45	−0.09	−2.64
Davydenko	Male	3.15	5.90	7.11	−2.75	1.21
Djokovic	Male	9.70	8.32	7.93	1.38	−0.39
Federer	Male	8.82	10.18	13.74	−1.36	3.56
Ferrer	Male	9.09	9.46	10.87	−0.36	1.41
Gasquet	Male	10.67	13.48	14.29	−2.81	0.81
Ginepri	Male	11.32	11.60	6.25	−0.28	−5.35
Gonzalez	Male	11.59	7.95	12.44	3.65	4.50
Haas	Male	13.11	12.95	17.86	0.16	4.91
Hewitt	Male	3.03	7.22	9.65	−4.19	2.43
Lee	Male	16.13	8.70	8.77	7.43	0.08
Lopez	Male	8.82	13.73	7.69	−4.90	−6.03
Monaco	Male	3.85	4.42	4.55	−0.58	0.12
Moya	Male	10.53	12.94	16.07	−2.41	3.14
Murray	Male	12.28	12.46	7.89	−0.18	−4.57
Nadal	Male	5.73	6.96	6.32	−1.23	−0.64
Nalbandian	Male	9.80	9.07	10.50	0.73	1.43
Robredo	Male	5.69	6.67	0.00	−0.98	−6.67
Roddick	Male	10.34	14.32	12.99	−3.98	−1.33
Verdasco	Male	6.94	8.37	9.68	−1.43	1.31
Wawrinka	Male	9.17	6.82	6.40	2.36	−0.42

difference variables 'Diff_BL' and 'Diff_AL' are transferred into the *Variable(s)* area. These two variables represent the difference in the percentage of points when the player goes to the net between being behind on service breaks and level on service breaks and between being ahead on service breaks and level on service breaks respectively. We use the **Plots** pop-up window to ensure a dendogram (Figure 17.2) is provided with the output.

316

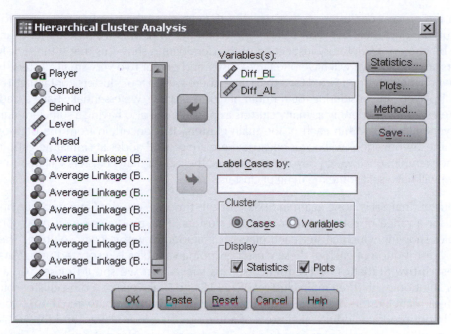

Figure 17.1 Hierarchical cluster analysis pop-up window.

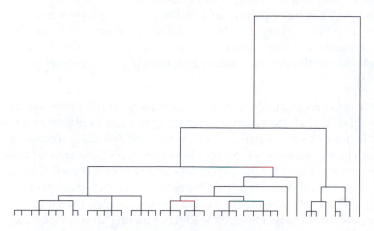

Figure 17.2 Dendogram showing the partitioning possibilities for the sample.

There are a variety of different outputs from SPSS including a table of average linkage between groups values and a vertical icicle plot which is also presented as a table. The dendogram is shown as an upside down tree in representing a hierarchy of clusters: the bottom most nodes represent the 43 players and root of the tree represents the full sample of 43 players. The dendogram can be shown with the root at the bottom and leaves at the top (Mardia *et al.*, 1994: 373) or rotated by 90° as in SPSS output. The lengths of vertical lines on the dendogram represent distance between different clusters as the sample is divided or agglomerated by the hierarchical clustering process. There is a human researcher decision within the process of cluster analysis which is the choice of the number of clusters to use. Any number of clusters between 1 and 43 (the number of players we have in our example) could be chosen. Exploratory inspection of the dendogram shows us that if three or fewer clusters are chosen, we will have one large cluster and one or two very small clusters. However, when the data are divided for a third time, there will be four clusters containing 18, 17, 7 and 1 players. In considering the number of clusters to use, we essentially have a trade off between wanting to identify as many clusters as possible but also having a sufficient number of players included within each of the main clusters. It is uncertain from the dendogram where the later divisions will occur because there are several nodes at similar heights. In the current example, once we go beyond 10 clusters, the size of many clusters will be too small and we will have split some meaningful clusters.

The cluster analysis is done again in SPSS, but this time we use the **Save** pop-up window to ask for a range of solutions from 4 to 10 clusters. This produces seven new nominal variables showing which cluster each player is included in within each solution. The frequency distribution of each of these cluster grouping variables is inspected using **Analyse → Descriptive Statistics → Frequencies**. This shows that we should either use a four cluster solution (with three main clusters of 17, 18 and 7 players) or a nine cluster solution (with four main clusters of 9, 9, 9 and 6 players). The divisions in between these two solutions do not produce additional clusters of any more than four players and the 10 cluster solution splits a cluster of nine into two smaller clusters which is not desirable. The question is do we wish to be able to identify an additional main cluster or do we wish to have as many players as possible included in the clusters to be used? The two key difference variables ('Diff_BL' and 'Diff_AL') can be displayed on a scatter plot with players in different clusters being shown by different marker symbols or in different colours. Inspection of the scatter plots for the four cluster and nine cluster solutions helps us make the decision to use the four cluster solution with three main clusters and a single player outside these (Figure 17.3).

The four clusters form a new grouping variable which can now be used the same way as any other grouping variable. Table 17.2 shows the percentage of net points played in different score-line states by the four clusters of players. A two-way ANOVA using score-line as a within-player effect and cluster as a between player effect reveals a significant interaction effect between cluster and score-line for the percentage of net points played (GG: $\varepsilon = 0.711$; $F_{6,78} = 21.1$, $p < 0.001$). Individual repeated measures ANOVA tests applied to the three main clusters reveal different significant score-line effects. Cluster 1 went to the net on a significantly lower percentage of points when leading or trailing on service breaks than when level on service breaks. Cluster 2 went to the net on a significantly greater percentage of points when trailing on service breaks than when level on service breaks. O'Donoghue (2003) interpreted this as players experimenting with strategies when trailing on breaks because if games within the set continued to be won by the server, the set would be lost. Cluster 3 went to the net on a significantly greater percentage of points when leading on service breaks than when level or trailing on service breaks. O'Donoghue (2003) interpreted this as players increasing their use of net play when it was effective. This could be due to how they got to a leading score-line rather than being due to the score-line itself.

318
cluster analysis

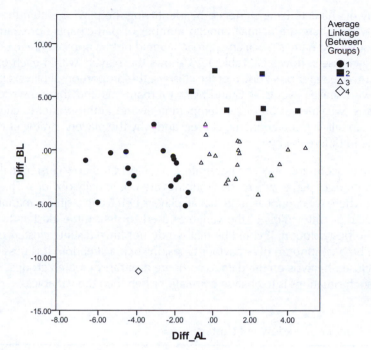

Figure 17.3 The four cluster solution.

Reporting results

Table 17.2 is an example of how the results of cluster analysis could be presented. The clusters are analysed using descriptive statistics supported by appropriate inferential statistics. Very often the inferential statistics compare the clusters in terms of the dependent variables used in the cluster analysis. However, in this example there is only a single conceptual variable (percentage of net points) which is compared between the different score-line conditions for each cluster separately.

Table 17.2 Percentage of net points played in different score line states

Score-line	Cluster			
	Cluster 1	Cluster 2	Cluster 3	Cluster 4
	(n = 17)	(n = 7)	(n = 18)	(n = 1)
Behind	8.1 ± 3.4 ^	10.1 ± 3.2 ^	7.1 ± 3.8	0.0
Level	10.0 ± 3.3	5.2 ± 2.9	7.9 ± 3.5	11.3
Ahead	6.6 ± 3.4 ^	6.9 ± 3.2	9.4 ± 4.1 ^ $	7.4
Repeated Measures ANOVA	GG: $\varepsilon = 0.733$ $F_{2,32} = 20.0$, $p < 0.001$	Sphericity Assumed: $F_{2,12} = 15.5$, $p = 0.004$	GG: $\varepsilon = 0.718$ $F_{2,34} = 12.1$, $p = 0.001$	N/A

Bonferroni adjusted post hoc tests: ^ Significantly different to level ($p < 0.01$), $ Significantly different to behind ($p < 0.01$).

Another way of characterizing clusters is by identifying the players within the clusters. This is possible where there are a small enough number of participants (for example, 43) and many are household names. Such an approach would not be appropriate in our next example of 840 gym users, however. Table 17.3 shows the players within each of the clusters. Consideration of gender breakdown of the clusters, the proportion of players in each cluster who have won singles events at Grand Slam tournaments and the known playing strategies, strengths, weaknesses and even temperament and anthropometric characteristics of the players can tell us more about the clusters and why the players' strategies are influenced by score-line in different ways.

The example of score-line effect on net strategy has illustrated how cluster analysis can identify different types of player where the groupings are not initially obvious. This is very useful in situations where we cannot assume that all players behave in a similar manner. It is always worth inspecting scatter plots of the variables used to determine clusters to see if cluster analysis is the best solution. It could be that we do not have distinct clusters of players, but there could be a continuum of behaviour types. In such a situation, it is possibly better not to enforce cluster analysis on the data to compare arbitrarily created groups. An alternative analysis in such situations is to analyse correlations between the variables.

Table 17.3 Gender breakdown of clusters

Gender	Cluster			
	Cluster 1 (n = 17)	Cluster 2 (n = 7)	Cluster 3 (n = 18)	Cluster 4 (n = 1)
Female	Bartoli	Chakvetadze	Henin &	Sugiama
	Hantuchova	Clijsters &	Ivanovic &	
	Jankovic	Hingis &	Li &	
	Kuznetsova &	Kudryatseva	Paszek	
	Peer	Williams S &	Zvonereva	
	Petrova			
	Sharapova &			
	Vaidisova			
	Williams V &			
Male	Berdych	Gonzales	Davydeno	
	Blake	Lee	Djokovic &	
	Chela		Federer &	
	Ginepri		Ferrer	
	Lopes		Gasquet	
	Murray		Haas	
	Robredo		Hewitt &	
	Roddick &		Monaco	
			Moya &	
			Nadal &	
			Nalbandian	
			Verdasco	
			Wawrinka	

& Grand Slam tournament singles winner.

320

cluster analysis

Second example: Customer preference for a flavouring ingredient in energy drinks

Market research is carried out to gauge customer preference for five versions of an energy drink that contain different concentrations of a flavouring ingredient. The products D1, D2, D3, D4 and D5 have concentrations of the flavouring ingredient of 0.02%, 0.03%, 0.04%, 0.05% and 0.06% respectively. Customer rating data are gathered at gyms belonging to a well-known gym company. Each participant tastes each of the five energy drinks, rating them on a 1 to 9 Likert scale. The participants also complete their gender, age group and other fields that allow socio-economic group to be estimated. Age group is classified as one of the following:

- Under 25 years
- 25 to 34 years
- 35 to 49 years
- 50 to 64 years
- 65 years or older.

Socio-economic groups are classed as A, B, C1, C2, D or E. Altogether, data are collected from 840 participants. Once the cluster analysis is completed, the rating of each product by the different clusters can be characterized before the demographic breakdown of those clusters is analysed. This can give very valuable marketing information about the five products.

SPSS

A cluster analysis is performed on the data which are found in the file 17-energy_drink.SAV, entering the variables D1, D2, D3, D4 and D5 which have an increasing concentration of flavouring from D1 (0.02%) to D5 (0.06%). We ask for a dendogram as before and **Save** a range of solutions from a two cluster solution to a 10 cluster solution. When we inspect the frequencies of the different clusters in each solution, we can see that the data were gradually partitioned into 10 clusters as shown in Table 17.4.

Table 17.4 Size of clusters within different solutions. The clusters shown in bold are partitioned as the next solution is produced

Clusters	Cluster sizes									
1	**840**									
2	**627**							213		
3	**443**					184		213		
4	401				42	184		**213**		
5	401				42	184		**183**		30
6	**401**				42	184		140	43	30
7	**299**			102	42	184		140	43	30
8	210	89		102	42	**184**		140	43	30
9	210	**89**		102	42	116	68	140	43	30
10	210	61	28	102	42	116	68	140	43	30

Why do we choose the cluster sizes that we do? In this example, we should think ahead to how the clusters will be used. The clusters will be cross-tabulated with gender, age group and socio-economic group with chi square tests of independence being performed. The largest of these categorical variables is socio-economic group which is measured at six levels. For the chi square test of independence to be valid, we need an expected frequency of at least five in at least 80 per cent of cells of the cross-tabulation. One of the socio-economic groups (E) only has 84 members, one 10th of the sample. Therefore, a cluster of size 50 would be expected to contain five Es. Of more concern is age group, even though there are five age groups. The problem here is that there are only 39 participants aged 65 years and over. Therefore, cluster sizes of over 100 would be needed to ensure at least five 65 year olds and over would be expected in each cluster. The 4, 5 and 6 cluster solutions only add small clusters of 42, 30 and 43 respectively. Therefore, we choose to use the three cluster solution.

We use **Analyse → Compare Means → Means** to determine how highly the three clusters rate the five products. Figure 17.4 shows the ratings of the different products by the three different clusters. Cluster 1 shows an increasing rating as flavour concentration increases up to D3 (0.04%) and the rating stays roughly the same. Cluster 3 shows a decreasing rating as flavour concentration increases up to D3 (0.04%) and the rating stays roughly the same. Cluster 2 gives the highest product rating to D3 (0.04%) with this flavour concentration appearing to be optimal.

Once the cluster variable has been produced, would use **Analyse → Descriptive Statistics → Crosstabs** requesting chi square tests and Cramer's V in the **Statistics** pop-up window. We would also request the percentage breakdown of each demographic variable among the different clusters in the **Cells** pop-up window.

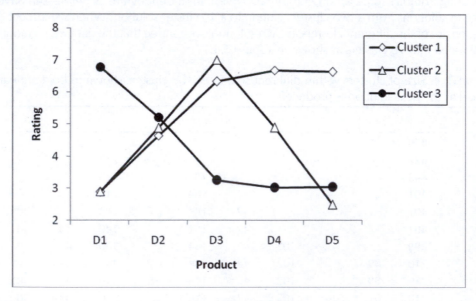

Figure 17.4 Product ratings by the three different clusters.

Reporting results

A line graph such as Figure 17.4 would be included to describe the product ratings by the different clusters. Table 17.5 shows the cross-tabulated frequencies of cluster members for different genders, age groups and socio-economic group. It is debatable as to whether the percentages should be of the rows or columns in Table 17.5. With a larger sample, the demographic variables could be looked at together rather than in isolation. For example, we could have a cluster break down for the under-25 females and the under-25 males. Currently, the gender breakdown conceals information about age groups within the gender and vice versa. The table would be supported by the following text. These types of results are very useful market research information.

Table 17.5 shows the distribution of each demographic group across the different clusters. There was a significantly different distribution of cluster membership for females and males ($\chi^2_2 = 11.6$, p = 0.003) with a small effect (V = 0.118). Cluster membership was independent of age group ($\chi^2_8 = 14.3$, p = 0.074, V = 0.092). Cluster membership was significantly influenced by socio-economic group ($\chi^2_{10} = 64.9$, p < 0.001) with a small effect (V = 0.197). More C1s and Es were members of cluster 1 than expected, more As were members of Cluster 2 than expected and more As and Bs were members of Cluster 3 than expected.

Table 17.5 Gender, age and socio-economic profile of the different clusters

Demographic group	Cluster			
	Cluster 1	Cluster 2	Cluster 3	Total
Gender				
Female	159 (46.9%)	93 (27.4%)	87 (25.7%)	339 (100.0%)
Male	284 (56.7%)	91 (18.2%)	126 (25.1%)	501 (100.0%)
Age				
Under 25 years	85 (54.5%)	33 (21.2%)	38 (24.4%)	156 (100.0%)
25 to 34 years	191 (56.5%)	63 (18.6%)	84 (24.9%)	338 (100.0%)
35 to 49 years	92 (43.6%)	63 (29.9%)	56 (26.5%)	211 (100.0%)
50 to 64 years	52 (54.2%)	20 (20.8%)	24 (25.0%)	96 (100.0%)
65 years and over	23 (59.0%)	5 (12.8%)	11 (28.2%)	39 (100.0%)
Socio-economic group				
A	51 (36.7%)	49 (35.3%)	39 (28.1%)	139 (100.0%)
B	103 (42.4%)	58 (23.9%)	82 (33.7%)	243 (100.0%)
C1	94 (71.8%)	14 (10.7%)	23 (17.6%)	131 (100.0%)
C2	71 (58.2%)	16 (13.1%)	35 (28.7%)	122 (100.0%)
D	71 (58.7%)	28 (23.1%)	22 (18.2%)	121 (100.0%)
E	53 (63.1%)	19 (22.6%)	12 (14.3%)	84 (100.0%)
Total	443 (52.7%)	184 (21.9%)	213 (25.4%)	840 (100.0%)

Discussion point

Consider Figure 17.4 and the mean ratings for each product by the different clusters. The standard deviations for the 3 × 5 (cluster × product) ratings vary from 1.3 to 1.7. Therefore, we cannot expect every individual within a particular cluster to follow the mean rating profile for the cluster. We might even consider if there is an alternative to cluster analysis for identifying different rating patterns according to flavour concentration. In Microsoft Excel, we could enter IF functions to determine if ratings tended to decrease, increase or have some optimum value. The limitation of such an approach is that the researcher imposes cluster types based on their opinion of the different rating responses to flavour concentrations. Cluster analysis also has a human input in that the solution chosen is based on subjective inspection of the dendogram and consideration of cluster sizes needed for further analysis. However, at least with cluster analysis, the partitioning or agglomeration of clusters is done by a statistics package based on the linkage algorithm chosen by the researcher.

SUMMARY

Cluster analysis is used to identify groups of participants that are not as obvious as gender, positional role, level of participation in sport and age group. The cluster variable produced can then be used in research studies in the same way as any other grouping variable. Cluster analysis is most useful where there are different types of people with respect to attitude or behaviour.

EXERCISE

Exercise 17.1. Athletes with different perceptions of anxiety

The file ex17.1-SAS2.SAV contains data for 175 fictitious student athletes who have completed a modified SAS2 questionnaire instrument that gives scores for:

- WI – worry intensity
- WD – worry direction
- SAI – somatic anxiety intensity
- SAD – somatic anxiety direction
- CDI – concentration distraction intensity
- CDD – concentration distraction direction.

The intensity scores are measured on a 5 to 20 scale while the direction scores are measured on a –15 to +15 scale. It is only the direction scores for competitive matches ('WD_COMP', 'SAD_COMP' and 'CDD_COMP') that are of interest to the current exercise. A negative direction score indicates that the given component of anxiety (worry, somatic anxiety or concentration distraction) is perceived as debilitative while a positive direction score indicates that it is perceived as facilitative.

Perform a cluster analysis using the three direction scores, saving cluster group variables for a range of solutions from a three cluster solution to a 10 cluster solution.

324

cluster analysis

a) Choose one of these, giving reasons for your choice of number of clusters to use.
b) Perform a series of one-way ANOVA tests to determine any significant differences between the clusters for each direction score. How would you characterize the clusters in terms of their perception of worry, somatic anxiety and concentration distraction?
c) Produce scatter plot(s) to help illustrate the differences between the clusters you have compared.

PROJECT EXERCISE

Exercise 17.2. Perceived performance in different modules

Ask members of your class to anonymously complete a form rating their perceived ability in research methods, biomechanics, physiology, psychology, sociology, sports development and practical sport. These should be rated on a 1 to 9 Likert scale with 1 meaning they think they will struggle to pass the module and 9 meaning they expect first-class marks in the module. Also ask them to tick their gender as male or female. Perform a cluster analysis on the perceived rating data. What different types of students are there in the group? Is the type of student independent of gender?

CHAPTER 18

DATA REDUCTION USING PRINCIPAL COMPONENTS ANALYSIS

INTRODUCTION

Sometimes when studying sport and exercise, researchers deal with multiple variables some of which may be related. This may happen in sports psychology and sports management where questionnaires contain numerous individual questions. There are other areas such as sports performance where multiple aspects of performance are represented by the set of variables used. Analysing these variables can lead to a cumbersome results section of a paper or thesis due to the volume of variables involved in the study. Factor analysis and principal components analysis are data reduction techniques that identify related variables and produce a smaller set of broad factors or components to represent different dimensions in the data. These factors or components can then be analysed leading to a more concise expression of the findings of a study. In this chapter, the use of principal components analysis for data reduction is covered. Those specifically interested in factor analysis are referred to other texts (Manly, 2005: 91–104; Tabachnick and Fidell, 2007: 607–75; Vincent, 2005: 234–6).

PRINCIPAL COMPONENTS ANALYSIS

Purpose of principal components analysis

The purpose of principal components analysis is to take a set of K variables and identify a smaller number of components that can be determined from the data while representing a large proportion of the variance in the data. This is done by identifying relationships between the K variables and producing a set of K uncorrelated components (new variables). Each component is a function of the original variables; this function is called an eigenvector. As each component is created, it maximizes the variance in component scores generated (Mardia *et al.*, 1994: 213). Subsequent components are created applying the same process to residual data leading to a set of K uncorrelated components. Because the amount of variance in the data differs between components, it is possible to represent more than half the variance with less than half the components. A small set of components can, therefore, be used instead of a large number of original variables.

Assumptions

Due to the very purpose of principal components analysis, it does not work well if the original variables are uncorrelated (Manly, 2005: 75). Tabachnick and Fidell (2007: 612–15) described the assumptions of factor analysis techniques in general. There should be 300 cases for a good factor analysis, although as few as 50 cases may be sufficient if several of the original variables have absolute correlations of over 0.8 with some of the components produced. The original variables should be free of outliers and any relationships between pairs of variables should be linear. Functional dependencies should also be removed from the set of variables being included. For example, body mass index (BMI) is calculated using height and body mass. Therefore, we would either include height and body mass in the principal components analysis or BMI, but we would not include all three of these variables.

Example: Perceptions of sports tourism

To illustrate principal components analysis, an example from sports tourism is used. This example is based on the research done by Hritz and Ross (2010) in Indianapolis but the data used here are fictitious. Questionnaires about the impact of sports tourism in the local area are distributed to members of the local community and there are 574 complete and valid responses. The questionnaire contains 14 statements that are answered using a five-point Likert scale with 1 representing 'strongly disagree' and 5 representing 'strongly agree':

1 Sport tourism has increased the crime rate in the area.
2 Local residents have suffered from living in a sport tourism destination area.
3 Sport tourism has encouraged a variety of cultural activities by local residents.
4 Meeting sport tourists from other regions is a valuable experience to understanding their culture and society.
5 Sports tourism has resulted in positive impacts on the cultural identity of the local area.
6 Sport tourism provides more parks and other recreation areas.
7 Our roads and public facilities are kept at a high standard due to sports tourism.
8 Sport tourism has provided an incentive for the restoration of historical buildings and the conservation of natural resources.
9 Sport tourism has resulted in traffic congestion, noise and pollution.
10 Construction of sports facilities has destroyed the natural environment.
11 Sport tourism has created more jobs for the local area.
12 Sport tourism has given economic benefits to local people and small businesses.
13 The cost of developing sport tourism facilities is too much.
14 I support sport tourism in the local area.

There were two additional statements included in Hritz and Ross's (2010) survey which are not being used in the current example. It is quite typical for this type of Likert scale data to be used in factor analysis and other advanced multivariate analysis techniques in sports psychology, sports management and other disciplines of sport and exercise science. The validity of using principal components analysis with such data is a matter of debate. The view of this author is that as long as methods are transparent and replicable, the research is open to

327

public scrutiny and readers can take into account the methods used when making decisions based on the research findings.

One of the purposes of Hritz and Ross's study was to produce a model of support for sport tourism in terms of other concepts. Therefore, the 14th statement listed above is not to be included in any principal components analysis. This is because we don't wish to integrate it with other variables as it is a key variable in its own right that is to be used in the research once the principal components analysis has been completed. It would be possible to produce a regression equation for support for tourism in terms of the other 13 statements. However, responses to these statements introduce multicollinearity within the data and the model would be based on 13 quite detailed variables rather than broader concepts of perception of sports tourism. Thus the role of principal components analysis in this type of study is to determine broader independent concepts that exist within the data and to use these within a more concise and meaningful model of support for tourism.

SPSS

This example uses the file 18-sport_tourism_data.SAV which contains the 14 variables for 574 fictitious responses. Once the file is opened, we use **Analyse → Dimension Reduction → Factor** to obtain the factor analysis pop-up window shown in Figure 18.1. Principal components is the default extraction method. The variables 'Q1' to 'Q13' are transferred into the *Variables* list; the 'respondent' identity number and 'support for sport tourism' remain in our unused variables at this time.

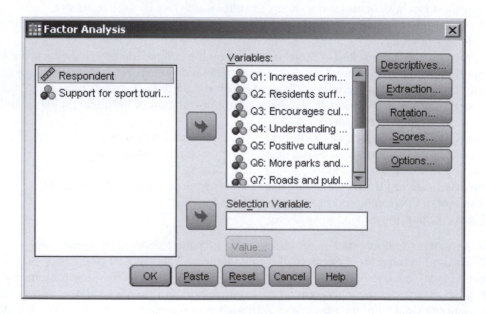

Figure 18.1 The Factor Analysis pop-up window.

The **Options** button is used to obtain a further pop-up window where we can specify how we wish to view any correlations between factors produced and our original 13 variables. By placing a tick in 'Suppress small coefficients' and requesting that absolute correlations below 0.5 are not displayed, we will be able to see the relationships between variables and components more easily. Click on **Continue** to close the Options pop-up window once this is done. The **Extraction** button provides a further pop-up window where we can request a scree plot to be provided. The meaning of the data shown in the scree plot will be discussed shortly. Once we close down this pop-up window and click on **OK** in the main factor analysis pop-up window, the computer will execute the principal components analysis and produce output including that shown in Figure 18.2. Principal components analysis takes our set of 13 variables and produces a new set of 13 variables. The original variables are each assumed to be equally important. The 13 new variables together represent 100 per cent of the variance in the data. However, some are more important than others representing more than one of the original variables. The eigenvalue of a component (new variable) is the number of original variables it is worth in terms of variance explained. An eigenvalue of one means that a component explains the same amount of variance in the data as an original variable does. There are five principal components that each represent more variance than a single original variable and which together explain 54.2 per cent of the variance in the data.

These components are constructed using eigenvectors which take the form shown in equation 18.1 where z_j is the jth component, x_i represents the ith original variable, K is the total number of original variables (and components), $b_{j,i}$ is a multiplying coefficient for the ith variable when computing values for the jth component. Therefore, each component is

Table 18.1 SPSS output for Principal Components Analysis showing eigenvalues

Total Variance Explained

Component	Initial Eigenvalues			Extraction Sums of Squared Loadings		
	Total	% of Variance	Cumulative %	Total	% of Variance	Cumulative %
1	1.923	14.789	14.789	1.923	14.789	14.789
2	1.533	11.792	26.581	1.533	11.792	26.581
3	1.412	10.862	37.443	1.412	10.862	37.443
4	1.168	8.985	46.428	1.168	8.985	46.428
5	1.004	7.726	54.153	1.004	7.726	54.153
6	.917	7.056	61.210			
7	.904	6.957	68.167			
8	.842	6.479	74.645			
9	.729	5.606	80.251			
10	.704	5.415	85.666			
11	.673	5.177	90.844			
12	.609	4.688	95.532			
13	.581	4.468	100.000			

Extraction Method: Principal Component Analysis.

data reduction using principal components analysis

computed using all the original variables with some original variables having higher loadings than others onto particular components.

$$z_j = b_{j,1}.x_1 + b_{j,2}.x_2 + \ldots + b_{j,K}.x_K \qquad (18.1)$$

Figure 18.2 is a scree plot which shows the eigenvalues for the new variables (components) produced by principal components analysis. Principal components analysis is not a black and white process of producing a given number of principal components or extracting only those principal components with an eigenvalue of over one. There are areas of intervention by the researcher that require knowledge of the research topic in order to decide how many components to extract and what broad concepts these represent. The scree plot can play a role in this although some researchers can make extraction decisions using the information in Table 18.1. Ultimately, there is a balance needed between maximizing the variance in the original data that is explained in the extracted components and extracting a small enough number of components to promote a concise analysis of the data. One approach is to use the 'elbow' of the scree plot (Thomas and Nelson, 1996: 185) while others might choose to use a number of components based on a minimum acceptable eigenvalue such as one.

In our initial exploration of the data, five principal components are extracted and the relationships that these have with the 13 original variables is shown in Table 18.2. These principal components currently do not have names and the SPSS package cannot be expected to have the human intelligence required to understand each application area in sport, exercise

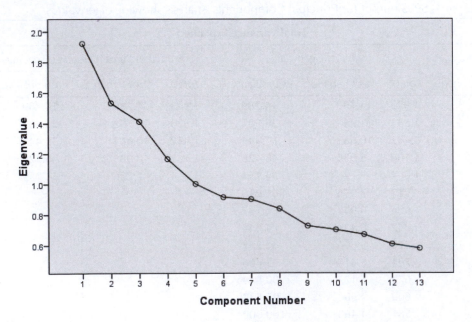

Figure 18.2 A scree plot.

330

data reduction using principal components analysis

Table 18.2 Component Matrix

Component Matrix[a]

	Component				
	1	2	3	4	5
Increased crime rate	.586				
Residents suffer	.577				
Encourages cultural activity		.681			
Understanding of culture and society		.649			
Positive cultural identity		.663			
More parks and recreation areas					−.548
Roads and public facilities are kept at a high standard				.614	
Restoration of historic buildings and conservation areas				.597	
Traffic congestion, noise and pollution	.583				
Construction destroys the environment	.617				
Job creation			.688		
Economic benefits for locals and small businesses			.678		
Development costs are too much	.508				

Extraction Method: Principal Component Analysis.

a 5 components extracted.

and other fields of study sufficiently to propose names for the components. It is down to the researcher to use their knowledge and experience to interpret the concepts represented by the components. The first principal component could be labelled 'negative impacts', the second 'social benefits', the third 'economic impact' and the fourth could be labelled 'environmental benefits' (Hritz and Ross, 2010). The fifth principal component has an absolute correlation of over 0.5 with only a single variable and so this could be used to label it. Note that when there is a negative correlation, the principal component represents the opposite concept to the particular variable. In this example, the fifth principal component may represent 'reduced parks and recreation areas'.

The researcher now has a decision to make: how many principal components will be extracted. The percentage of variance explained by the extracted principal components comes into this decision. We could extract four principal components representing 46.4 per cent of the variance in the data, five principal components representing 54.2 per cent of the variance in the data or six principal components representing 61.2 per cent of the variance in the data. When these principal components are extracted, there is room for manoeuvre that did not exist when the 19 original variables were used to produce an optimal set of 19 uncorrelated components. This allows the extracted principal components to be 'rotated' to elevate the loadings (correlations) with those variables used to interpret them and to make the extracted principal components as independent of each other (uncorrelated) as possible. Typically, rotation reduces the eigenvalue of the primary principal component while the eigenvalues of the other components tend to increase. This means that if, in our sport tourism example, we wished to extract six components, although the sixth component has an eigenvalue of less than one before rotation, the sixth component could well have an eigenvalue greater than one after rotation.

There are five different rotation techniques provided in SPSS. Varimax is used where the components are assumed to be independent while Direct Oblimin is used where relationships are assumed between components (Ntoumanis, 2001: 141). In this example, we will use a four component solution and apply the Varimax factor rotation technique. This involves running the principal component analysis again in SPSS using **Analyse → Dimension Reduction → Factor**. A good feature of the SPSS package is that when we do any analysis, if that type of analysis has been done before during the current session with SPSS, then the various options and variables used on the previous occasion are set up the next time that analysis is done. Therefore, all we need to do is enter the options that make this particular principal component analysis different to the previous one. To begin, we click on the **Extraction** button and use the pop-up window to state that we are using a fixed number of factors and that the number of factors to be extracted is four. This time when we click on **Continue** and **OK**, two component matrices are produced: one showing the correlations prior to rotation and one showing correlations after rotation. These are shown in Table 18.3 and 18.5 respectively. Table 18.4 shows that the rotation process slightly reduced the eigenvalues of the first two principal components and increased the eigenvalues of the other two extracted components.

As before, the four extracted principal components are labelled 'negative impacts', 'social benefits', 'economic impact' and 'environmental benefits' (Hritz and Ross, 2010). In this analysis, the author did request to see absolute correlations greater than 0.35 due to 'more

Table 18.3 Unrotated extracted components

Component Matrix[a]				
	Component			
	1	*2*	*3*	*4*
Increased crime rate	.586			
Residents suffer	.577			
Encourages cultural activity		.681		
Understanding of culture and society		.649		
Positive cultural identity		.663		
More parks and recreation areas				
Roads and public facilities are kept at a high standard				.614
Restoration of historic buildings and conservation areas				.597
Traffic congestion, noise and pollution	.583			
Construction destroys the environment	.617			
Job creation			.688	
Economic benefits for locals and small businesses			.678	
Development costs are too much	.508			

Extraction Method: Principal Component Analysis.
a 4 components extracted.

data reduction using principal components analysis

Table 18.4 Eigenvalues of extracted components

Total Variance Explained

Component	Initial Eigenvalues			Extraction Sums of Squared Loadings			Rotation Sums of Squared Loadings		
	Total	% of Variance	Cumulative %	Total	% of Variance	Cumulative %	Total	% of Variance	Cumulative %
1	1.923	14.789	14.789	1.923	14.789	14.789	1.890	14.536	14.536
2	1.533	11.792	26.581	1.533	11.792	26.581	1.499	11.529	26.066
3	1.412	10.862	37.443	1.412	10.862	37.443	1.422	10.938	37.003
4	1.168	8.985	46.428	1.168	8.985	46.428	1.225	9.424	46.428
5	1.004	7.726	54.153						
6	.917	7.056	61.210						
7	.904	6.957	68.167						
8	.842	6.479	74.645						
9	.729	5.606	80.251						
10	.704	5.415	85.666						
11	.673	5.177	90.844						
12	.609	4.688	95.532						
13	.581	4.468	100.000						

Extraction Method: Principal Component Analysis.

Table 18.5 Rotated extracted components

Rotated Component Matrix[a]

	Component			
	1	2	3	4
Increased crime rate	.596			
Residents suffer	.630			
Encourages cultural activity		.686		
Understanding of culture and society		.692		
Positive cultural identity		.718		
More parks and recreation areas			.430	.333
Roads and public facilities are kept at a high standard				.734
Restoration of historic buildings and conservation areas				.618
Traffic congestion, noise and pollution	.612			.309
Construction destroys the environment	.599			
Job creation			.762	
Economic benefits for locals and small businesses			.735	
Development costs are too much	.510			

Extraction Method: Principal Component Analysis.
Rotation Method: Varimax with Kaiser Normalization.

a Rotation converged in 5 iterations.

parks and recreation areas' not loading onto any factor when 0.5 was set as the minimum absolute correlation to be displayed. Very small absolute correlations (< 0.3) should not be used to interpret components (Ntoumanis, 2001: 138).

At this stage, the components have not been saved as new variables in the datasheet. The reason for this being that SPSS can execute the principal components analysis so quickly that we can look at several alternatives based on differing numbers of components to extract and alternative rotation algorithms to be applied. This flexible and efficient use of SPSS combined with the researcher's knowledge and experience in their field allows the most suitable solution for their research to be selected. Once we have decided which solution to use, we run the analysis one more time entering the number of components we wish to extract and the rotation algorithm to be used. On this occasion, we click on the **Scores** button and when the score pop-up window appears we place a tick in the option to save the components as variables. Four new variables will appear in our datasheet when we do this and they can be analysed the same way as any other variables. Factor (component) scores outside the range -2.5 to $+ 2.5$ are possible outliers and may need to be removed (Ntoumanis, 2001: 138). This may require the factor analysis to be redone without these cases as the factors would have been computed using the cases that turned out to be outliers.

These new variables are basically z-scores and have no units and we will have difficulty finding other literature that uses the same variables. Indeed, the eigenvectors used to create the factors are derived purely from the data in the given study and are not universal. Studies that have used factor analysis in sports performance (O'Donoghue, 2008; Csatjalay et al., 2008; Choi et al., 2008) end up using the original variables anyway. This is because the original variables in such cases are valid, understood and recognized variables allowing comparison of results with findings in other research studies. Principal components analysis still has a role to play in such studies as redundant variables can be identified and excluded from the analysis. Principal components analysis is probably best used with psychometric type data where instruments are being developed to measure concepts that are not directly observable.

Another issue to beware of is that some relationships between variables might be specific to the particular population of interest. For example, one study of fire fighter fitness found a positive correlation between measures of strength and endurance (McTrustry and O'Donoghue, 2002) that we might not see in some athletic populations where these variables might be negatively correlated.

Using the component scores

The component scores have been saved as new variables in our datasheet and can now be used like any other variables. In this example, they are used to form a model for 'support for sport tourism' using linear regression. We use **Analyse → Regression → Linear** to form a model of 'support for sport tourism' in terms of the four components. The four new variables ('REGR factor score 1' to 'REGR factor score 4') are transferred into the *Independent Variables* list with 'support for sport tourism' being transferred into the *Dependent Variable* area. Table 18.6 shows that all the components except the fourth have a negative effect on support for tourism.

Table 18.6 SPSS output for linear regression using the component scores as independent variables

Coefficients[a]

Model		Unstandardized Coefficients B	Std. Error	Standardized Coefficients Beta	t	Sig.
1	(Constant)	4.509	.029		153.714	.000
	REGR factor score 1 for analysis 1	−.265	.029	−.331	−9.032	.000
	REGR factor score 2 for analysis 1	−.226	.029	−.283	−7.712	.000
	REGR factor score 3 for analysis 1	−.021	.029	−.027	−.730	.466
	REGR factor score 4 for analysis 1	.170	.029	.212	5.788	.000

a Dependent Variable: Support for tourism

Equation 18.2 shows the regression model for support for sport tourism, y, where z_1 = 'negative impact', z_2 = 'social benefits', z_3 = 'economic impact' and z_4 = 'environmental benefits'.

$$y = 4.509 - 0.265\, z_1 - 0.266\, z_2 - 0.021\, z_3 + 0.170\, z_4 \qquad (18.2)$$

Reporting results

There are different ways of reporting the results of factor analysis including loading plots that display the correlations of original variables with pairs of components. The author prefers to use a table to show correlations between extracted components and original variables because a single table can present the results for all the components whereas the number of loading plots required would be the number of extracted components divided by two and rounded up. Once the components are saved, they can be analysed and presented like any other variables.

SUMMARY

Principal components analysis is a data reduction technique that allows us to reduce a large set of correlated variables to a smaller set of broader components. These components facilitate more concise data analysis. The components are dimensionless z-scores and are not suitable replacement variables for well-understood variables in sport and exercise science. However, principal components analysis has an important role to play in the study of concepts that are not directly observable such as constructs in sports psychology.

EXERCISES

Exercise 18.1. Six component solution

Use SPSS with the data in the file 'ex-18.1-sport_tourism_data.SAV' selecting a six component solution and applying the Varimax rotation technique. How would you interpret the extracted components in this case.

PROJECT EXERCISE

Exercise 18.2. Sports tourism

Undertake a research project to investigate sport tourism in your local city, especially if that city has a football club that plays regularly and other sports events that attract regular spectators or participants. Adapt the questionnaire developed by Hritz and Ross (2010) to the particular scenario and include questions on demographic variables such as sex and age group. Once you have gathered the responses, use principal components analysis to produce broad components that can be used within a model of tourism support. Compare the different dimensions of sports tourism perception between different demographic groups identified in the questionnaire.

data reduction using principal components analysis

CHAPTER 19

RELIABILITY

INTRODUCTION

Reliability can have different meanings when applied to human performance, equipment and measurement (Crowder et al., 2000: 1). Reliability of equipment and systems is concerned with the probability that the system is working successfully. Typical measures used in the evaluation of this type of reliability are mean time between failure and rate of failure occurrence (Sommerville, 1992: 394–5). The reliability of a human performer could be the ability of the person to consistently perform specified tasks to a required standard. This chapter is concerned with the reliability of measurement within scientific research. The chapter commences by covering the measurement issues of validity, objectivity and reliability and their inter-relationships. The different types of reliability study done in sport and exercise science are then classified. There are many different reliability statistics that can be used appropriately or inappropriately. Therefore, the chapter includes a section on selecting the appropriate reliability statistic to use in different situations. Finally, the chapter describes how to produce reliability statistics.

MEASUREMENT ISSUES

Validity

In any scientific investigation, the correct use of the most appropriate descriptive and inferential statistical techniques allows results to be presented effectively and conclusions to be drawn. However, the analysis is meaningless if the raw data that have been gathered and analysed are invalid or inaccurate. When data are used to support important decisions, decision makers often question the relevance and accuracy of variables used as well as whether the data are up to date (O'Donoghue, 2010: 149). Therefore, the measurement issues of validity, objectivity and reliability are crucially important in sport and exercise research.

The validity of a variable depends on its relevance and its reliability (Morrow Jr. et al., 2005: 82). The relevance of a variable is how well the variable represents an important concept. Consider the example of sit-and-reach test result as a measure of flexibility. The validity of this variable to a study first depends on the importance and relevance of the concept of flexibility to the research problem. Validity also depends on how well the sit-and-reach test result measures the particular type(s) of flexibility of concern to the research study. Linacre

(2000) distinguished the term 'accuracy' from validity with accuracy being how well a variable fairly represents an individual's true level of the ability of interest.

Objectivity

During the development of a research problem, vague concepts are transformed into operationalized variables. The use of operational definitions adds objectivity to a study. A variable is objective if a measurement procedure is used that is independent of the subjective opinion of an individual data gatherer. Objectivity and reliability are related concepts because objective measurement procedures help independent operators apply measurement procedures consistently.

Reliability

Reliability is the consistency of the values obtained when a measurement is used. Some measurement methods are more reliable than others due to the different difficulties involved in applying the methods. For example, measuring the height and width of a sheet of paper with a ruler is a straightforward process and one would expect independent people applying this process to achieve consistent results when measuring the same sheets of paper. Using a stopwatch to time the amount of time a field hockey player is engaging in moderate to vigorous activity during a match is not so straightforward. The measurement process requires an element of human judgement to decide if activity is of a moderate or vigorous intensity. This will be inevitable no-matter what operational definitions are provided. Other sources of inaccuracy are operator reaction time using the stopwatch, possible misidentification of the player and using incorrect buttons on the stopwatch. Therefore, a greater amount of inconsistency would be expected in this example than in the example of measuring the height and width of a sheet of paper. Very often it is not possible to determine 'accuracy' as described by Linacre (2000) because the true score for an individual may not be known and we simply have different measurements, none of which can be taken as 100 per cent accurate.

There are relationships between reliability, objectivity and validity. Objectivity can improve reliability through the consistent use of well-defined measurement procedures. A variable that is not reliable cannot be valid, no matter how relevant the variable is. This is because a high degree of inconsistency in measurement leads to values that may be a poor reflection of the variable in the particular case being measured.

TYPES OF RELIABILITY STUDY

What is reliability?

At the beginning of this chapter it was mentioned that the word 'reliability' has different uses in different contexts. Even within sport and exercise science, there are different types of reliability with different disciplines using the word 'reliability' to mean different things. The reliability of a variable has been described as the consistency with which values for the same

338

reliability

performance can be measured by independent observers (O'Donoghue, 2010: 150). This is a type of reliability which is used within performance analysis of sport. There are other forms of reliability used in other disciplines of sport and exercise science. The type of reliability described by O'Donoghue (2010: 150) is inter-rater reliability which is one of the four types of reliability classified by Trochim (2000) which are:

- Inter-rater reliability
- Test-retest reliability
- Parallel forms reliability
- Internal consistency.

Performance analysts would view parallel forms reliability as criterion validity, because different measurement procedures are being compared. Furthermore, internal consistency would not be viewed as reliability by performance analysts because different variables are being compared. However, in some fields of sport and exercise science, internal consistency has been described under the heading of reliability (Patterson, 2000). Test–retest reliability has been used extensively in fitness testing but could not be used where variables are inherently unstable. Indeed, where questionnaire items may vary between repeated tests, some recognize this as a stability issue rather than a reliability issue (Freedman and Miller, 2000). The four different types of reliability will now be discussed.

Inter-rater reliability

Inter-rater reliability is where the same performance(s) are independently measured by different trained personnel. This can be done live while the participant(s) are performing or it can be done after the performance has occurred using video recordings of the performance. This form of reliability is used in performance analysis of sport but can also be used for independent assessments of other measurements such as MRI (Magnetic Resonance Imaging) scans and skill competency tests.

Test–retest reliability

A fundamental difference between test–retest reliability and inter-rater reliability is that the performers are tested or observed more than once. The reliability of a variable tested in this way depends on the consistency with which performers can perform the test as well as the consistency of measurement by an observer and/or any equipment involved. Test–retest reliability is, therefore, unsuitable for inherently unstable variables such as sports performance indicators where values vary due to opposition effect (McGarry and Franks, 1994; O'Donoghue, 2004; Taylor et al., 2008) and other situational variables such as venue, importance of the match and environmental factors. Test–retest reliability studies are applied to stable characteristics of performers such as anthropometric variables and elements of fitness. A set of participants is typically tested or measured and then the same measurement process is repeated some time later under identical circumstances to ensure measurements are taken under the same controlled conditions. The time between the test and retest of the measurement should be long enough to avoid agreement purely due to recall and short enough to avoid changes in values due to participant maturation, improved fitness, etc. Test-retest reliability tests are used in areas such as physiology and quantitative biomechanics.

Parallel forms reliability

Parallel forms reliability is where some concept of interest is measured using two alternative techniques and the consistency of those different techniques is being evaluated. Where one of the techniques is a gold standard measurement of the concept of interest, many in sport and exercise science would actually refer to this as validation testing rather than reliability testing. Parallel forms reliability involves applying the two measurement procedures to the same set of participants. The participants may be required to perform a single test with different measurement processes being applied to their performance. An example of this is where the Prozone3® player tracking system (Prozone Sports Ltd., Leeds, UK) was evaluated using measurements taken by timing gates (Di Salvo et al., 2006). The players participating in the study performed a series of straight line and arced runs at different speeds. The Prozone3® system installed at the stadia used in the study and the timing gates were used in parallel to determine the speeds of the runs made by the players. Another type of parallel forms reliability involves the participants having to perform two different tests. For example, we might wish to compare the results of a multistage fitness test (Ramsbottom et al., 1988) with a laboratory treadmill test of $\dot{V}O_2$ max. The stage of the multistage fitness test at which a participant can no longer run at the required speed is used to predict $\dot{V}O_2$ max. This way, the two tests are at least producing the same variable using the same measurement scale even if the protocol of measurement is different. However, when a test such as the multistage fitness test is originally devised, it is necessary to determine a mapping process from the stage at which a participant completed the test and the gold standard laboratory $\dot{V}O_2$ max test result. This can be done through cross-validation. A cross-validation study is where participants perform both tests, but half of the participants are used to statistically determine a model of the gold standard variable using the other variable being tested. An example of this is a badminton-specific test of $\dot{V}O_2$ max that was validated against a laboratory treadmill test (Henderson, 1998). The badminton-specific test required the participants to make movements from the centre of one side of the badminton court to six different locations on the perimeter of the court. This was controlled by playing an audio file indicating where the player had to run next. This gradually increased the pace of movement and eventually the player would have to terminate the test. The point of the test at which the test terminated was recorded. A set of 30 badminton players participated in the study with half being used to create a regression equation for $\dot{V}O_2$ max in terms of the point at which players dropped out of the court test. This equation was then used to predict the laboratory $\dot{V}O_2$ max test results for the other half of the set of participants. These predicted values were compared with the values the players actually achieved during the laboratory treadmill test.

Internal consistency

Internal consistency is a type of reliability used to test the consistency of components that make up some overall construct. Constructs are used in areas such as sports psychology to represent some ability or characteristic that cannot be directly observed. These include anxiety, mood and intelligence. A questionnaire instrument is used to record raw data, typically using Likert scales, which are used to form dimensions of the construct. For example, a construct for wellbeing might be composed of emotional wellbeing, intellectual wellbeing, physical wellbeing, social wellbeing and spiritual wellbeing (Corbin *et al.*, 2000: 17–18). Internal consistency is how consistent these individual dimensions (or components) of wellbeing are with the overall construct. This can be assessed using correlation techniques. As with parallel forms reliability, there are some in sport and exercise science who would not regard internal consistency of construct components as reliability. The components could be consistent with the wider construct, but that does not necessarily mean they were each measured in a consistent manner. Furthermore, inconsistency of components would not necessarily be viewed as poor reliability by scientists in certain areas of sport and exercise. For example, a performance profile may consist of indicators of different abilities all of which are of interest. These may individually be measured reliably, but for reasons of positional role within a team there may be low correlations between these components.

RELIABILITY STUDIES

There are other factors involved in the classification of reliability studies besides the type of reliability study. These include how widely the measurement process will be used, the number of participants (or matches) in the reliability study, how many repeated tests are involved and the role of human operation in the measurement process. There are standard instruments used in sports psychology, standard laboratory tests used in biomechanics, physiology and motor control as well as standard field tests used in fitness testing (Carling *et al.*, 2009). Measures such as these are widely used and are not specific to one single study. Therefore, research papers have been written with the exclusive purpose of establishing the reliability and validity of such measures so that these measures can be used in other studies that reference published evidence of reliability. The wide use of such measurement procedures is a motivation for extensive reliability assessment involving many participants. Indeed, the validity and reliability of some instruments has been tested for different population types in different published reliability studies (Zhu, 2000).

Performance analysis of sport is an area where systems are often developed for use in one specific study. It is, therefore, difficult to justify the publication of a stand-alone reliability study for the system. Instead, the reliability of the system would be reported within the methods section of a paper with a wider purpose investigating some aspect of sports performance. The use of the system within a single study typically means that the reliability of the system will not be tested as extensive as that of more general instruments. Therefore, single match reliability studies are commonly undertaken to establish the reliability of systems used in performance analysis investigations.

We have described test–retest reliability. It is possible to do reliability studies where participants are tested on more than two occasions. It is also possible that measurement procedures are made up of steps and reliability studies may be applied to each of the steps in the measurement process. For example, an observational method may involve verbal coding of behaviour and entering the recorded verbal codes into a computerized system. A participant could be independently observed by two verbal coders and then the two sets of verbal codes could be independently entered into a computerized system by different operators.

MEANINGFUL RELIABILITY ASSESSMENT

In general, the reliability of a measurement can be classified as good, acceptable or poor. There are many different types of reliability statistic that are used with the different types of reliability study that have been described in this chapter. These statistics have values that need to be interpreted during the assessment of reliability of a given measurement. For example, the kappa statistic has a value between −1.0 and +1.0, but we really need to know what kappa values indicate acceptable and good levels of reliability for a given measurement. Altman (1991: 404) stated that kappa values of 0.8 or above indicated a very good strength of agreement and that kappa values between 0.6 and 0.8 represented a good strength of agreement. This was based on an example of xeromammograms. O'Donoghue (2007) showed that kappa values of over 0.6 were possible in time–motion analysis when activity patterns associated with completely different energy system mixes were compared. It is, therefore, essential to establish the values of any reliability statistic that would be associated with good, acceptable and poor reliability for whatever measurement the statistic is evaluating. Choi *et al.* (2008) described a process of using synthetic data to intentionally represent different severities of disagreement in basketball performance data. For example, one pair of synthetic observations represented good agreement where there were minor differences and other pairs represented agreements that would be considered borderline acceptable/good, acceptable, borderline acceptable/poor and poor. The kappa values produced from these synthetic data allowed threshold values to be set for good and acceptable levels of agreement. O'Donoghue (2007) used a similar approach to propose kappa values of 0.8 and greater for good strength of agreement and 0.5 to 0.8 for acceptable strength of agreement in a work-rate analysis system. The use of such values is evidenced and better than using arbitrary values.

Another issue that should be considered is the type of study that a measurement will be used for. Atkinson and Nevill (1998) have proposed relating reliability statistics to the analytical goals of investigations. The same system may require completely different levels of reliability in different studies. The finer the difference being tested, the greater the level of reliability required. Investigations using repeated measures are particularly sensitive to measurement error.

SELECTING RELIABILITY STATISTICS

Factors

There is a great choice of reliability statistics that can be used and in many specific situations there is no one correct answer as to which is the most suitable. The factors involved in choosing a reliability statistic are:

- The scale of measurement of the given variable.
- The type of reliability study.

- Whether different samples of the same variable are being compared or if a relationship is being established between different variables (relevant in parallel forms reliability such as cross-validation).
- Whether the reliability study involves a single participant (performance) or multiple participants (performances).
- Whether the reliability of outcome variables or raw gathered data are being evaluated.
- The number of measurements (repeated tests or trials) being made on each participant.

Multiple participant reliability studies

The main factor that determines the reliability statistic to be used is the scale of measurement. When using multiple participants in test–retest studies, inter-rater reliability studies and parallel forms reliability studies, the same reliability statistics are recommended. Table 19.1 shows the reliability statistics recommended when the variable of concern is measured on different measurement scales in such studies. There is a wide choice of statistics for absolute reliability of ratio scale measures. The 95 per cent limits of agreement and 95 per cent ratio limits of agreement require at least 40 participants. Where we cannot gather this volume of reliability data, the other techniques such as mean absolute error, root mean squared error and percentage error could be used. Very often the choice may come down to the statistic that is commonly used in the particular discipline of sport and exercise science. For example, percentage error might be used in performance analysis while root mean squared error might be used in biomechanics.

Errors are expressed as ratios of the values recorded in 95 per cent ratio limits of agreement. Such ratios are only meaningful where the variables are measured on a ratio scale and values can be validly divided. The percentage error method proposed by Hughes *et al.* (2004) should also be avoided with interval scale measures to avoid division by zero errors and meaningless negative values for percentage errors. There is an alternative version of percentage error that expresses errors as a percentage of the meaningful range of values (O'Donoghue, 2010: 170–1). This can be used with either ratio or interval scale measures.

Table 19.1 Reliability statistics recommended in inter-rater, test retest and parallel forms (same variable) reliability studies

Measurement Scale	Reliability Statistics
Nominal	Kappa
Ordinal	Weighted kappa
Ratio	Relative reliability: Pearson's r. Absolute reliability: root mean square error, absolute error, percentage error, 95 per cent limits of agreement, change of the mean, standard error of measurement (typical error), 95 per cent ratio limits of agreement
Interval	Same methods as for ratio reliability except don't use percentage error, 95 per cent ratio limits of agreement

Multiple retest reliability studies

Where test–retest studies involve more than two trials, there are reliability statistics that could be used in addition to those identified in Table 19.1. We could simply use a method in Table 19.1 such as 95 per cent limits of agreement to compare pairs of trials. However, sometimes it is more efficient to have a single value indicating the reliability of the given measure. Coefficient of variation can be used as a measure of relative reliability to express the standard deviation of trial values as a percentage of their mean. This can be done for each participant in a multiple retest reliability study with the mean or maximum coefficient of variation being reported. Coefficient of variation should not be used with interval scale measures where the standard deviation of values (always positive) could be divided by a zero or negative mean value. Another reliability statistic used in multiple retest studies is the intra-class correlation coefficient.

Single case reliability studies

There are measurement techniques that are specific to particular studies and which may be time consuming to apply. For example, in performance analysis of sport, a system may be developed for use in a single study. It is most unlikely that a multiple performance reliability study for the system would warrant publication as the system might not be used in any other study. Therefore, a reliability assessment must be reported within any paper where research has been done using the system. A multiple performance reliability investigation could be larger than the main investigation of the research! This could easily be the case if 95 per cent limits of agreement were being used where a minimum of 40 matches would be needed in the reliability study. There are many high-profile performance analysis investigations where fewer performances than this have been used in the main investigation. Therefore, a single performance reliability study may be justified.

An inter-operator agreement study comparing a single measured value is not recommended (irrespective of how long the measurement process took). This is because the random error component of error could cancel out the systematic bias component leading to an unrealistically low total error value. Similarly, if a single nominal or ordinal value was being compared between independent observers, there would be an unknown chance of agreeing by guessing. Therefore, single case reliability studies typically use multiple measurements from the case. For example, multiple 100m section times could be measured from the same 1,500m race or categorical and/or numerical data could be gathered from many points from within the same tennis match. This introduces a further issue in reliability assessment which is whether we should be analysing the reliability of the observation or the reliability of the particular variables to be used in a given study. There are some who would argue that the reliability of each variable to be used in an investigation should be assessed. However, there are many studies where the reliability of a system has been tested as a whole and then multiple performance variables have been derived from the system and used within research studies. Consider the tennis example illustrated in Table 19.2. Here, there are only two raw observed variables point type and point outcome. However, these are used to generate a set of 11 performance indicators. If one additional raw variable such as whether the point emanated from first or second serve was added, we could almost triple the number of

Table 19.2 Raw variables and performance indicators generated by an example tennis observational instrument

Raw observed variables (and values)	Performance indicators
Point type (ace, double fault, baseline rally, server to net, receiver to net)	%Points that are aces
	%Points that are double faults
	%Service points that are won
Outcome of point (server wins, receiver wins)	%Points that are baseline rallies
	%Points where server goes to the net
	%Points where receiver goes to the net
	%Service points won on serve
	%Baseline rallies won by server
	%Baseline rallies won by receiver
	%Points won by server when server goes to the net
	%Points won by receiver when receiver goes to the net

performance indicators generated by using performance indicators for all service points, points where the first serve was in and points where a second serve was required (we would not require percentage of first serve points that are double faults). If we were to add a fourth raw variable such as the set in which the point was played, we would multiply the number of performance indicators generated again. Reporting the results of a reliability study of a complete set of performance indicators would be word-intensive and would suffer from errors in some points being cancelled out by other types of errors in other points. For example, if one observer classified a point as a server to net and the other classified it as a baseline rally and the two observers made these errors the other way round in another point, we could report 100 per cent agreement for the two performance indicators concerned. Therefore, in such situations, the reliability of the observed events could be evaluated with inductive reasoning used to discuss the reliability of the performance indicators generated by the system.

Once the decision has been made to assess reliability at the individual point level, the reliability statistics shown in Table 19.1 can be used. For example, nominal variables such as point type and point outcome can be assessed using kappa. The use of such techniques requires the system to record a chronological list of values for each variable. In situations where a manual form is used to record a tally (total frequency) for different values, it is not possible to identify individual errors or determine the kappa statistic. In such cases, percentage error has been applied to the recorded frequencies.

Cross-validation

Earlier in the chapter, the cross-validation of an on-court test of $\dot{V}O_2$ max for badminton players was described (Henderson, 1998). All participants were tested using the on-court test and a gold standard laboratory test of $\dot{V}O_2$ max. Data for half of the participants were used to determine a regression equation for $\dot{V}O_2$ max from the drop-out time from the on-court test. The data for the other half of the participants were used to test the validity of the $\dot{V}O_2$ max predicted from the on-court test against the actual measured $\dot{V}O_2$ max from the laboratory test. The combined use of linear regression for modelling and then Pearson's correlation for relative reliability and mean absolute error for absolute reliability can be used in parallel forms reliability tests as well as validation studies. Where the second part of such a study involves at least 40 participants, it is possible to use 95 per cent limits of agreement.

Internal consistency of components

Internal consistency of components of some construct differs from the other three types of reliability to the extent that the reliability statistics listed in Table 19.1 can be used for the other three types but not for internal consistency. Correlations could be done between each pair of components this could be done using Pearson's r. Once the correlations are determined, there is a need to make an assessment based on the range of correlation values or average correlation produced. A statistic for internal consistency reliability that produces a single value for the construct under consideration is Cronbach's α. Now that various reliability statistics have been identified and guidance has been given for which ones to use in different situations, the following section of this chapter covers the calculation of reliability statistics.

PRODUCING RELIABILITY STATISTICS

Kappa

The kappa statistic (Cohen, 1960) will be illustrated using an example of netball player assessment. Joll (2011) developed a system for assessing netball players during trials. The system required assessors to observe and score players on a 1 to 9 scale for each of the following 13 aspects of the game (1 being awarded for a poor performance and 9 being awarded for an excellent performance):

- Game fitness
- Shoulder pass
- Chest pass
- Bounce pass
- Overhead pass
- Speed
- Balance
- 3 foot mark
- Taking their feet to the ball
- Turn fully on receipt of ball to face direction of play
- Finding space with attacking drives (using angled runs)
- Sprinting without stepping back
- Basic man to man defence.

The file 19-netball-selection.XLS contains a sheet 'Raw data' which is fictitious data for six assessors independently observing a set of 30 players over a trial weekend at the end of which a 12-player squad would be selected. Three of the assessors (Q1, Q2 and Q3) are qualified coaches while three others are unqualified (U1, U2 and U3). With 30 players being assessed on 13 items, each assessor makes a total of 390 assessments using the 1 to 9 scale. Table 19.3 compares the assessments made by two of the assessors (Q1 and Q2).

Table 19.3 Agreements (bold) and disagreements for ratings made by assessors Q1 and Q2

Assessor Q1	Assessor Q2									
	1	2	3	4	5	6	7	8	9	Total
1										0
2										0
3			**8**	4						12
4			3	**51**	16					70
5				19	**79**	15				113
6					11	**84**	13			108
7						16	**33**	10		59
8							3	7	**14**	24
9								1		3	4
Total	0	0	11	74	106	118	54	24	3	390

The proportion of agreements, P_0, is given by equation 19.1 where N is the total number of assessments made by each assessor, V is the maximum score of 9, k is a variable representing score between 1 and 9 and A is the 9×9 table shown in Table 19.3 without the totals, row and column headings. P_0 is the sum of the values shown in bold on the diagonal divided by the total number of assessments, $P_0 = 272 / 390 = 0.70$.

$$P_0 = \frac{\sum_{k=1}^{V} A_{k,k}}{N}$$ (19.1)

The number of agreements we would expect by chance is determined by assuming that an assessor who was guessing would use the same distribution of scores but in a random fashion that is if assessor 'Q2' was trying to guess scores for each aspect for each player without observing the performances of the players, 'Q2' would still make 106 scores of 5 and 118 scores of 6, for example, to give herself a chance of agreeing by guessing. Therefore, whenever 'Q1' makes a score, there is a 106/390 chance that 'Q2' will award a score of 5. This means that on this fraction (106/390) of the 113 occasions when 'Q1' scored a 5 could be agreed if 'Q2' was guessing. This comes to 30.71 expected agreements by guessing for the value 5. This is done for all nine scores using equation 19.2 to determine the total number of agreements that could be expected by guessing (86.69). P_C expresses this as a proportion of the 390 pairs of scores ($P_C = 0.22$).

$$P_C = \frac{\sum_{k=1}^{V} \left(\sum_{i=1}^{V} A_{i,k} \right) \left(\sum_{j=1}^{V} A_{k,j} \right)}{N^2}$$ (19.2)

Consider P_0 as being itself divided by one. Kappa simply takes P_C away from the numerator and the denominator of this division in order to address the probability of agreeing by guessing as shown in equation 19.3 which gives κ.

$$\kappa = (P_0 - P_C) / (1 - P_C)$$ (19.3)

This gives a kappa value, $\kappa = 0.61$ which represents a good strength of agreement between assessors Q1 and Q2. Altman (1991: 404) stated that values of kappa were very good if they were over 0.8, good if between 0.6 and 0.8, moderate if between 0.4 and 0.6, fair if between 0.2 and 0.4 and poor if less than 0.2.

The remainder of this section describes the process of determining the kappa statistic in the 19-netball-selection.XLS spreadsheet. Columns D to I of the 'Raw data' sheet represent the six different assessors with 'Q1', 'Q2', 'Q3', 'U1', 'U2' and 'U3'. In Excel, a cross-tabulation of frequencies such as that shown in Table 19.3 can be done using pivot tables. For example, to compare assessors 'Q1' and 'Q2', the columns with their scores in are selected and then **Insert → Pivot Table** is used. We choose to place the pivot table in new worksheet ('Solution Q1 v Q2') as shown in Figure 19.1. We then drag 'Q1' and 'Q2' into the row label and column label area respectively and then drag 'Q1' into the values area and we see the cross-tabulation of the two variables populate with frequencies.

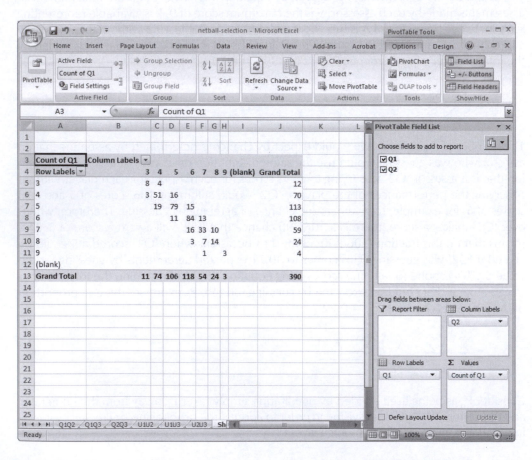

Figure 19.1 The pivot table cross-tabulating the scores of assessors 'Q1' and 'Q2'.

reliability

Figure 19.2 shows the P_0, P_C and κ values determined for assessors Q1 and Q2. There are additional sheets in the netball selection file showing the calculation of kappa for the remaining pairs of assessors.

Weighted kappa

Kappa may not be the best reliability statistic to use in this example because a situation where the assessors disagree by one is counted as a total disagreement exactly the same as

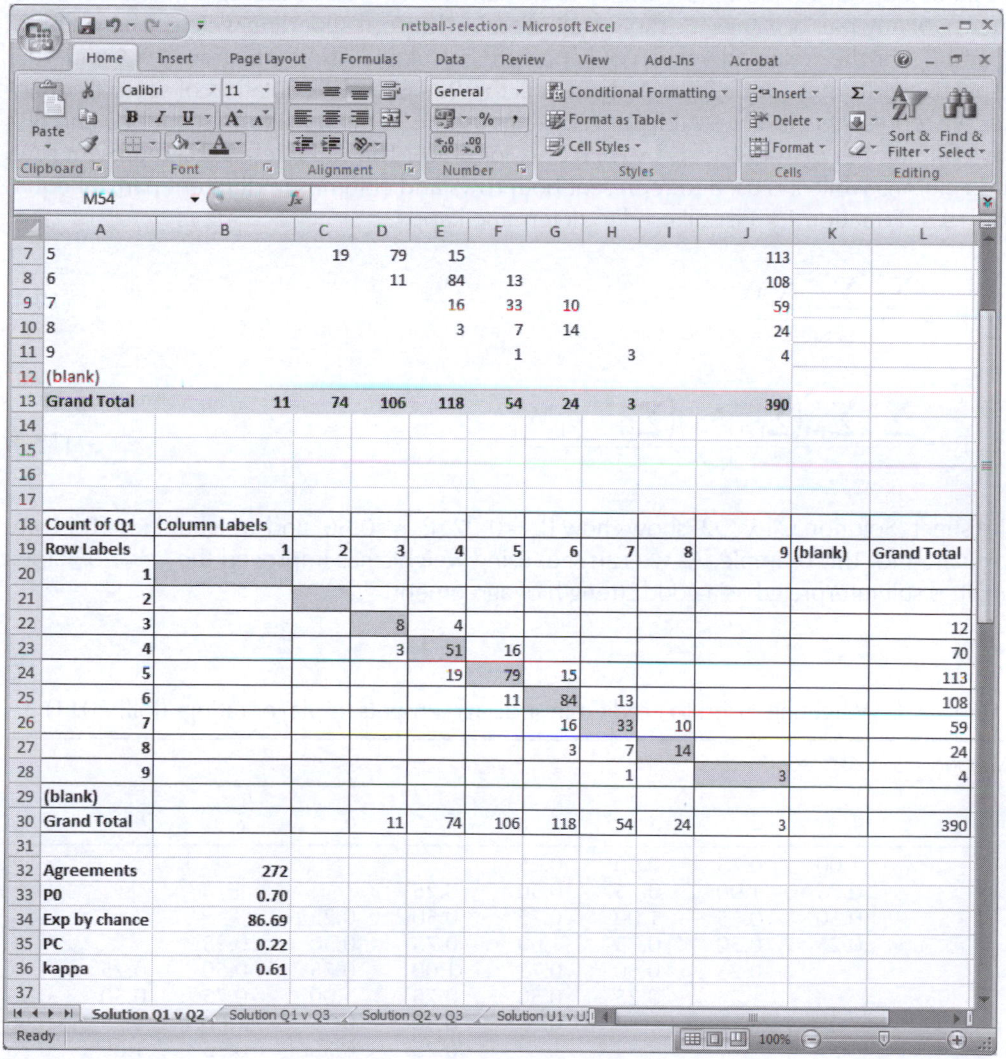

	A	B	C	D	E	F	G	H	I	J	K	L
7	5		19	79	15					113		
8	6		11	84	13					108		
9	7			16	33	10				59		
10	8			3	7	14				24		
11	9				1		3			4		
12	(blank)											
13	Grand Total		11	74	106	118	54	24	3	390		
14												
15												
16												
17												
18	Count of Q1	Column Labels										
19	Row Labels	1	2	3	4	5	6	7	8	9	(blank)	Grand Total
20	1											
21	2											
22	3			8	4							12
23	4			3	51	16						70
24	5				19	79	15					113
25	6					11	84	13				108
26	7						16	33	10			59
27	8							3	7	14		24
28	9								1		3	4
29	(blank)											
30	Grand Total			11	74	106	118	54	24	3		390
31												
32	Agreements	272										
33	P0	0.70										
34	Exp by chance	86.69										
35	PC	0.22										
36	kappa	0.61										
37												

Figure 19.2 Spreadsheet to determine kappa.

when one assessor scores 9 while the other scores 1. Other statistics have been proposed for assessing the stability of questionnaire items that also fail to give any credit for minor disagreements in items (Wilson and Batterham, 1999). Nevill et al. (2001) proposed an alternative approach for assessing test–retest stability of five-point scale items within psychometric questionnaires where absolute differences of zero or one were credited. A much earlier development to addressing minor disagreements is Cohen's (1968) weighted version of kappa. The weights suggested by Joll (2011) for the netball player assessment system are shown in Table 19.4. These give credit to disagreements of less than four points on the nine-point rating scale.

P_0 and P_C are calculated differently for the weighted version of kappa than they were for kappa as both of P_0 and P_C use minor disagreements as well as the agreements. The frequency of any pair of values is simply multiplied by the corresponding weight. The values of P_0 and P_C for the weighted version of kappa are calculated using equations 19.4 and 19.5 respectively. In these equations, A represents the 9×9 cross-tabulation of score frequencies, W represents the 9×9 table of weightings used, V is the maximum score of 9 and N is the total number of assessments by each assessor (390). The variables i and j represent rows and columns while k is used to represent both rows and columns in different parts of equation 19.5. Equation 19.3 is still used to determine kappa.

$$P_0 = \frac{\sum_{i=1}^{V}\sum_{j=1}^{V} A_{i,j}W_{i,j}}{N} \tag{19.4}$$

$$P_C = \frac{\sum_{i=1}^{V}\sum_{j=1}^{V}\left(\sum_{k=1}^{V} A_{i,k}\right)\left(\sum_{k=1}^{V} A_{k,j}\right)W_{i,j}}{N^2} \tag{19.5}$$

The sheet 'Solution Q1 v Q2' shows how $P_0 = 0.92$, $P_C = 0.66$, and weighted $\kappa = 0.77$ are calculated for this example. As we can see, using weights has improved the value of kappa which is still interpreted as a good strength of agreement.

Table 19.4 Weightings used to credit minor disagreements in player ratings (Joll, 2011)

Assessor A	Assessor B								
	1	2	3	4	5	6	7	8	9
1	1.00	0.75	0.50	0.25					
2	0.75	1.00	0.75	0.50	0.25				
3	0.50	0.75	1.00	0.75	0.50	0.25			
4	0.25	0.50	0.75	1.00	0.75	0.50	0.25		
5		0.25	0.50	0.75	1.00	0.75	0.50	0.25	
6			0.25	0.50	0.75	1.00	0.75	0.50	0.25
7				0.25	0.50	0.75	1.00	0.75	0.50
8					0.25	0.50	0.75	1.00	0.75
9						0.25	0.50	0.75	1.00

Correlation coefficients

An example of a test–retest reliability investigation of the Y-Balance Test Kit™ is used to describe the reliability of ratio scale measures. The data for 60 fictitious participants who performed the test with the dominant leg and with their eyes open are shown in Table

Table 19.5 Test-retest reliability study for the Y-Balance test performed with the eyes open and with the dominant leg (fictitious data)

Participant	Test	Retest	Error	Abs Error	Squared error	Mean value	% Error (Hughes et al., 2004)	% Error (O'Donoghue, 2010)
1	83.88	81.91	−1.97	1.97	3.88	82.90	2.38	1.31
2	103.40	108.36	4.96	4.96	24.60	105.88	4.68	3.31
3	117.40	116.21	−1.19	1.19	1.42	116.81	1.02	0.79
4	80.40	80.56	0.16	0.16	0.03	80.48	0.20	0.11
5	116.60	121.25	4.65	4.65	21.62	118.93	3.91	3.10
6	92.80	96.96	4.16	4.16	17.31	94.88	4.38	2.77
7	98.30	102.34	4.04	4.04	16.32	100.32	4.03	2.69
8	100.00	105.28	5.28	5.28	27.88	102.64	5.14	3.52
9	97.30	98.22	0.92	0.92	0.85	97.76	0.94	0.61
10	96.60	101.49	4.89	4.89	23.91	99.05	4.94	3.26
11	100.40	100.60	0.2	0.2	0.04	100.50	0.20	0.13
12	100.50	106.15	5.65	5.65	31.92	103.33	5.47	3.77
13	80.39	75.56	−4.83	4.83	23.33	77.98	6.19	3.22
14	102.70	106.04	3.34	3.34	11.16	104.37	3.20	2.23
15	85.90	92.69	6.79	6.79	46.10	89.30	7.60	4.53
16	99.35	101.68	2.33	2.33	5.43	100.52	2.32	1.55
17	106.00	105.63	−0.37	0.37	0.14	105.82	0.35	0.25
18	115.00	114.96	−0.04	0.04	0.00	114.98	0.03	0.03
19	102.30	105.71	3.41	3.41	11.63	104.01	3.28	2.27
20	99.80	107.90	8.1	8.1	65.61	103.85	7.80	5.40
21	82.89	90.98	8.09	8.09	65.45	86.94	9.31	5.39
22	113.30	112.03	−1.27	1.27	1.61	112.67	1.13	0.85
23	86.91	88.34	1.43	1.43	2.04	87.63	1.63	0.95
24	88.70	90.98	2.28	2.28	5.20	89.84	2.54	1.52
25	91.00	95.64	4.64	4.64	21.53	93.32	4.97	3.09
26	106.70	111.71	5.01	5.01	25.10	109.21	4.59	3.34
27	99.50	100.90	1.4	1.4	1.96	100.20	1.40	0.93
28	100.50	101.10	0.6	0.6	0.36	100.80	0.60	0.40
29	107.29	112.83	5.54	5.54	30.69	110.06	5.03	3.69
30	107.70	116.04	8.34	8.34	69.56	111.87	7.46	5.56
31	82.45	84.04	1.59	1.59	2.53	83.25	1.91	1.06
32	114.30	126.27	11.97	11.97	143.28	120.29	9.95	7.98
33	98.30	99.90	1.6	1.6	2.56	99.10	1.61	1.07
34	118.00	118.71	0.71	0.71	0.50	118.36	0.60	0.47
35	102.60	105.05	2.45	2.45	6.00	103.83	2.36	1.63
36	110.59	112.42	1.83	1.83	3.35	111.51	1.64	1.22
37	97.20	99.50	2.3	2.3	5.29	98.35	2.34	1.53
38	105.50	119.95	14.45	14.45	208.80	112.73	12.82	9.63
39	79.71	77.19	−2.52	2.52	6.35	78.45	3.21	1.68
40	102.90	112.93	10.03	10.03	100.60	107.92	9.29	6.69

41	95.80	96.76	0.96	0.96	0.92	96.28	1.00	0.64
42	89.18	92.36	3.18	3.18	10.11	90.77	3.50	2.12
43	94.60	101.59	6.99	6.99	48.86	98.10	7.13	4.66
44	90.90	93.26	2.36	2.36	5.57	92.08	2.56	1.57
45	105.90	107.36	1.46	1.46	2.13	106.63	1.37	0.97
46	81.50	83.52	2.02	2.02	4.08	82.51	2.45	1.35
47	90.10	94.37	4.27	4.27	18.23	92.24	4.63	2.85
48	112.00	115.62	3.62	3.62	13.10	113.81	3.18	2.41
49	90.00	96.94	6.94	6.94	48.16	93.47	7.42	4.63
50	81.07	84.83	3.76	3.76	14.14	82.95	4.53	2.51
51	95.46	103.02	7.56	7.56	57.15	99.24	7.62	5.04
52	83.85	84.38	0.53	0.53	0.28	84.12	0.63	0.35
53	104.40	107.32	2.92	2.92	8.53	105.86	2.76	1.95
54	107.90	109.18	1.28	1.28	1.64	108.54	1.18	0.85
55	99.30	103.95	4.65	4.65	21.62	101.63	4.58	3.10
56	122.20	124.40	2.2	2.2	4.84	123.30	1.78	1.47
57	97.70	96.98	−0.72	0.72	0.52	97.34	0.74	0.48
58	109.20	104.18	−5.02	5.02	25.20	106.69	4.71	3.35
59	103.90	104.69	0.79	0.79	0.62	104.30	0.76	0.53
60	93.41	93.37	−0.04	0.04	0.00	93.39	0.04	0.03
Mean	98.72	101.73	3.01	3.61	22.03	100.23	3.58	2.41
SD	10.75	11.70	3.63	3.02	36.31	11.09	2.85	2.02

19.5. These data are also provided in the 'Absolute and Relative Rel' worksheet of the 19-Y-Balance-Test.XLS spreadsheet. First, we will distinguish between relative and absolute reliability. Relative reliability is how well participants in a reliability study maintain their rank between a test and a retest (or between independent observations of performances). Pearson's coefficient of correlation, r, can be used as a relative reliability statistic. In this test–retest reliability study of the Y-Balance test, the correlation between test and retest scores is high (r = 0.951). It must be noted, however, that perfect positive correlations (r = 1.000) are possible when there are large differences between test and retest scores; for example, each retest score could be 10 times the corresponding test score and we would have a perfect positive correlation (Patterson, 2000). Therefore, relative reliability statistics should always be used in conjunction with absolute reliability statistics. The remainder of this section describes absolute reliability statistics.

Absolute error

The fifth column of Table 19.5 shows absolute errors between test and retest scores. Absolute error is the magnitude of error that is the error with the sign (+ or −) removed. Using absolute errors avoids positive and negative errors cancelling each other out when determining an average error within the sample. The mean absolute error in this case is 3.61 per cent. Some may prefer to use a more stringent measure of absolute reliability which covers 95 per cent of absolute errors. Therefore, O'Donoghue (2010: 171–2) proposed using the 95th percentile of absolute errors which in this example is 8.42 per cent. In Microsoft Excel, the function PERCENTILE(range of cells, 95%) is used to determine the 95th percentile of a series of values. We could also use maximum absolute error (14.45 per cent).

352

Percentage error

Percentage error has been used by performance analysts since the mid-1980s, often with single case reliability studies. Hughes et al. (2004) described a percentage error statistic that expressed absolute error as a percentage of the value recorded. This version of percentage error is shown in equation 19.6 where T1 is the test score and T2 is the retest score. For test–retest studies of performance measures, there is a case for dividing by the better value of T1 and T2 as the participant would have shown this ability over the two tests. This better value could be the minimum of the two values if the measure was a sprint time where lower values are better or it could be the maximum of the two values if the measure was a jump distance where higher scores are better. In observational analysis studies, T1 and T2 represent values recorded by two independent observers. The absolute error between the two is divided by the mean of T1 and T2 because neither would be known to be the correct value for the participant.

$$\%Error = 100 \times |T1-T2| / ((T1+T2)/2) \tag{19.6}$$

O'Donoghue (2010: 170–1) criticized the percentage error statistic proposed by Hughes et al. (2004) because dividing by the mean recorded value could lead to a misleading representation of error which could either underestimate or overestimate measurement error for the given variable. An example of underestimation of error is two independent measures of 100m sprint time producing values of 10.2s and 10.6s. The absolute error of 0.4s is 3.8 per cent of the mean value of 10.4s. However, values of under 9s are not realistic (at the time of writing!) and dividing by the whole of the mean value gives low percentage error values. Performances of 10.2s and 10.6s represent different standards with an even paced 10.6s performance meaning the athlete would still have very noticeable 3.77m to go as the 10.2s athlete finished. An example of overestimation of error is where two independent observers counting the number of incomplete passes by a basketball player record frequencies of one and zero. According to equation 19.6 this is a percentage error of 200 per cent and yet both observers are in broad agreement that the player did not make many incomplete performances. O'Donoghue (2010: 171) proposed an alternative method of calculating percentage error by dividing absolute error by the meaningful range of values for the given variable as shown in equation 19.7 where Max and Min are the maximum and minimum realistic values for the given variable. In Table 19.5, Min and Max values for the Y-Balance test of 0 per cent and 150 per cent are used in the calculation of O'Donoghue's (2010) percentage error.

$$\%Error = 100 \times |T1-T2| / (Max - Min) \tag{19.7}$$

In multiple participant reliability studies, the percentage error statistics of Hughes et al. (2004) and O'Donoghue (2010) can be summarized using mean percentage error or 95th percentile of percentage error. The penultimate row of Table 19.5 shows that mean percentage error is 3.58 per cent and 2.41 per cent when using the methods of Hughes et al. (2004) and O'Donoghue (2010) respectively. The 95th percentile of the percentage errors using these two techniques are 9.29 per cent and 5.62 per cent respectively.

Root mean squared error

Root mean squared error is an absolute reliability statistic that has been used in biomechanics. This statistic removes negative signs from the error values by squaring them. However, the mean squared error will not be in the correct units. For example, if our unit of measurement was m, then the differences would also be in m but the squared differences would be in m². We are dealing with length, height, distance or displacement rather than area and, therefore, the square root is taken of the mean squared error to ensure the reliability statistic is in the correct units. Table 19.5 shows that the mean squared error is 22.03 for the Y-Balance test example (and the %² units would be meaningless) and so the root mean squared error is 4.69 per cent.

Limits of agreement

A criticism of mean percentage error, mean absolute error and root mean square error is that each gives a single value describing an average error concealing important information about range and consistency of error values. An error between test–retest scores (or between values recorded by independent raters) is made up of two components: systematic bias and random error. The systematic bias is an average difference between test and retest scores. An example of systematic bias could be a tendency for retest scores to show an average improvement on test scores. Another example of systematic bias could be for one observer in a time–motion study to record higher values on average for the amount of time spent running than another observer. In a reliability study involving 40 participants, it is very unlikely that all test–retest increases in score will be the same. The consistency or inconsistency of error values about the systematic bias is of interest because an evaluation of reliability should give some indication of how large errors can be. Random error is this variability in errors due to factors other than the systematic difference between tests or between raters. Factors causing random error include inherent biological factors, mechanical factors or inconsistent measurement procedure (Atkinson and Nevill, 1998). A reliability statistic describing both systematic bias and random error is the 95 per cent limits of agreement.

The reliability study of the Y-balance test will also be used as an example of 95 per cent limits of agreement. The data for this example are found in the 19-Y-Balance-Test.XLS spreadsheet and the 'Dominant Leg' worksheet represents the test performed on the dominant leg. The data are also shown in Table 19.5. The error column of Table 19.5 contains the values used in the calculation of the 95 per cent limits of agreement. The mean difference is 3.01 per cent and this is the systematic bias component of the error between test and retest. There is a general tendency for participants to improve their performance between the test and retest. This is not a consistent increase from test to retest performance with 10 participants having lower retest scores than their test score, while others had an above average increase from test to retest performance, the highest increase being 14.45 per cent. Assuming that the test–retest increases are normally distributed, the standard deviation of 3.63 per cent means that 68 per cent of errors should be within 3.63 per cent of the mean error of 3.01 per cent. The 95 per cent limits of agreement represent a range of errors such that 95 per cent of all errors would be expected to be within them. If the test–retest increases are normally distributed, then 95 per cent of these increases would be expected to lie within 1.96 standard deviations of the mean increase. Since $1.96 \times 3.63 = 7.12$, the 95 percent limits of agreement are 3.01 ± 7.12 per cent. Therefore, 95 per cent of retest scores would be expected to lie between the test score – 4.10 per cent and the test score + 10.13 per cent. The reliability data and the 95 per cent limits of agreement can be summarized on a Bland–Altman (1986) plot which plots errors against mean values as shown in Figure 19.3.

Ninety-five per cent ratio limits of agreement

There are situations where the difference between observer ratings (or test–retest scores) increases as the ratings increase. An example of this is valuation of soccer players. One rater might value an international player at £12,000,000 while another rater might value the same player at £11,000,000. The £1,000,000 difference between their valuations is actually more than some non-international players are worth. For example, the two observers might rate a non-international player as being worth £250,000 and £240,000. This tendency for differences to be related to the valuations made means that the differences are better expressed as ratios rather than differences. Ratio limits of agreement have been used to assess the between-trials and between-days reliability of power output during repeated treadmill sprinting (Tong et al., 2001).

The Excel spreadsheet 19-footballers.XLS contains two raters' valuations of 50 international soccer players. This example resulted from Ireland's non-qualification for the 2010 FIFA World Cup at the hand of Thierry Henry. The author picked a squad of 23 players who

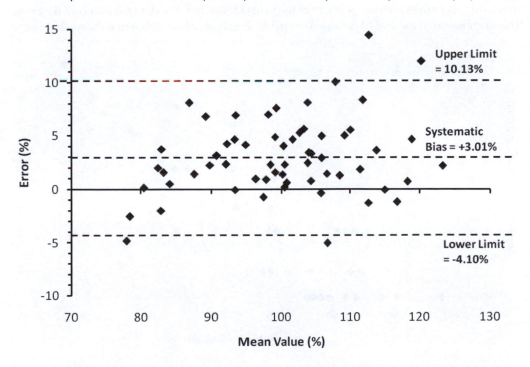

Figure 19.3 Bland–Altman plot for test–retest reliability of the Y-Balance test performed on the dominant leg.

were from nations other than the 32 qualifying nations. Some of the author's colleagues picked their own 23 player squads from players from non-qualifying nations. Naturally with soccer supporters, the author and his friends had differences of opinion as to who should be included in the squad. Fifty of the players considered are included in the spreadsheet and these are shown in Table 19.6. The worksheet 'Calculation 95 percent limits of agreement' shows that the 95 per cent limits of agreement between the two raters is −£0.282M ± £1.684M. That is, according to 95 per cent limits of agreement, the first rater had a tendency to value players £282,000 lower than the second rater. This difference is not uniform with the first rater valuing 17 of the players higher than the second rater. The random error of ±£1,684,000 is greater than 42 of the 100 valuations made between the two raters. The differences between the raters for the most valuable players have had a disproportionate effect on the mean difference in ratings made by the two raters – the difference in the ratings of Stefan Jovetić, for example, is £3,800,000. The worksheet 'Calculation 95% Ratio LOA' shows a very strong correlation between absolute differences in the valuations and mean valuations between the two raters (r = 0.920). Figure 19.4 shows this positive relationship between mean value and absolute difference, indicating that the disagreements are more of a ratio (or percentage) than a difference (plus or minus) disagreement. Figure 19.4 was best shown using logarithmically scaled axes (in this case $\log_2$) to avoid large numbers of players being clustered in the bottom left of the chart.

The data in this example are heteroscedastistic. Heteroscedasticity is where there is a positive relationship between the magnitude of error and value recorded. Where no such relationship exists and errors are independent of the values recorded, the data are homoscedastistic. This correlation of r = 0.920 between mean value and absolute difference shows that there

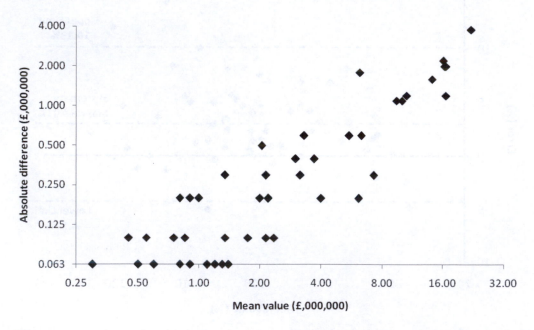

Figure 19.4 Relationship between absolute inter-rater difference and the mean value recorded by raters.

Table 19.6 Soccer player valuations

Player	Nationality	Position	Valuation (A) £,000,000	Valuation (B) £,000,000	Mean	Abs diff	ln A	ln B	Diff (lns)
Adebayor, E	Togo	Forward	10.6	9.5	10.1	1.1	2.36	2.25	0.11
Bale, G	Wales	Defender	11.2	10.0	10.6	1.2	2.42	2.30	0.11
Ben Haim, T	Israel	Defender	1.4	1.4	1.4	0.0	0.34	0.34	0.00
Benayoun, Y	Israel	Midfielder	2.4	2.3	2.4	0.1	0.88	0.83	0.04
Berbatov, D	Bulgaria	Forward	16.0	17.2	16.6	1.2	2.77	2.84	−0.07
Bogdani, E	Albania	Forward	0.2	0.2	0.2	0.0	−1.61	−1.61	0.00
Chamakh, M	Morocco	Forward	3.9	3.5	3.7	0.4	1.36	1.25	0.11
Chivu, C	Romania	Defender	2.2	2.1	2.2	0.1	0.79	0.74	0.05
Da Silva, E	Croatia	Forward	8.9	10.0	9.5	1.1	2.19	2.30	−0.12
Diawara, S	Senegal	Defender	0.6	0.6	0.6	0.0	−0.51	−0.51	0.00
Dunne, R	Ireland	Defender	0.9	0.9	0.9	0.0	−0.11	−0.11	0.00
Edel, A	Armenia	Keeper	0.3	0.3	0.3	0.0	−1.20	−1.20	0.00
Elmander, J	Sweden	Forward	6.2	6.0	6.1	0.2	1.82	1.79	0.03
Evans, J	N. Ireland	Defender	3.3	3.0	3.2	0.3	1.19	1.10	0.10
Farfan, J	Peru	Forward	13.4	15.0	14.2	1.6	2.60	2.71	−0.11
Fletcher, D	Scotland	Midfielder	2.0	2.3	2.2	0.3	0.69	0.83	−0.14
Ghezzal, J	Algeria	Forward	0.5	0.5	0.5	0.0	−0.69	−0.69	0.00
Giggs, R	Wales	Midfielder	0.8	0.8	0.8	0.0	−0.22	−0.22	0.00
Gillet, J	Belgium	Keeper	0.2	0.2	0.2	0.0	−1.61	−1.61	0.00
Given, S	Ireland	Keeper	2.1	1.9	2.0	0.2	0.74	0.64	0.10
Gordon, C	Scotland	Keeper	2.3	2.1	2.2	0.2	0.83	0.74	0.09
Gudjohnsen, E	Iceland	Forward	1.3	1.3	1.3	0.0	0.26	0.26	0.00
Hangeland, B	Norway	Defender	2.3	2.1	2.2	0.2	0.83	0.74	0.09
Hleb, A	Belarus	Midfielder	6.2	6.0	6.1	0.2	1.82	1.79	0.03
Ibertsberger, A	Austria	Defender	0.9	0.8	0.9	0.1	−0.11	−0.22	0.12
Ibrahimovic, Z	Sweden	Forward	15.6	17.6	16.6	2.0	2.75	2.87	−0.12
Jovetić, S	Montenegro	Forward	20.2	24.0	22.1	3.8	3.01	3.18	−0.17
Kanoute, F	Mali	Forward	1.8	2.3	2.1	0.5	0.59	0.83	−0.25
Keita, S	Mali	Midfielder	1.2	1.2	1.2	0.0	0.18	0.18	0.00
Kladec, M	Czech Rep	Defender	1.8	1.7	1.8	0.1	0.59	0.53	0.06
Makiada, C	DR Congo	Forward	0.7	0.8	0.8	0.1	−0.36	−0.22	−0.13
Modric, L	Croatia	Midfielder	5.3	7.1	6.2	1.8	1.67	1.96	−0.29
Nekounan, J	Iran	Midfielder	0.5	0.6	0.6	0.1	−0.69	−0.51	−0.18
Nikolov, O	Macedonia	Keeper	0.2	0.2	0.2	0.0	−1.61	−1.61	0.00
Oliech, D	Kenya	Forward	0.7	0.9	0.8	0.2	−0.36	−0.11	−0.25
Ospina, D	Columbia	Keeper	0.4	0.5	0.5	0.1	−0.92	−0.69	−0.22
Petrov, S	Bulgaria	Midfielder	0.9	1.1	1.0	0.2	−0.11	0.10	−0.20
Pizarro, C	Peru	Forward	4.1	3.9	4.0	0.2	1.41	1.36	0.05
Pjanic, M	Bosnia Herz	Midfielder	15.4	17.4	16.4	2.0	2.73	2.86	−0.12
Radu, S	Romania	Defender	0.8	1.0	0.9	0.2	−0.22	0.00	−0.22
Raketic, I	Croatia	Midfielder	6.0	6.6	6.3	0.6	1.79	1.89	−0.10
Misimovic, Z	Bosnia Herz	Midfielder	15.0	17.2	16.1	2.2	2.71	2.84	−0.14
Riise, JA	Norway	Midfielder	1.3	1.4	1.4	0.1	0.26	0.34	−0.07
Rosicky, T	Czech Rep	Midfielder	5.2	5.8	5.5	0.6	1.65	1.76	−0.11
Sahin, N	Turkey	Midfielder	3.0	3.6	3.3	0.6	1.10	1.28	−0.18
Samba, C	Congo	Defender	1.2	1.5	1.4	0.3	0.18	0.41	−0.22
Sessegnon, S	Benin	Midfielder	2.3	2.1	2.2	0.2	0.83	0.74	0.09
Valencia, L	Ecuador	Midfielder	3.2	2.8	3.0	0.4	1.16	1.03	0.13
Vargas, J	Peru	Midfielder	7.4	7.1	7.3	0.3	2.00	1.96	0.04
Vermaelen, T	Belgium	Defender	1.1	1.1	1.1	0.0	0.10	0.10	0.00
Mean			4.31	4.59	4.45	0.50	0.77	0.81	−0.04
SD			5.06	5.67	5.36	0.75	1.24	1.25	0.12

is severe heteroscedasticity in the data. Correlations of r > 0.25 are sufficient to justify that there is at least mild heteroscedasticity in the data. Where data are heteroscedastistic, it is better to express inter-rater disagreement as a ratio rather than a difference. For example, when two estate agents value properties, one might have a tendency to value properties 5 per cent higher than the other. This would be the systematic bias between the estate agents and there would be some additional random error between them.

Returning to our example of valuation of soccer players, the calculation of 95 per cent ratio limits of agreement is done on the worksheet 'Calculation 95% Ratio LOA'. One way of dealing with ratios is to take logarithms of values recorded by the two raters and subtract them. The antilog of the difference is the value one would get if dividing the original values (Atkinson and Nevill, 1998). Columns H and I are the natural logarithms (ln) of the valuations while column J is the difference between the ln transformed values. The average of these differences is −0.04 and the antilog of this average ($e^{-0.04}$) is 0.96. The systematic bias is actually the geometric mean of the ratio differences between the raters' values rather than the arithmetic mean. The geometric mean of the 50 ratio differences is the product of the 50 ratio differences raised to the power of 1/50. Whether using ratios or differences between ln transformed values, values of zero cannot be included in data to be dealt with using 95 per cent ratio limits of agreement. When we take the arithmetic mean of ln transformed values and then take the antilog (EXP in Excel) of that average we compute the geometric mean. In this case the systematic bias is 0.96 which means there is a tendency for the first rater to give a valuation 4 per cent less than that of the second rater. The standard deviation of the differences between the ln transformed values is 0.12 and represents 68 per cent of the spread of these differences. Therefore, the standard deviation of the ln transformed differences is multiplied by 1.96 in order to represent 95 per cent of the differences this assumes that the difference between ln transformed values is normally distributed. The random error component of 95 per cent ratio limits of agreement is $e^{1.96 \times 0.12} = 1.26$. Therefore, 95 per cent ratio limits of agreement are expressed as 0.96 × /÷1.26. Therefore, the first rater tends to value players at 96 per cent of the value reached by the second rater. There is a 95 per cent chance that the rating given by the first rater is within 26 per cent of this expected value (96 per cent of that given by the second rater). For example, if we consider a player who is rated at £6,000,000 by the second rater, we are 95 per cent confident that the first rater would value the player between £6,000,000 × 0.96 / 1.26 = £4,571,000 and £6,000,000 × 0.96 × 1.26 = £7,258,000. Figure 19.5 shows the Bland–Altman Plot for this example illustrating a clear 'shotgun' effect resulting from heteroscedasticity in the data.

Change of the mean and standard error of measurement (typical error)

Hopkins (2000a) described the use of change of the mean and typical error for test–retest reliability studies. Consider the Y-Balance test data shown in Table 19.5. The mean of the test and retest scores are 98.72 per cent and 101.73 per cent respectively. Therefore, the change in the mean is 3.01 per cent. The change in the mean described by Hopkins (2000a) is always the same as the systematic bias component of 95 per cent limits of agreement. Hopkins (2000a) expressed a preference for 'typical error' as a measure of random error, describing 95 per cent limits of agreement as too stringent. Typical error is the standard

358

reliability

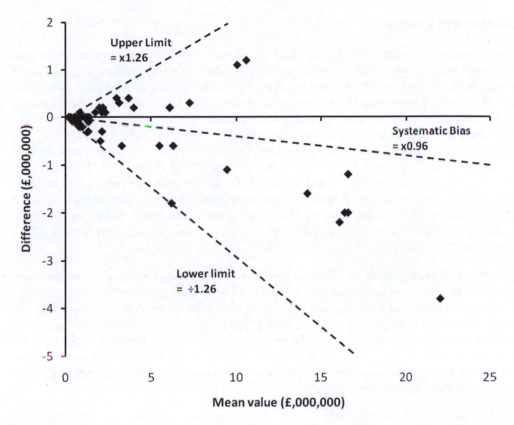

Figure 19.5 Bland–Altman Plot for soccer player valuations.

deviation of the test–retest differences divided by the square root of two. In Table 19.5, we can see that the standard deviation of the test–retest differences is 3.63 per cent. Therefore, the typical error is 3.63% / $\sqrt{2}$ = 2.57%. Dividing the standard deviation by $\sqrt{2}$ (the same as multiplying it by 0.71) means that typical error covers 52 per cent of test–retest differences if the differences are normally distributed.

The relative merits of 95 per cent limits of agreement and 'typical error' have been debated by Atkinson and Nevill (2000) and Hopkins (2000a, 2000b). Both statistics are useful but it is essential that those reporting them, or indeed reading reliability studies using them, understand what they mean. The 95 per cent limits of agreement represent a range of errors such that only 5 per cent of errors are outside this range. Typical error, however, represents 52 per cent of errors. These interpretations of limits of agreement and typical error assume test–retest differences are normally distributed. If researchers doing a test–retest reliability study wished to express a range that covered 50 per cent of errors, they could simply multiply the standard deviation of the test–retest differences by a z score of 0.67. Hopkins' (2000a) point about the use of the t distribution for small samples is valid and Atkinson and Nevill (2000) recommended the use of 50+ participants if using 95 per cent limits of agreement.

Coefficient of variation

The coefficient of variation is the standard deviation of a sample of numbers expressed as a percentage of the mean of those numbers. Within reliability studies, the coefficient of variation, CV, can be used where the reliability study consists of multiple trials. Variation in test values is represented by the standard deviation of the values. In a performance analysis investigation, a group of six independent observers may analyse a single performance leading to one value of CV for each numerical performance indicator of interest. The fact that CV is a dimensionless percentage does allow the reliability of different performance indicators to be compared. However, O'Donoghue's (2010) criticism of Hughes *et al.*'s (2004) percentage error statistic is also relevant to CV as division by the mean may be using an unrealistic range of values. Where multiple participants perform three or more repeated trials within a reliability study, there will be a CV for each participant. The researcher must decide how to summarize this data with options including the presentation of mean CV or maximum CV.

CV will be illustrated using an example based on computerized scoring in amateur boxing. The computerized scoring system consists of five button pressing devices connected to a personal computer that executes the scoring system software. A bout is scored by five judges, each operating a button pressing device. The judges watch the bout entering scoring punches they see the boxers making. There are two buttons used to score punches: a red button used to enter a score for the boxer wearing red kit and a blue button used to enter a score for the boxer wearing blue kit. Within the rules of amateur boxing, a scoring punch must satisfy the following three criteria:

- The punch must be accurately placed on the opponent's target area. The target area consists of the front and both sides of the head and also the front and sides of the body above the waist.
- The punch must be made with a clenched fist striking with the white frontal knuckle padding of the glove.
- The punch must be clean, fair, unguarded and have the proper weight of the boxer's body or shoulder behind the blow.

Where three or more of the five judges press a button to register a scoring punch for the same boxer within 1s, the computerized scoring system will record a score. All scoring punches are scored equally regardless of the quality of the blow or whether three, four or five judges recognized it. The judges are strategically placed around the ring so as to provide as much coverage of the bout as possible. Where a contest goes the distance, the winner is the boxer who has the most points recorded by the computerized scoring system. In this example, the button pressing activity of five individual judges is analysed irrespective of whether the button pressed resulted in a point for a boxer (if there are at least two other judges pressing a button of the same colour) or not. Table 19.7 shows the button pressing activity for five trained judges who observed six bouts. Table 19.7 does not actually reveal the outcome of the bout which depends not only on the number of times the red and blue buttons are pressed, but also how many times three or more judges press a button of the same colour within 1s. The outcomes in the rightmost five columns of Table 19.7 are the outcomes according to individual judges rather than the single outcome produced by the

360

Table 19.7 Button presses by 5 trained judges (A to E) observing six amateur boxing matches

Bout	Red Button Presses					Blue Button					Outcome (R = Red win, B = Blue win)				
	A	B	C	D	E	A	B	C	D	E	A	B	C	D	E
1	82	78	74	81	62	71	66	67	69	65	R	R	R	R	B
2	43	39	54	54	57	63	57	57	64	62	B	B	B	B	B
3	36	31	29	37	34	50	46	46	54	54	B	B	B	B	B
4	49	56	51	54	52	50	40	50	48	56	B	R	R	R	B
5	41	39	50	38	46	58	59	53	53	43	B	B	B	B	R
6	83	71	72	69	74	54	51	49	49	41	R	R	R	R	R

computerized system. The question of interest in the current example is whether the process of a judge observing and scoring a contest is reliable.

The Microsoft Excel spreadsheet 19-boxing_judging.XLS contains a worksheet 'CV' which analyses the data in Table 19.7. Table 19.8 shows the coefficients of variation. There is a range of coefficients of variation in this example meaning there is a choice of ways in which to present the results of the reliability study. We could report that coefficients of variation had a mean of 9.3 per cent with individual CVs ranging from 3.6 per cent to 16.0 per cent over the 12 boxing performances (there are two performances within each bout – the red boxer and the blue boxer). Alternatively we might report the same information for the scoring of the red boxer (mean = 10.2%, min = 5.2%, max = 16.0%) and the blue boxer separately (mean = 8.5%, min = 3.6%, max = 11.9%). We can see in Table 19.8 that the CVs of over 10 per cent occur in some of the performances for which the lowest numbers of button presses were made. Uniform coefficients of variation would require the standard deviations in button pressing activity to be proportional to mean button pressing frequency. Note that in this example the judge is a condition and the bouts (or more accurately the performances within the bouts) are the subjects of the analysis.

Table 19.8 Coefficients of variation for red and blue button pressing in the six bouts

Bout	Red Button Presses			Blue Button		
	Mean	SD	CV	Mean	SD	CV
1	75.4	8.1	10.8%	67.6	2.4	3.6%
2	49.4	7.9	16.0%	60.6	3.4	5.5%
3	33.4	3.4	10.1%	50.0	4.0	8.0%
4	52.4	2.7	5.2%	48.8	5.8	11.8%
5	42.8	5.1	11.8%	53.2	6.3	11.9%
6	73.8	5.4	7.4%	48.8	4.8	9.9%

Intraclass correlation coefficient

The intraclass correlation coefficient, ICC, like the coefficient of variation, can be used when reliability studies consist of repeated trials performed by a sample of participants. ICC has been used to assess inter-observer agreement during observation of sports performance (Richard et al., 2000) and between trials reliability of a 5m multiple shuttle run test (Boddington et al., 2001). There are at least six different methods of calculating ICC which produce different values (Atkinson and Nevill, 1998). In this section, the technique described by Vincent (1999: 182–5) is used. A repeated measures ANOVA test can be applied to the reliability data the same way as it can be applied to any other repeated measures data as described in Chapter 11. Information provided by the repeated measures ANOVA test is used in the calculation of ICC shown in equation 19.8. In equation 19.8, MS_R is the mean square for the subjects of the investigation which in this case are the six bouts. The 'Between-subjects effects table' for the red boxer shows as the mean square for the error term which is 1421.8.

$$ICC = (MS_R - MS_{C+E}) / MS_R \qquad (19.8)$$

In equation 19.9, SS_C is sum of squares 44.5 for our condition or repeated measure and df_C is the four degrees of freedom for judge. The SS_E term is the error term for the repeated measure 763.9 while df_E is the 20 degrees of freedom for the error term. All four of these values can be seen in the within-subjects effects table shown in Table 19.9. MS_{C+E} is the mean square for the judge condition (repeated measure) and the error effect combined.

$$MS_{C+E} = (SS_C + SS_E) / (df_C + df_E) \qquad (19.9)$$

The SPSS datasheet 19-boxing_judging.SAV contains the boxing scoring data shown in Table 19.7. To obtain the information we need to determine ICC, we perform a repeated measures ANOVA test on the data. The ANOVA includes judge as a repeated measure of five levels (judges A to E). The ANOVA is done separately for the scoring of the red and the blue boxer. Tables 19.9 to 19.12 show the output of the two repeated measures ANOVA tests.

Table 19.9 SPSS output for judge effect on red boxer's scoring

Tests of Within-Subjects Effects – Scoring of the Red boxer

Measure:MEASURE_1

Source		Type III Sum of Squares	df	Mean Square	F	Sig.
judge	Sphericity Assumed	44.467	4	11.117	.291	.880
	Greenhouse-Geisser	44.467	2.301	19.323	.291	.781
	Huynh-Feldt	44.467	4.000	11.117	.291	.880
	Lower-bound	44.467	1.000	44.467	.291	.613
Error(judge)	Sphericity Assumed	763.933	20	38.197		
	Greenhouse-Geisser	763.933	11.506	66.395		
	Huynh-Feldt	763.933	20.000	38.197		
	Lower-bound	763.933	5.000	152.787		

Table 19.10 SPSS output for bout effect on red boxer's scoring

Tests of Between-Subjects Effects

Measure: MEASURE_1
Transformed Variable:Average

Source	Type III Sum of Squares	Df	Mean Square	F	Sig.
Intercept	89216.533	1	89216.533	62.748	.001
Error	7109.067	5	1421.813		

Table 19.11 SPSS output for judge effect on blue boxer's scoring

Tests of Within-Subjects Effects Scoring of the Blue boxer

Measure: MEASURE_1

Source		Type III Sum of Squares	df	Mean Square	F	Sig.
judge	Sphericity Assumed	94.333	4	23.583	1.111	.379
	Greenhouse-Geisser	94.333	1.246	75.731	1.111	.350
	Huynh-Feldt	94.333	1.458	64.701	1.111	.356
	Lower-bound	94.333	1.000	94.333	1.111	.340
Error(judge)	Sphericity Assumed	424.467	20	21.223		
	Greenhouse-Geisser	424.467	6.228	68.152		
	Huynh-Feldt	424.467	7.290	58.226		
	Lower-bound	424.467	5.000	84.893		

Table 19.12 SPSS output for bout effect on blue boxer's scoring

Tests of Between-Subjects Effects

Measure: MEASURE_1
Transformed Variable:Average

Source	Type III Sum of Squares	df	Mean Square	F	Sig.
Intercept	90200.833	1	90200.833	305.690	.000
Error	1475.367	5	295.073		

Mauchly's test revealed that sphericity can be assumed for the scoring of the red boxer (p > .05) but not for the scoring of the blue boxer (p = .002). In both cases we use the sphericity assumed results. The reason for this is that there is no significant judge effect on button pressing frequency. Violation of sphericity increases the chance of making a Type I error, but in the absence of a significant difference the only error that could have been made in such an analysis was a Type II error.

For the scoring of the red boxer:

$$MS_{C+E} = (44.5 + 763.9) / (4 + 20) = 33.7$$

$$ICC = (1421.8 - 33.7) / 1421.8 = .976$$

For the scoring of the blue boxer:

$$MS_{C+E} = (94.3 + 424.5) / (4 + 20) = 21.6$$

$$ICC = (295.1 - 21.6)/ 295.1 = .927$$

ICC values of 0.9 or greater represent a high level of agreement, values between 0.8 and 0.89 represent moderate agreement and values of less than 0.8 represent questionable agreement for physiological data (Vincent, 1999: 184). For behavioural sciences, values of 0.7 to 0.8 might be acceptable (Vincent, 1999: 184).

The ICC values of 0.976 and 0.927 do not mean that the boxing scores shown have an acceptable agreement for decision making in competition. Three of the six bouts were unanimous between the judges, the others would have been split decisions. The real question is whether the computerized scoring system provides a result that is consistent with actual scoring punches made. That study would make an interesting research project.

Cronbach's alpha

Cronbach's alpha, α, is a measure of internal consistency between different elements that contribute to some overall construct score. Those who use psychometric tests often refer to this internal consistency as reliability. The statistic is used widely in social sciences, business and sports sciences. Cronbach's α is not robust to missing values and this is something that needs to be considered during data collection. The example of netball squad selection (Joll, 2011) that was used to illustrate kappa and the weighted version of kappa is also used as an example of the calculation of Cronbach's α. The data for this example are found in the SPSS datasheet 19-Netballselection.SAV. There are different versions of Cronbach's α, the one used by SPSS is $\alpha_{standardized}$ which is given by equation 19.10 where K =13 is the number of items and $\bar{r}$ is the mean of the K(K–1)/2 correlation coefficients from the K items.

$$\alpha = \frac{K.\bar{r}}{1 + (K - 1).\bar{r}} \tag{19.10}$$

We determine Cronbach's α in SPSS using **Analyse → Scale → Reliability Analysis**. Figure 19.6 shows the pop-up window for reliability analysis. The 13 items are entered into the *Items* area and the model to be used is 'Alpha'. When we click **OK**, Cronbach's α is computed as shown in Table 19.13. The result is .926 which is well above the 0.7 typically required to conclude acceptable internal consistency.

Table 19.13 SPSS output for Cronbach's α

Reliability Statistics	
Cronbach's Alpha	N of Items
.926	13

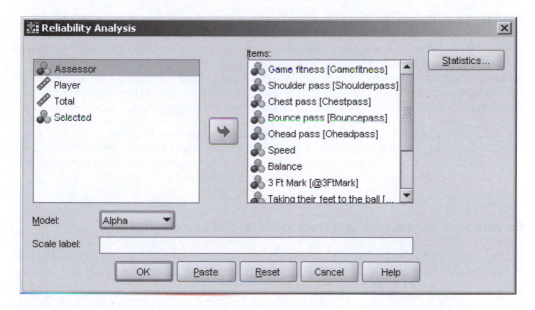

Figure 19.6 Reliability Analysis pop-up window in SPSS.

SUMMARY

There are four different types of reliability: inter-rater agreement, test–retest reliability, parallel forms reliability and internal consistency. Depending on the nature of the data, there are different statistics that can be used to assess each form of reliability. Kappa can be used with nominal variables and weighted kappa can be used with ordinal variables provided we have a chronological list of events for each observation or test being compared. If our system or method only provides the accumulated totals, then percentage error of frequencies could be used instead. Interval scale variables can be assessed using correlation techniques for relative reliability supported by measures of absolute reliability. There are two main types of absolute reliability statistic that provide measures of systematic bias and random error. These are 95 per cent limits of agreement as well as change of the mean and typical error. Where there is heteroscedasticity within reliability data, 95 per cent ratio limits of agreement should be used instead of 95 per cent limits of agreement. Multiple trial studies can use coefficient of variation or intra-class correlation coefficient (ICC) to assess reliability. Cronbach's α is used as a measure of internal consistency of questionnaire items and other data that are constructed from a set of related items.

EXERCISES

Exercise 19.1. Kappa for agreement in netball performance assessment

Use pivot tables and the steps described in the previous section to determine kappa and the weighted version of kappa for the following assessor pairs Q1 vs Q3, Q2 vs Q3, U1 vs U2, U1 vs U3 and U2 vs U3. The data are found in the 'Raw data' sheet of the ex19.1-netball_selection.XLS spreadsheet.

365

reliability

Exercise 19.2. Kappa for decision accuracy in netball performance assessment

There is a form of reliability referred to as 'decision accuracy' where a measurement process is considered sufficiently reliable if it results in correct decisions being made (Thomas and Nelson, 1996: 219). The worksheet 'Selection' of the ex19.2-netball_selection.XLS spreadsheet summarizes the data for each of the 30 players with columns for the total score awarded by each assessor and the selection recommendation made by each assessor (1 = Selected, 2 = Borderline, 3 = Not Selected).

a) Compare each pair of assessors applying kappa to the selection recommended.
b) Compare each pair of assessors applying the weighted version of kappa to the selection recommended. Use weights of 0.5 for disagreements where one assessor has classed a player as borderline and the other has recommended selection or non-selection.

Exercise 19.3. Reliability of split times in middle distance athletics

The file ex19.3-1,500m.XLS contains data for a fictitious observational study of 1,500m running. Two independent observers watched a race video 12 times, once for each athlete taking part and recorded the split times for each 100m point of the race for each athlete. The worksheet 'data' contains the 180 values (12 athletes × 15 sections of 100m) recorded by observers A and B. In this example we are not concerned with elapsed time as the time taken to run the first 100m and the time taken to run the first 700m are essentially different variables. We are concerned with 100m section time which is stored in columns E and F. Determine the following for the test–retest errors:

a) Relative reliability using Pearson's r.
b) Mean absolute error and 95th percentile for absolute error.
c) Root mean squared error.
d) Mean percentage error and the 95th percentile for percentage error according to Hughes *et al.* (2004).
e) Mean percentage error and the 95th percentile for percentage error according to O'Donoghue (2010) assuming a minimum possible value for 100m section of 10s and a maximum expected time of 20s for athletes at this level.
f) Mean absolute error for the fifteen 100m sections individually. What might this tell us about the reliability of the data.

Exercise 19.4. Y-Balance test performed with eyes open and using the non-dominant leg

a) Use the worksheet 'Non-Dominant Leg' in the ex19.4-Y-Balance-Test.XLS spreadsheet to determine if the data are homoscedastic or heteroscedastistic and then decide whether 95 per cent limits of agreement or 95 per cent ratio limits of agreement are appropriate, before determining the chosen reliability statistic.
b) Determine the change of the mean and typical error for the data.

Exercise 19.5. Scoring in amateur boxing

Determine ICC combining the red and blue scorings in the file ex19.5-boxing_judging.SAV into 12 rows of data.

reliability

Exercise 19.6. Internal consistency of the Behavioural Regulation instrument (BRSQ)

The file ex19.6-BRSQ.SAV contains data for 175 fictitious student athletes who have completed a behavioural regulation in sport questionnaire (Lonsdale *et al.*, 2008). The behavioural regulation in sport instrument (BRSQ) is composed of nine subscales that constitute the concept of self-determination theory
with four of the subscales being concerned with intrinsic motivation. The nine subscales are scored on
a 4 to 28 scale and are listed below:

- Intrinsic Motivation
- Intrinsic Motivation for Accomplishment
- Intrinsic Motivation to Gain Knowledge
- Intrinsic Motivation to Experience Stimulation
- Integration
- Identified Regulation
- Introjected Regulation
- External Regulation
- Amotivation.

Use Cronbach's α to assess the internal consistency of the BRSQ instrument.

Exercise 19.7. Internal consistency of a wellbeing construct

A total of 123 students complete a wellbeing questionnaire instrument (Corbin *et al.*, 2000: 17–18) that
produces scores from 3 to 12 for emotional wellbeing, intellectual wellbeing, physical wellbeing, social
wellbeing and physical wellbeing. These all add up to a total wellbeing score between 15 and 60. Use
Cronbach's α to assess the internal consistency of the instrument based on the five component scores
(emotional wellbeing, intellectual wellbeing, physical wellbeing, social wellbeing and physical wellbeing).
The data for this exercise are found in the file ex19.7-wellbeing.SAV.

CHAPTER 20

STATISTICAL POWER

INTRODUCTION

Statistical power analysis is an area of statistics that warrants a book in its own right. For a full coverage of statistical power analysis, the author recommends Murphy *et al.*'s (2009) textbook on the area. This chapter will briefly cover statistical power, Murphy *et al.*'s (2009) model and the four applications of power analysis.

WHAT IS STATISTICAL POWER?

Statistical power is the conditional probability of finding a significant difference or relationship in a sample when such a difference or relationship exists in the population from which the sample is drawn. A power of 0.2 indicates that a study is four times as likely to accept the null hypothesis than to reject it when the null is actually false in reality. A power value of 0.5 gives a study a 50:50 chance of rejecting the null hypothesis when it is actually true. Therefore, the desired power for a research study is typically 0.8 or higher.

This is an important area of statistics that has been neglected in much sport and exercise research as investigations are primarily concerned with the chance that a Type I Error has been made. The p value that is reported for the inferential statistical tests done in much published research is an estimate of the probability that a Type I Error has been made when statistical significance has been concluded. Where α has been set of 0.05, p values of under 0.05 indicate significant differences or relationships. Even when a significant difference is not found, it is this p value that is reported. For example, a p value of 0.25 being determined when α is 0.05 is not a significant result, but p is still the probability that we would be making a mistake if we claimed any difference or relationship observed in our sample existed in the population from which the sample was drawn. Statistical power is concerned with Type II Errors and situations where there may be a difference or relationship in the population of interest which has not been concluded by the study performed on a sample. There may be a real effect, but the study has not provided sufficient evidence to say that there is any effect.

As we have already seen in Chapter 7, the power of a statistical test is the probability that the test will reject the null hypothesis when the null hypothesis really is false. That sounds simple enough, but after a short time thinking about statistical power, many questions will

be raised. If we consider the example that we used for 2,000m rowing in Chapter 7, there are a number of limitations to what was done. The main limitation was that we used one idealized population where the null hypothesis was false and one where the null hypothesis was true. The latter was set at the hypothesized value of seven minutes with which our one tailed test would be comparing rowing performances. However, this is not the only potential population where the null hypothesis would be true. Similarly, there are a range of populations where the mean 2,000m rowing time would be faster than 2,000m. Indeed, these times could go right up to but not equalling our hypothesized value of seven minutes. In a quite critical paper on power analysis, O'Keefe (2007) stated that it was misleading to refer to 'the' power of a statistical test for the simple reason that there are a full range of statistical powers that are possible for the range of situations possible in the population of interest. As well as using assumed population means, the rowing example in Chapter 7 used assumed population standard deviations which is an added limitation of power analysis.

There is a question about two tailed tests for dependent variables measured on a continuous numerical scale. Can the null hypothesis ever be true? If height is measured to the nearest nanometre, the probability of two group means within a population being exactly equal is negligible. Murphy et al. (2009: 4–5) referred to the null hypotheses used in many research studies as nil hypotheses. A nil hypothesis is a commonly used null hypothesis where we are testing whether the difference between groups or conditions is zero or not or whether the correlation between two variables is zero or not. Rejecting the null hypothesis may simply mean that the difference or relationship is not negligibly small. Therefore, Murphy et al. (2009: 4–5) discussed other types of null hypotheses where some meaningful difference between samples is being tested for or where some meaningful strength of relationship is being tested for. Nil hypotheses do have an important role in research especially where exploratory studies are being conducted on new concepts, constructs and populations where there may be no previous research evidence supporting some minimum effect.

On first being introduced to power analysis many years ago, the author felt that it could be misused to undertake biased research. The author works in performance analysis, which is still a young discipline within sport, and exercise science, which itself is a young discipline. Research studies done by the author are typically to determine if differences between groups or conditions are significant or not. The answer to the research question would be based purely on the results of statistical tests performed on the data which would be independent of the author's personal opinion. Coming from this culture, the author was concerned when researchers in other disciplines were using statistical power to determine what sample size they required to make sure they would achieve a significant difference. This could be seen as biased research to provide evidence supporting a particular hypothesis. The good thing is that effect size can show whether significant effects are meaningful or not.

The power of a statistical test is a function of three other factors: the sensitivity of the data, the population effect size and decision criterion set (Murphy et al., 2009: 7–8). Sensitivity is concerned with how well studies have been designed to control unwanted sources of variability. Variability in the data can come from measurement error and sampling error. The higher the unwanted variability is, the lower the power of the test. Variance is accounted for within the sensitivity of a method, but has such an impact on power that Atkinson and Nevill (2001) include it as an additional factor influencing statistical power. Earlier, Atkinson *et al*. (1999) had produced a nomogram for statistical power of 0.9 to show the impact of variability due to measurement error on studies with different expected population effects when different sample sizes were used. The effect size is the percentage of variance in the dependent variable for the population that can be explained by the groupings or conditions being tested. With correlations, the effect is the percentage of the variability in the data that can be explained by the relation between two variables. Larger effect sizes lead to increased statistical power. The decision criterion is the α level used to reject the null hypothesis; this is typically 0.10, 0.05 or 0.01 in sport and exercise research. Statistical tests are more likely to find significant results when using higher α values such as 0.10 than lower α values such as 0.01.

Power analysis can be done *a priori* or *post hoc*. The planning of a research project may include *a priori* power analysis where population parameters can be estimated, sample sizes can be decided and an α level can be justified (Hinton, 2004: 99). Sometimes the purpose of power analysis is not to determine power prior to the study but to decide on a required statistical power and then determine the sample sized needed for the required statistical power, the α level determined and the anticipated population parameters. Low statistical power may be due to too stringent decision criteria being used (Murphy *et al*., 2009: 18–19). Researchers should consider why they are using the α levels they are using in research studies.

Power analysis is often an important part of research proposals when researchers are applying for funding or ethical clearance to undertake clinical experiments. Funding bodies will want to know the likelihood of the experiment leading to actionable results. Ethics committees may take the view that invasive research studies with little chance of finding a significant result may waste participants' time. Indeed, Murphy *et al*. (2009: 17–18) expressed 'shock' at how low the power of many published studies in the behavioural and social sciences have been.

Post hoc power analysis is done after a study has been carried out to explain why the study did not produce significant results. Given that power is a function of a population effect, a chosen α level and a given sample size, there is no difference between the power of a test when calculated before or after the study has been done (O'Keefe, 2007). Post hoc power analysis simply estimates power after the study rather than before it. The term 'observed power' may give a different value to power than that determined prior to a study. This is because observed power assumes a population effect that is assumed to be the same as that observed in the sample analysed. O'Keefe (2007) referred to three different published research papers on statistical power that stated that observed power was uninformative where a study has not produced significant results. The reasons given for this were that non-significant results correspond to low observed powers and that the power would not be sufficient for detecting a population effect equivalent to the non-significant sample effect. While there are criticisms of some uses of observed power, there are certainly useful forms of *post hoc* power analysis that can be done where effects shown in previous research are considered. This type of power analysis may still be different to that which could be performed before a study if measures of dispersion from the study are used that could not have been known prior to the study. Another way in which post hoc power can be used is to apply the power considered in the study to a range of theoretical population effects. This can provide useful information where it can be demonstrated that the study would have been likely to detect such effects if they actually existed.

MURPHY *ET AL.*'S (2009) MODEL OF STATISTICAL POWER

The main purpose of Murphy et al.'s (2009) book on statistical power analysis was to propose a general model of power analysis that could be used with traditional nil hypothesis tests as well as hypothesis testing where some minimum effect is being tested. The model is based on the F-statistic due to its familiarity and its suitability for nil hypothesis testing and other null hypothesis testing (Murphy et al., 2009: 26). The F-statistic is used within ANOVA tests and the General Linear Model (GLM). The test statistics used in other statistical tests can be mapped onto the F-statistic. These include the t-statistic, chi square, Pearson's r, R^2 and Cohen's d (Murphy et al., 2009: 34). Other models of statistical power have also been based on single test statistics (Kraemer and Thiemann, 1987; Lipsey, 1990) while Cohen (1988) produced tables to be used in power analysis for several statistical tests.

When a traditional nil hypothesis is being tested, the model uses the traditional F-distribution. When other null hypotheses are being tested, the model uses a non-central F-distribution. The population effect size is represented by the percentage of the variance (PV) of the dependent variable that is based on the effects in the general linear model on which the test is based (Murphy et al., 2009: 27). A PV value of 0.01 represents a small effect, 0.10 represents a medium effect and 0.25 represents a large effect (Murphy et al., 2009: 38–9). Where the non-central F-distribution is used, a non-centrality parameter λ_{est} is used in the analysis (see equation 20.1) where df_{err} is the error degrees of freedom given by n - k where n is the number of participants and k is the number of terms in the general linear model.

$$\lambda_{est} = df_{err}(PV / (1-PV)) \tag{20.1}$$

Given α, the F value needed to reject the null hypothesis and PV, statistical power, $1 - \beta$, can be determined (Murphy et al., 2009: 31). Murphy et al. (2009) have developed a 'one stop' F-calculator and in their appendices provide tables for F and PV. Interpolation may need to be done where these tables do not cover the specific degrees of freedom for a test being done by a user. However, the author found the use of the FDIST and FINV functions in Microsoft Excel satisfactory for determining areas of the F-distribution and critical F-values respectively. Appendix A of Murphy et al.'s (2009) book illustrates an analytical approach to using their power analysis model. Equation 20.2 shows the relation between the F-test and the effect of the factors in the general linear model. The error degrees of freedom ensure sample size is accounted for and α is used with the F-distribution.

$$PV = df_{hyp}.F / (df_{hyp}.F + df_{err}) \tag{20.2}$$

Appendix A of Murphy et al.'s (2009: 163) book shows how to determine the degrees of freedom, g, to use so that the non-central F-distribution maps onto the traditional F-distribution. This is shown in equation 20.3.

$$g = (df_{hyp} + \lambda)^2 / (df_{hyp} + 2\lambda) \tag{20.3}$$

The λ value in equation 20.3 is given by the ratio of sum of squares for the treatment to the mean squares for the error effect from the ANOVA table used with the F-test as shown in equation 20.4 (Murphy et al., 2009: 30).

$$\lambda = SS_{treatment} / MS_{err} = df_{err}.PV.(1-PV) \qquad (20.4)$$

The critical F value, $F_{critical}$, is determined from the F distribution for our chosen α level with df_{hyp} main effect degrees of freedom and df_{err} error effect degrees of freedom. F* is the F value used to determine the power of the test and is shown in equation 20.5.

$$F* = F_{critical} / ((df_{hyp} + \lambda) / df_{hyp}) \qquad (20.5)$$

The percentage of the F distribution that is greater than F* with g main effect degrees of freedom and df_{err} error degrees of freedom gives us a value for $1-\beta$.

The author used this analytical approach for the ANOVA table shown in Table 11.8 in Chapter 11 where SPSS had reported an 'observed power' of 0.729. The non-centrality parameter, λ, reported in this table was 10.358 which matched perfectly the value computed from equation 20.4 using the $SS_{treatment}$ and MS_{err} values in the ANOVA table. Furthermore, the partial η^2 value of 0.214 reported in the ANOVA table agreed with the PV value calculated using equation 20.2. After this point there was some divergence from what was reported by SPSS and what the author calculated using Murphy et al.'s (2009) model. It is not certain where the disagreement occurred due to the SPSS output not showing the steps between the calculation of λ and observed power. The author calculated the value of g to be 7.52 using equation 20.3 which was rounded down to 7 for the purposes of determining F*. $F_{critical}$ was 2.852 with an α level of 0.05, three main effect degrees of freedom and 38 error effect degrees of freedom. Therefore F* was 0.640 based on equation 20.5 and the area of the F distribution above F* was 0.720 when seven main effect degrees of freedom and 38 error effect degrees of freedom were used. The disagreement between this power value of 0.720 and the value of 0.729 reported by SPSS could be due to definitional differences or table look up between the analytical approach described by Murphy et al. (2009) and the way in which observed power is calculated by SPSS.

FOUR APPLICATIONS OF POWER ANALYSIS

Determining power levels

If we have a population effect size, an α level and a sample size then we can determine power. In Chapter 7, an a priori power analysis was conducted for 2,000m rowing time assuming a population mean and standard deviation of 6 minutes 52s and 25s respectively. The example in Chapter 7 was limited because it used a single sample size (n=30), the standard deviation of 25s was not supported by evidence and one particular assumed population where the null hypothesis would be false.

In this chapter, we show how the power of the one-tailed one-sample t-test can be used with this indoor rowing example a priori; that is, before data are gathered during the planning stage of the study. Our null hypothesis is that mean 2,000m rowing time will be seven minutes or slower. Statistical power is the conditional probability that we achieve a significant result when the null hypothesis is false. Therefore, we are interested in situations where the mean 2,000m rowing time is actually less than 420s. It is more beneficial to examine a range of potential 2,000m rowing times below 420s for the population rather than using a single fictitious situation as we did in Chapter 7. We can decide on an α level (say 0.05), look at a range of sample sizes and consider a range of population means where the null hypothesis is false. This leaves us with one problem, which is the effect of any differences. The difference between the hypothesized 420s and the range of population means we are considering will not give us effect size without some indication of standard deviation. Where power analysis is done prior to clinical studies, there may have been some exploratory study used to give an indication of standard deviation. We will use the study in Chapters 7 where the standard deviation was 23.2s to provide an estimate of the standard deviation in the population. Some may prefer to multiply this by $\sqrt{(n-1)/n}$ to correct for the difference in equations for the standard deviation of a population and the standard deviation for a sample which is what the value of 23.2s is. However, using this larger value for standard deviation is a more cautious approach to estimating power. Power curves can be used to plot power against a range of population means (Anderson et al., 1994: 320–1; Newell et al., 2010: 395–407).

Figure 20.1 shows power curves for this example covering a range of population means from 400s to 420s, sample sizes of 5, 10, 15, 20, 25 and 30 using an α value of 0.05 and a standard deviation of 23.2s. A series of these curves could be produced for different α levels and different standard deviations. These could then be used to determine the power of alternative designs and assist in decisions where there is a trade off between the cost of the study and the statistical power of the study.

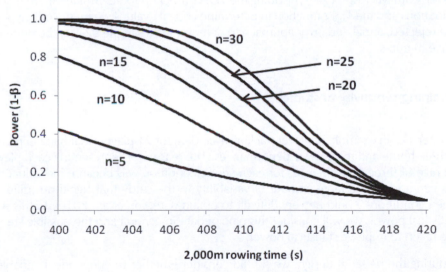

Figure 20.1 Power curves for a one-tailed one-sample t-test comparing mean rowing time with a hypothesized upper limit of seven minutes (420s) The assumed standard deviation is 23.2 and α = 0.05.

The author produced these power curves in Microsoft Excel using the TDIST function to determine the relevant area of the sampling distribution for each population mean that was above the critical value for t where the null hypothesis would be rejected. The critical values were different for each power curve due to the differing sample sizes and hence differing degrees of freedom involved. The power values do not decrease below 0.05 when the population mean is 420s because the α value set gives a 0.05 chance of a Type I Error meaning that a significant result could be produced even when the null hypothesis is true.

Determining sample size

The power curves shown in Figure 20.1 can also be used to estimate the sample size required for a study. For example, if we wanted have a conditional probability of concluding a significant difference of 0.8 if the population mean $\pm$ SD is 410 ± 23.2s and the α level set is 0.05, we can see from the power curves in Figure 20.1 that we need more than 30 participants. Effect size curves plot sample size against power for given values of α and effect size (Thomas and Nelson, 1996: 110–11). Effect size curves have an advantage of being applicable to many different studies using different number ranges as long as effect sizes can be predicted. The effect size charts shown by Thomas and Nelson (1996: 110–11) each contained curves for different effect sizes showing that larger sample sizes were required to achieve a given statistical power where effect sizes were low.

Analytical approaches to determining sample size involve taking equations for t scores, expanding the SEM in the equation for t and changing the subject of the equation to n (Vincent, 1999: 141–3). Changing the subject of the equation to n is problematic because n is used in two places; the calculation of the SEM and the number of degrees of freedom. However, with Microsoft Excel providing the TDIST function to determine different areas of t distributions and the TINV function to determine critical t values where the null hypothesis can be rejected, a trial and error approach to determining a sample size can be done in less than five minutes.

Determining sensitivity of studies

In Chapter 12, an example of player tracking used data for 21 players that had been derived from three home and three away performances. The reason for representing each player by an average of three performances for each venue condition was because individual match effects can add such a large amount of variability to the study that significant differences between conditions would be very difficult to obtain (Gregson et al., 2010). In this section on statistical power, we will consider this type of data in general and the possible steps that can be taken to improve statistical power.

Early during the research design, we should consider whether the dependent variable (the percentage of time spent moving at 4m.s⁻¹ or faster) is appropriate. This variable can be influenced by abnormally high or low amounts of stoppage time during matches. In Chapter 12, we were concerned with 15 minute periods during the match. The percentage of time spent moving at 4m.s⁻¹ or faster during such short periods would certainly be influenced by any prolonged stoppages during those periods. We should consider using an alternative dependent variable such as the percentage of ball-in-play time spent moving at $4m.s^{-1}$ or faster. We should also seriously consider whether it is necessary to compare 15 minute periods. If we are trying to show changing movement patterns over the course of matches, comparing the first and second halves might be sufficient for this purpose. These steps of choosing appropriate dependent and independent variables are an attempt to reduce unwanted variability in the data. Lower variability increases the effect size and the chance of finding a significant difference.

There are further steps that can be taken to reduce unwanted variability by improving reliability and using typical match performances. The commercial player tracking systems used by professional soccer clubs are not fully automated and require manual verification by quality control personnel who have been reported to change between 38 per cent and 97 per cent of locations recorded by the player tracking algorithms (Di Salvo et al., 2009). A reliability study has shown some inter- and intra-operator variance in the data entered by quality control personnel. Although 100 per cent agreement is impossible to achieve in such a system, end-user training can improve intra- and inter-operator agreement. Such training should be systematic and supported by user performance metrics (O'Donoghue et al., 1996). For example, a set of player performances could be used for which their movement data are recorded by experienced and expert users of the system. The data recorded by trainee operators could be compared with those of experienced operators showing percentage agreement levels. The levels of agreement that one would see between trainee, experienced and expert users of the system should be understood to allow interpretation of trainee performance and feedback to the operators during training. Once trainees have demonstrated the required level of agreement with experts, they can commence work collecting data to be used by soccer clubs as well as in research projects. A reduction in variability due to measurement error will increase the effect size of research done with the data. However, we must also consider the costs involved in end-user training. The example of end-user training described here requires data sets to be created by expert users for matches to be used to train operators. There is a time and financial cost to this exercise before the end-user training can actually commence.

Even if player movement could be tracked with 100 per cent accuracy, player movement during a soccer match is not a stable characteristic of a player such as height or body mass would be. One study of work-rate revealed that David Beckham exhibited a greater variability in work-rate over 11 matches than the variability between 115 different midfield players in the same study (O'Donoghue, 2004). Gregson et al. (2010) found coefficients of variation averaging 17.7 per cent and 30.8 per cent for total high-speed running distance and total sprint distance respectively in English FA Premier League soccer players. The unstable nature of sports performance, particularly where there are direct opposition effects, adds a great deal of variance that can cripple any attempt to find significant within-subjects effects. Therefore, each player to be included in a study should be represented by a series of performances to allow an average (or typical) performance to be determined for that player. This will reduce variability due to individual match effects. Other potential ideas for reducing variability are to use similar sets of matches to gather data for each player. If one player is represented by 10 matches played on pitches that are wider than those in the 10 matches used to represent another player, this could add unwanted variability to the study. Even factors such as time of the season may effect performance and so all players should be represented by sets of matches spanning a similar spread of times of the season.

Determining criteria for statistical significance

Given that statistical power can be determined if α, sample size and population effect sizes are known, it is also possible to determine α if we have a target for statistical power, an upper limit for the number of participants and a good estimate of effect size. However, such an approach does look like a case of setting the goal posts based on what can be achieved with the resources available to do the study. Criteria for significance need to take into account many important factors that go beyond statistical power calculation. For a given sample size and population effect size, as α is decreased, β is increased leading to a reduction in power. Similarly, for a given sample size and population effect size, as α is increased, β is decreased leading to an increase in power. It is necessary to consider the financial and human consequences of decisions that may result from unknowingly making Type I and Type II Errors (Fallowfield et al., 2005: 47–8). Murphy et al. (2009: 85–6) defined desired relative seriousness (DRS) as the ratio of the seriousness of a Type I Error to the seriousness of a Type II Error. If a Type I Error is considered to be three times as serious as a Type II Error, then DRS = 3.0. If we have a target power of 0.8, then β will be 0.2. If p is an estimate of the probability of the alternative hypothesis being true, then the desired α level is given by equation 19.6 according to Murphy et al. (2009: 85). This approach could lead to α levels that are not typically used in research. The researcher needs to decide whether $\alpha_{desired}$ is to be used literally or to assist in the selection of a conventional a level such as 0.10, 0.05 or 0.01.

$$\alpha_{desired} = \frac{p.\beta}{(1-p).DRS} \tag{20.6}$$

Often it is not possible to isolate the decision about the α level to be used from the sample size used in the study. The DRS is more important in some types of studies than others, especially where their may be some risks to participants that need to be managed. Where risks are not so high, there will be a trade off between the α level to be set and the cost of the study which depends on the number of participants involved. A further issue with equation 20.6 is that it can calculate a desired α level of greater than one which is impossible. In such cases, researchers need to carefully consider their beliefs about the seriousness of Type I and Type II Errors and the evidence supporting the belief in the alternative hypothesis.

SUMMARY

The statistical power of a study is the probability that the study will reject the null hypothesis and find a significant effect when the null hypothesis is indeed false for the wider population. Statistical power is, therefore, the conditional probability of finding a significant difference where there is such a difference in reality, 1 - β, where β is the probability of making a Type II Error. Statistical power can be determined where an α level has been set, the sample size for the study is known and the effect size within the population is known. Models of statistical power such as that of Murphy *et al.* (2009) have been developed and observed power can be determined by statistical package such as SPSS. Any of statistical power, α, sample size and population effect size can be determined if the other three elements are known. Usually population effects are unknown but there may be theoretical or previous research evidence to allow population effect sizes to be estimated. Alternatively, power curves can be used to show the statistical power of a study for a range of potential population effects. Power analysis can be done prior to a study assisting design decisions and the justification of resources required for the study. Power analysis can also be conducted after a study has been completed if new information from the study can be used in the discussion of a range of potential population effects.

EXERCISES

Exercise 20.1. Desired relative seriousness

In a planned experiment, there is an estimated probability of 0.75 for the alternative hypothesis. The desired relative seriousness is four; that is, a Type I Error is four times as serious as a Type II Error. We wish to have a statistical power of 0.8 for the study. What α level should we use?

Exercise 20.2. Determining power for a training study

Use the 'General z-test – 1-tailed' sheet of the ex20.2-power.XLS spreadsheet to undertake this exercise. Previous research suggests that a 10-week training programme will improve the estimated $\dot{V}O_2$ max of games players by 0.5 mL.kg^{-1}.min^{-1}while control subjects undertaking alternative skill based training will maintain their estimated $\dot{V}O_2$ max. A study is planned to test the training programme with an experimental group of 30 participants and a control group of 30 participants. It is assumed that the standard deviation of each group will be 1.5 mL.kg^{-1}.min^{-1}. Let α be 0.1.

a) What will the statistical power of the study be?
b) If the power is less than 0.8, how many participants are needed for it to increase to 0.8.

Exercise 20.3. Verifying observed power

Consider Table 11.13 in Chapter 11 that shows the observed power (retrospectively determined) of 0.979 for the effect of quarter on the percentage of time spent performing high-intensity activity in netball. Using the analytical model of Murphy *et al.* (2009) described in this chapter, check the value for observed power. If there is a difference, explain why this difference has occurred.

Exercise 20.4. Power curves

The '2-tailed t-test' sheet of the ex20.2-power.XLS spreadsheet is used in this example. Consider a study to compare the vertical jump performance of male club level basketball players with the vertical jump performance of a control group of male club level games players from a selection of different sports. The club level games players are expected to achieve a mean vertical jump score of 45cm with a standard deviation of 5cm. We assume that the standard deviation of the basketball players will also be 5cm. The null hypothesis of the independent samples t-test to be done is that there is no difference between vertical jump performance of the control group and the basketball players. An α level of 0.05 is to be used. The curves on the chart in this spreadsheet are for different numbers of participants, assuming an equal split between the two groups. What sample size is needed to achieve a statistical power of 0.8 if the basketball players have a true population mean of (a) 47cm, (b) 48cm and (c) 49cm. It is possible to change the sample size for each curve in the cells G2:L2. The idealized control group mean and standard deviation should be entered into the cells B4 and C4 respectively.

PROJECT EXERCISE

Exercise 20.5. Research project planning

When preparing a future research project, take the opportunity to consider the null hypothesis, potential effects of independent variables and the standard deviation of dependent variables of interest. Consider the α level you wish to use and the desired statistical power. Determine the sample size needed to achieve the desired level of statistical power. Consider whether such a sample size is feasible given the nature of the study.

REFERENCES

ACSM (2005) *ACSM's Health-related physical fitness assessment manual,* Philadelphia, PA: Wolters Kluwer/Lippincott Williams and Wilkins.

ACSM (2010) *ACSM's Health-related physical fitness assessment manual*, 3rd edn, Philadelphia, PA: Wolters Kluwer/Lippincott Williams and Wilkins.

Alexandros, L. and Athanasios, M. (2011) 'The setting pass and performance indices in volleyball', *International Journal of Performance Analysis in Sport*, 11: 34–9.

Allison, P.D. (1999) *Multiple regression: a primer*, Thousand Oaks, CA: Pine Forge Press.

Altman, D.G. (1991) *Practical statistics for medical research*, London: Chapman & Hall.

Anderson, D.R., Sweeney, D.J. and Williams, T.A. (1994) *Introduction to statistics: concepts and applications*, 3rd edn, Minneapolis/St Paul: West Publishing Company.

Atkinson, G. and Nevill, A.M. (1998) 'Statistical methods for assessing measurement error (reliability) in variables relevant to sports medicine', *Sports Medicine*, 26: 217–38.

Atkinson, G. and Nevill, A.M. (2000) 'Typical error versus limits of agreement', *Sports Medicine*, 30: 375–7.

Atkinson, G. and Nevill, A.M. (2001) 'Selected issues in the design and analysis of sport performance research', *Journal of Sports Sciences*, 19: 811–27.

Atkinson, G., Nevill, A.M. and Edward, B. (1999) 'What is an acceptable amount of measurement error? The application of meaningful "analytical goals" to the reliability analysis of sports science measurements made on a ratio scale', *Journal of Sports Sciences*, 17: 18.

Baker, J., Ramsbottom, R. and Hazeldine, R. (1993) 'Maximal shuttle performance over 40m as a measure of anaerobic performance', *British Journal of Sports Medicine*, 27: 228–32.

Bland, J.M. and Altman, D.G. (1986) 'Statistical methods for assessing the agreement between two methods of clinical measurement', *Lancet*, I: 307–10.

Bloomfield, J., Polman, R.C.J., Butterly, R. and O'Donoghue P.G. (2005) 'An analysis of quality and body composition of four European soccer leagues', *Journal of Sport Medicine and Physical Fitness*, 45: 58–67.

Brown, L.E., Murray, D. and Hagerman, P. (2008) 'Test administration and interpretation', in T.J. Chandler and L.E. Brown (eds) *Conditioning for strength and human performance* (pp. 147–63), Philadelphia, PA: Lippincott, Williams and Wilkins.

Boddington, M.K., Lambert, M.I., Gibson, A. St. and Noakes, T.D. (2001) 'Reliability of a 5-m shuttle test', *Journal of Sports Sciences*, 19: 223–8.

Boyle, P.M., Mulligan, D. and O'Donoghue, P.G. (2002) 'Fitness profile of junior level county female Gaelic football players', *Journal of Sports Sciences*, 20: 32–3.

Carling, C., Reilly, T. and Williams, A.M. (2009) *Performance assessment for field sports*, London: Routledge.

Carron, A.V., Loughhead, T.M. and Bray, S.R. (2005) 'The home advantage in sport competitions: Courneya and Carron's (1992) conceptual framework a decade later', *Journal of Sports Sciences*, 23: 395–407.

Choi, H., O'Donoghue, P.G. and Hughes, M. (2008) 'The identification of an optimal set of performance indicators for real-time analysis using principle components analysis', in A. Hokelmann, K. Witte and P. O'Donoghue (eds) *Current trends in performance analysis – World Congress of Performance Analysis of Sport VIII; selected proceedings* (pp. 295–301), Aachen: Shaker.

Cohen, J. (1960) 'A coefficient of agreement for nominal scales', *Educational and Psychological Measurement*, 20: 37–46.

Cohen, J. (1968) 'Weighted kappa: Nominal scale agreement with provision for scaled agreement or partial credit', *Psychological Bulletin*, 70: 213–20.

Cohen, J. (1988) *Statistical power analysis for the behavioural sciences*, 2nd edn, Hillside, NJ: Lawrence Erlbaum Associates.

Cohen, L., Manion, L. and Morrison, K. (2007) *Research methods in education*, 6th edn, London: Routledge.

Conway, P. and O'Donoghue, P.G. (2001) 'Factors influencing the location of Europe's major soccer cities', poster presented at Exercise and Sports Science Association of Ireland Annual Conference, Carlow, Ireland, April.

Corbin, C.B., Welk, G.J., Corbin W.R. and Welk K.A. (2000) *Concepts of fitness and wellness: a comprehensive lifestyle approach*, 3rd edn, New York: McGraw-Hill.

Costa, G.C., Caetano, R.C.J., Ferreira, N.N., Junqueira, G., Alfonso, J., Costa, P. and Mesquita, I. (2011) 'Determinants of attacking tactics in youth male elite volleyball', *International Journal of Performance Analysis in Sport*, 11: 96–104.

Courneya, K.S. and Carron, A.V. (1992) 'The home advantage in sport competitions: A literature review', *Journal of Sport and Exercise Psychology*, 14: 28–39.

Cramer, H. (1999) *Mathematical methods of statistics*, Princeton, NJ: Princeton University Press.

Croucher, J.S. (1986) 'The conditional probability of winning games of tennis', *Research Quarterly in Exercise and Sport*, 57: 23–6.

Crowder, M.J., Kimber, A.C., Smith, R.L. and Sweeting, T.J. (2000) *Statistical analysis of reliability data*, Boca Raton, FL: Chapman and Hall/CRC.

Csataljay, G., O'Donoghue, P.G., Hughes, M. and Dancs, H. (2008) 'Principal components analysis of basketball performance indicators', in A. Hokelmann, K. Witte and P. O'Donoghue (eds) *Current trends in performance analysis – World Congress of Performance Analysis of Sport VIII; selected proceedings* (pp. 278–83), Aachen: Shaker.

Di Salvo, V., Collins, A., McNeill, B. and Cardinale, M. (2006) 'Validation of Prozone®: A new video-based performance analysis system', *International Journal of Performance Analysis of Sport*, 6(1): 108–19.

Di Salvo, V., Gregson, W., Atkinson, G., Tordoff, P. and Drust, B. (2009) 'Analysis of high intensity activity in Premier League soccer', *International Journal of Sports Medicine*, 30: 205–12.

Diamantopoulos, A. and Schlegelmilch, B.B. (1997) *Taking the fear out of data analysis*, London: The Dryden Press.

Dorado, C., Sanchis Moysi, J., Vicente, G., Serrano, J.A., Rodriguez, L.P. and Calbet, J.A.L. (2002) 'Bone mass, bone mineral density and muscle mass in professional golfers', *Journal of Sports Sciences*, 20: 591–7.

Durnin, J.G.V.R. and Wormersley, J. (1974) 'Body fat assessed from total body density and its estimation from skinfold thickness: measurements on 481 men and women aged from 16 to 72 years', *British Journal of Nutrition*, 32: 77–97.

Fallowfield, J.L., Hale, B.J. and Wilkinson, D.M. (2005) *Using statistics in sport and exercise science research*, Chichester: Lotus Publishing.

Freedman, P.S. and Miller, K. (2000) 'Objective monitoring of physical activity using motion sensors and heart rate', *Research Quarterly for Exercise and Sport*, 71: 21–9.

Gale, D. (1971) 'Optimal strategy for serving in tennis', *Mathematics Magazine*, 5: 197–9.

Graham, A. (2006) *Developing thinking in statistics*, London: Sage.

Gratton, C. and Jones, I. (2004) *Research methods for sports studies*, London: Routledge.

Greene, D., Leyshon, W. and O'Donoghue, P.G. (2008) 'Elite male 400m hurdle tactics are influenced by race leader', in A. Hokelmann, K. Witte and P. O'Donoghue (eds) *Current trends in performance analysis – World Congress of Performance Analysis of Sport VIII; selected proceedings* (pp. 730–6), Aachen: Shaker.

Gregson, W., Drust, B., Atkinson, G. and Di Salvo, V. (2010) 'Match-to-match variability of high speed activities in Premier League soccer', *International Journal of Sports Medicine*, 31: 237–42.

Groebner, D.F., Shannon, P.W., Fry, P.C. and Smith, K.D. (2005) *Business statistics: a decision making approach*, Upper Saddle River, NJ: Pearson Education International.

Henderson, C. (1998) 'An on-court test of aerobic fitness for badminton players', unpublished thesis, University of Ulster.

Hinton, P.R. (2004) *Statistics Explained*, 2nd edn, London: Routledge.

Hinton, P.R., Brownlow, C., McMurray, I. and Cozens, B. (2004) *SPSS Explained*, London: Routledge.

Hoffman, J. (2006) *Norms for fitness, performance and health*, Champaign, IL: Human Kinetics Publishers.

Hopkins, W.G. (2000a) 'Measures of reliability in sports medicine and science', *Sports Medicine*, 30: 1–15.

Hopkins, W.G. (2000b) 'Typical error versus limits of agreement: author's reply', *Sports Medicine*, 30: 377–81.

Horwill, F. (1982) 'Statistics for runners', in D. Watts, H. Wilson and F. Horwill (eds) *The complete middle distance runner* (pp. 63–7), London: Stanley Paul.

Hritz, N. and Ross, C. (2010) 'The perceived impacts of sports tourism: an urban host community perspective', *Journal of Sports Management*, 24: 119–38.

Hughes, M., Cooper, S-M. and Nevill, A.M. (2004) 'Analysis of notation data: reliability', in M. Hughes and I.M. Franks (eds) *Notational Analysis of Sport, 2nd Edition, Systems for better coaching and performance in sport* (pp. 189–204), London: Routledge.

Hulens, M., Beunen, G., Claessens, A.L., Lefevre, J., Thomis, M., Philippaerts, R., Borms, J., Vrijens, J., Lysens, R. and Vansant, G. (2001) 'Trends in BMI among Belgian children, adolescents and adults from 1969 to 1996', *International Journal of Obesity*, 25: 395–9.

Impellizzeri, F.M., Rampinini, E., Maffiuletti, N. and Marcora, S.M. (2007) 'A vertical jump force test for assessing bilateral strength asymmetry in athletes', *Medicine and Science in Sport and Exercise*, 39: 2044–50.

Joll, C. (2011) 'To test the reliability of player grading by qualified and unqualified coaches', unpublished thesis, University of Wales Institute Cardiff.

Joll, A. and O'Donoghue, P.G. (2007) 'Relative age distribution of Welsh netball players', *International Journal of Coaching Science*, 1(2): 3–30.

Jones, G. and Swain, A.B.J. (1992) 'Intensity and direction dimensions of competitive state anxiety and relationships with competitiveness', *Perceptual and Motor Skills*, 74: 467–72.

King, S. and O'Donoghue, P. (2003) 'Specific high intensity training based on activity profile', *International Journal of Performance Analysis in Sport*, 3(2): 130–44.

Kirk-Smith, M. (1998) 'Psychological issues in questionnaire based research', *Journal of the Market Research Society*, 40: 223–6.

Klaasen, F.J.G.M. and Magnus, J.R. (2001) 'Are points in tennis independent and identically distributed? Evidence from a dynamic binary panel data model', *Journal of the American Statistical Association*, 96: 500–9.

Kraemer, H.C. and Thiemann, S. (1987) *How many subjects?*, Newbury Park, CA: Sage.

Levine, D.M. and Stephan, D.F. (2005) *Even you can learn statistics: a guide for everyone who has ever been afraid of statistics*, Upper Saddle River, NJ: Pearson Education Inc.

Linacre, J.M. (2000) 'New approaches to determining and validity', *Research Quarterly for Exercise and Sport*, 71: 129–36.

Lipsey, M.W. (1990) *Design sensitivity*, Newbury Park, CA: Sage.

Lonsdale, C., Hodge, K. and Rose, E.A. (2008) 'The Behavioral Regulation in Sport Questionnaire (BRSQ): Instrument Development and Initial Validity Evidence', *Journal of Sport & Exercise Psychology*, 30: 323–55.

Manly, B.F.J. (2005) *Multivariate statistical methods: a primer*, 3rd edn, Boca Raton: Chapman Hall.

Mardia, K.V., Kent, J.T. and Bibby, J.M. (1994) *Multivariate analysis*, London: Academic Press Ltd.

McGarry, T. and Franks, I.M. (1994) 'A stochastic approach to predicting competition squash match-play', *Journal of Sports Sciences*, 12: 573–84.

McLaughlin, E. and O'Donoghue, P.G. (2002) 'Activity profile of primary school children in the playground', *Journal of Human Movement Studies*, 42: 91–108.

McTrustry, T. and O'Donoghue, P.G. (2002) 'Lifestyle and fitness of new recruits and experienced fire-fighters in Northern Ireland', *Fire Safety, Technology and Management*, 7(2): 11–6.

Morris, C. (1977) 'The most important points in tennis', in S.P. Ladany and R.E. Machol (eds) *Optimal Strategies in Sport* (pp. 131–40), New York: North Holland.

Morrow Jr., J.R., Jackson, A.W., Disch, J.G. and Mood, D.P. (2005) *Measurement and evaluation in human performance*, 3rd edn, Champaign, IL: Human Kinetics.

Mulligan, D. (2001), *Fitness of women's Gaelic footballers*, M.Sc dissertation, University of Ulster, School of Applied Medical Sciences and sports Studies.

Murphy, K.R., Myors, B. And Wolach, A. (2009) *Statistical power analysis: a simple and general model for traditional and modern hypothesis tests*, 3rd edn, New York: Routledge.

Nevill, A. (2000) 'Just how confident are you when publishing the results of your research?', *Journal of Sports Sciences*, 18: 569–70.

Nevill, A.M., Lane, A.M., Kilgour, L.J., Bowes, N. and Whyte, G.P. (2001) 'Stability of psychometric questionnaires', *Journal of Sports Sciences*, 19: 273–8.

Nevill, A.M., Atkinson, G., Hughes, M.D. and Cooper, S-M. (2002) 'Statistical methods for analysing discrete and categorical data recorded in performance analysis', *Journal of Sports Sciences*, 20: 829–44.

Newell, J., Aitchison, T. and Grant, S. (2010) *Statistics for sport and exercise science: a practical approach*, Harlow: Prentice Hall.

Newton, P.K. and Aslam, K. (2006) 'Monte Carlo tennis', *SIAM Review*, 48: 722–42.

Ntoumanis, N. (2001) *A step-by-step guide to SPSS for sport and exercise studies*, London: Routledge.

O'Donoghue, P.G. (2001) 'The most important points in Grand Slam singles tennis', *Research Quarterly for Exercise and Sport*, 72: 125–31.

O'Donoghue, P.G. (2003) 'The effect of scoreline on elite tennis strategy: a cluster analysis', *Journal of Sports Sciences*, 21: 284–5.

O'Donoghue, P.G. (2004) 'Sources of variability in time-motion data: measurement error and within player variability in work-rate', *International Journal of Performance Analysis of Sport*, 4(2): 42–9.

O'Donoghue, P.G. (2006) 'The effectiveness of satisfying the assumptions of predictive modeling techniques: an exercise in predicting the FIFA World Cup 2006', *International Journal of Computing Science in Sport*, 5(2): 5–16.

O'Donoghue, P.G. (2007) 'Reliability issues in performance analysis', *International Journal of Performance Analysis of Sport*, 7(1): 35–48.

O'Donoghue, P.G. (2008) 'Principal components analysis in the selection of key performance indicators in sport', *International Journal of Performance Analysis in Sport*, 8(3): 145–55.

O'Donoghue, P.G. (2009) 'Variability in men's singles tennis strategy at the US Open', in A. Lees, D. Cabello and G. Torres (eds) *Science and Racket Sports IV* (pp. 232–8), London: Routledge.

O'Donoghue, P.G. (2010) *Research methods for sports performance analysis*, London: Routledge.

O'Donoghue, P.G., Martin, G.D. and Murphy, M.H. (1996) 'Systematic evaluation of end-user training for time and motion analysis applications' paper presented at the 2nd International Conference of Technical Informatics, Timisoara, Romania, November.

O'Donoghue, P.G., Dubitzky, W., Lopes, P., Berrar, D., Lagan, K., Hassan, D., Bairner, A. and Darby, P. (2004) 'An evaluation of quantitative and qualitative methods of predicting the 2002 FIFA World Cup', *Journal of Sports Sciences*, 22: 513–14.

O'Donoghue, P.G., Mayes, A., Edwards, K.M. and Garland, J. (2008) 'Performance norms for British National Super League netball', *International Journal of Sports Science and Coaching*, 3: 501–11.

O'Keefe, D.J. (2007) 'Post hoc power, observed power, a priori power, prospective power, achieved power: sorting out appropriate uses of statistical power analyses', *Communication Methods and Measures*, 1: 291–9.

Patterson, P. (2000) 'Reliability, validity and methodological response to the assessment of physical activity via self-report', *Research Quarterly for Exercise and Sport*, 71: 15–20.

Raedeke, T.D. and Smith, A.L. (2001) 'Development and preliminary validation of an athlete burnout measure', *Journal of Sport and Exercise Psychology*, 23: 281–306.

Ramsbottom, R., Brewer, J. and Williams, C. (1988) 'A progressive shuttle run test to estimate maximal oxygen uptake', *British Journal of Sports Medicine*, 22: 141–4.

Rehner, N.J., (1994) 'The maintenance of fluid balance during exercise', *International Journal of Sports Medicine*, 15: 122–5.

Richard, J-F., Godbout, P. and Gréhaigne, J-F. (2000) 'Students' precision and inter-observer reliability of performance assessment in team sports', *Research Quarterly for Exercise and Sport*, 71(1): 85–91.

Robinson,G., O'Donoghue, P.G. and Nielson, P. (2011) 'Path changes and injury risk in English FA Premier League soccer', *International Journal of Performance Analysis of Sport*, 11: 40–56.

Rowntree, D. (2004) *Statistics without tears: a primer for non-mathematicians*, Boston, MA: Pearson.

Rutherford, A. (2001) *Introducing ANOVA and ANCOVA: a GLM approach*, London: Sage.

Salkind, N.J. (2004) *Statistics for people who (think they) hate statistics*, 2nd edn, Thousand Oaks, CA: Sage.

Singer, R.N. and Janelle, C.M. (1999) 'Determining sport expertise: From genes to supremes', *International Journal of Sport Psychology*, 30: 117–50.

Smith, R.E., Smoll, F.L., Cumming, S.P. and Grossbard, J.R. (2006) 'Measurement of multidimensional sport performance anxiety in children and adults: The Sport Anxiety Scale-2', *Journal of Sport and Exercise Psychology*, 28: 479–501.

Sommerville, I. (1992) *Software engineering*, 4th edn, Wokingham: Addison-Wesley.

Soric, M., Misigoj-Durakovic, M. and Pedisic, Z. (2008) 'Dietary intake and body composition of prepubescent female aesthetic athletes', *International Journal of Sports Nutrition and Exercise Metabolism*, 18: 343–54.

Swalgin, K.L. (1998) 'The basketball evaluation system, a computer assisted factor weighted model with means of validity', *International Scientific Journal of Kinesiology and Sport*, 30: 31–7.

Swalgin, K. and Knjaz, D. (2006) 'Euro-BES: a computerised performance evaluation model in the Euro-League', in H. Dancs, M.D. Hughes and P.G. O'Donoghue (eds) *Performance Analysis of Sport VII* (pp. 162–72), Cardiff: UWIC CPA Press.

Swalgin, K. and Knjaz, D. (2007) 'The Euro-basketball evaluation system, a computerised seemless model to grade player performance', paper presented at the 6th International Symposium on Computer Science in Sport, Calgary, Canada, June.

Tabachnick, B.G. and Fidell, L.S. (1996) *Using multivariate statistics*, 3rd edn, NY: Harper Collins.

Tabachnick, B.G. and Fidell, L.S. (2007) *Using multivariate statistics*, 5th edn, New York: Harper Collins.

Taylor, J., Mellalieu, S., James, N. and Shearer, D. (2008) 'The influence of match location, qualify of opposition and match status on technical performance in professional association football', *Journal of Sports Sciences*, 26: 885–95.

Theodorakis, Y., Weinberg, R., Douma, I. and Panagiotis, K. (2000) 'The effects of motivational versus instructional self-talk on improving motor performance', *The Sport Psychologist*, 14: 253–72.

Thomas, J.R. and Nelson, J.K. (1996) *Research Methods in Physical Activity*, 3rd edn, Champaign, IL: Human Kinetics Publishers.

Tong, R.J., Bell, W., Ball, G. and Winter, E.M. (2001) 'Reliability of power output measurements during repeated treadmill sprinting in rugby players', *Journal of Sports Sciences*, 19: 289–97.

Trochim, W. (2000) *The Research Methods Knowledge Base*, 2nd edn, Cincinnati, OH: Atomic Dog Publishing.

Ulijaszek, S.J., Johnston, F.E. and Preece, M.A. (1998) *The Cambridge encyclopedia of human growth and development*, Cambridge: Cambridge University Press.

Vincent, W.J. (1999) *Statistics in Kinesiology*, 2nd edn, Champaign, IL: Human Kinetics Publishers.

Vincent, W.J. (2005) *Statistics in Kinesiology*, 3rd edn, Champaign, IL: Human Kinetics.

Waterhouse, J., Drust, B., Weinert, D., Edwards, B., Gregson, W., Atkinson, G., Kao, S., Aizawa, S. and Reilly, T. (2005) 'The circadian rhythm of core temperature: origin and some implications for exercise performance', *Chronobiology International, 22*: 207–25.

Wildt, A.R. and Ahtola, O.T. (1978) *Analysis of covariance, quantitative applications in the social sciences series*, Newbury Park, CA: Sage.

Wilson, K. and Batterham, A. (1999) 'Stability of questionnaire items in sport and exercise psychology: bootstrap limits of agreement', *Journal of Sports Sciences*, 17: 725–34.

Wood, M. (2003) *Making sense of statistics: a non-mathematical approach*, Basingstoke: Palgrave MacMillan.

Wright, D.B. (2002) *First steps in statistics*, London: Sage.

YMCA of the USA (2000) *YMCA fitness testing and assessment manual*, 4th edn, Champaign, IL: Human Kinetics.

Zhu, W. (2000) 'Score equivalence is at the heart of international measures of physical activity', *Research Quarterly for Exercise and Sport*, 71: 121–8.

INDEX

388

390